锐捷ICT认证系列丛书

锐捷认证互联网专家
RCIE&RS 实验指南

黄君羡　汪双顶　江　政　卢金莲　著
正月十六工作室　组编

电子工业出版社
Publishing House of Electronics Industry
北京·BEIJING

内 容 简 介

本书是锐捷认证互联网专家（RCIE）的配套教材，由锐捷金牌讲师、国家教学名师、锐捷 RCIE 认证讲师等共同完成，全面融入 RCIE&RS 认证最新标准，通过 40 个实战型案例，详细讲解了 RCIE 核心知识与应用，实现了书证融通，有助于读者快捷通过 RCIE&RS 认证。

本书适合具备一定网络基础的读者提升路由与交换技能。无论是准备参加 RCIE&RS 认证的考生，还是希望提升数通技能的网络工程师，都能从本书中获得帮助。

未经许可，不得以任何方式复制或抄袭本书之部分或全部内容。
版权所有，侵权必究。

图书在版编目（CIP）数据

锐捷认证互联网专家 RCIE&RS 实验指南 / 黄君羡等著；正月十六工作室组编. -- 北京：电子工业出版社，2024.6. -- ISBN 978-7-121-48130-7
Ⅰ．TP393
中国国家版本馆 CIP 数据核字第 20244050ZJ 号

责任编辑：孙　伟
印　　刷：三河市良远印务有限公司
装　　订：三河市良远印务有限公司
出版发行：电子工业出版社
　　　　　北京市海淀区万寿路 173 信箱　邮编：100036
开　　本：787×1092　1/16　印张：18.75　字数：480 千字
版　　次：2024 年 6 月第 1 版
印　　次：2024 年 6 月第 1 次印刷
定　　价：198.00 元

凡所购买电子工业出版社图书有缺损问题，请向购买书店调换。若书店售缺，请与本社发行部联系，联系及邮购电话：(010) 88254888, 88258888。
质量投诉请发邮件至 zlts@phei.com.cn，盗版侵权举报请发邮件至 dbqq@phei.com.cn。
本书咨询联系方式：(010) 88254608，sunw@phei.com.cn。

前　　言

随着数字经济的迅猛发展，网络技术的作用和价值受到前所未有的关注和重视。锐捷网络股份有限公司（以下简称"锐捷"）作为成立于 2003 年的老牌 ICT 民族企业，是行业领先的网络基础设施及解决方案提供商，始终致力于将技术与应用充分融合，创造性地解决客户问题。锐捷在全球拥有 8 大研发中心、8000 余名员工，业务范围覆盖 80 多个国家和地区，服务于各行业客户的数字化转型。锐捷认证体系不仅代表了网络技术、数据通信技术的行业至高标准，更是网络工程师们打开职业生涯晋升通道的黄金钥匙。

本书的著者亲历、研究锐捷认证体系（从早先的 RCNA 到之后的 RCNP，再到如今的 RCIE）超过 10 年，以国育产教融合教育科技（海南）有限公司为代表的锐捷体系的人才供应商和技术服务商也为本书提供了大力支持。

本书旨在通过一系列的实验案例，帮助读者深入理解锐捷网络技术的原理和应用，掌握锐捷网络设备的配置和管理技能，为其顺利通过 RCIE&RS 认证打下坚实的基础。本书具有以下特色：

1. 金牌团队，赋能人才培养

本书由锐捷金牌讲师、国家教学名师、锐捷 RCIE 认证讲师等共同完成，全面融入 RCIE&RS 认证最新标准，理实结合，力求学以致用。

2. 书证融通，快速提高水平

本书通过 40 个实战型案例，让读者在实践中领悟理论和快速提高水平。

3. 配套资源丰富

本书配套锐捷官方模拟器、实验拓扑、配置手册、模拟题库等电子资源，并动态同步 RCIE&RS 认证。

本书由正月十六工作室组编，著者信息如下：

著者单位	著　者
锐捷网络股份有限公司	汪双顶
广东交通职业技术学院	黄君羡
正月十六工作室	卢金莲
国育产教融合教育科技（海南）有限公司	江　政

由于著者学术水平有限，书中难免存在不足之处，衷心希望读者提出宝贵的意见和建议，以便我们不断完善本书内容。

目 录

第 1 章 BGP 路由协议 ... 1

1.1 BGP 路径属性 ... 1

1.1.1 Origin 属性 ... 1
- 原理 ... 1
- 任务拓扑 ... 1
- 实施步骤 ... 1
- 任务验证 ... 4
- 问题与思考 ... 5

1.1.2 AS-Path 属性 ... 5
- 原理 ... 5
- 任务拓扑 ... 5
- 实施步骤 ... 5
- 任务验证 ... 7
- 问题与思考 ... 8

1.1.3 Next-Hop 属性 ... 8
- 原理 ... 8
- 任务拓扑 ... 9
- 实施步骤 ... 9
- 任务验证 ... 11
- 问题与思考 ... 13

1.1.4 Local Preference 属性 ... 13
- 原理 ... 13
- 任务拓扑 ... 14
- 实施步骤 ... 14
- 任务验证 ... 16
- 问题与思考 ... 17

1.1.5 MED 属性 ... 17
- 原理 ... 17
- 任务拓扑 ... 17
- 实施步骤 ... 17
- 任务验证 ... 19
- 问题与思考 ... 21

1.1.6 Community 属性 21
  ➢ 原理 21
  ➢ 任务拓扑 22
  ➢ 实施步骤 22
  ➢ 任务验证 24
  ➢ 问题与思考 26

1.2 BGP 选路原则 26

1.2.1 选路原则 1——优选 Weight 值大 26
  ➢ 原理 26
  ➢ 任务拓扑 26
  ➢ 实施步骤 26
  ➢ 任务验证 29
  ➢ 问题与思考 30

1.2.2 选路原则 2——优选 Local Preference 值大 31
  ➢ 原理 31
  ➢ 任务拓扑 31
  ➢ 实施步骤 31
  ➢ 任务验证 36
  ➢ 问题与思考 42

1.2.3 选路原则 3——优选 AS-Path 长度短 42
  ➢ 原理 42
  ➢ 任务拓扑 42
  ➢ 实施步骤 42
  ➢ 任务验证 47
  ➢ 问题与思考 51

1.2.4 选路原则 4——Origin 优选 i>e>? 51
  ➢ 原理 51
  ➢ 任务拓扑 52
  ➢ 实施步骤 52
  ➢ 任务验证 54
  ➢ 问题与思考 55

1.2.5 选路原则 5——优选 MED 值小 55
  ➢ 原理 55
  ➢ 任务拓扑 56
  ➢ 实施步骤 56
  ➢ 任务验证 60
  ➢ 问题与思考 62

1.2.6 选路原则 6——优选 EBGP 路由 63
  ➢ 原理 63
  ➢ 任务拓扑 63

　　　　　　➢ 实施步骤 ··· 63
　　　　　　➢ 任务验证 ··· 66
　　　　　　➢ 问题与思考 ·· 67
　　　1.2.7　选路原则 7——优选最近的 IGP 邻居通告 ·· 67
　　　　　　➢ 原理 ··· 67
　　　　　　➢ 任务拓扑 ··· 68
　　　　　　➢ 实施步骤 ··· 68
　　　　　　➢ 任务验证 ··· 71
　　　　　　➢ 问题与思考 ·· 73
　　　1.2.8　选路原则 8——等价负载 ·· 73
　　　　　　➢ 原理 ··· 73
　　　　　　➢ 任务拓扑 ··· 73
　　　　　　➢ 实施步骤 ··· 73
　　　　　　➢ 任务验证 ··· 76
　　　　　　➢ 问题与思考 ·· 78
　　　1.2.9　选路原则 9——优选 Router ID 值小 ·· 78
　　　　　　➢ 原理 ··· 78
　　　　　　➢ 任务拓扑 ··· 79
　　　　　　➢ 实施步骤 ··· 79
　　　　　　➢ 任务验证 ··· 81
　　　　　　➢ 问题与思考 ·· 82
　　　1.2.10　选路原则 10——优选 Cluster List 短 ··· 83
　　　　　　➢ 原理 ··· 83
　　　　　　➢ 任务拓扑 ··· 83
　　　　　　➢ 实施步骤 ··· 83
　　　　　　➢ 任务验证 ··· 86
　　　　　　➢ 问题与思考 ·· 88
　　　1.2.11　选路原则 11——优选邻居 IP 地址小 ·· 88
　　　　　　➢ 原理 ··· 88
　　　　　　➢ 任务拓扑 ··· 88
　　　　　　➢ 实施步骤 ··· 88
　　　　　　➢ 任务验证 ··· 90
　　　　　　➢ 问题与思考 ·· 92
　1.3　BGP 高级特性 ··· 92
　　　1.3.1　BGP 反射器 ··· 92
　　　　　　➢ 原理 ··· 92
　　　　　　➢ 任务拓扑 ··· 93
　　　　　　➢ 实施步骤 ··· 93
　　　　　　➢ 任务验证 ··· 96

- ➢ 问题与思考 ·········· 98
- 1.3.2 BGP 联盟 ·········· 98
  - ➢ 原理 ·········· 98
  - ➢ 任务拓扑 ·········· 99
  - ➢ 实施步骤 ·········· 99
  - ➢ 任务验证 ·········· 103
  - ➢ 问题与思考 ·········· 105
- 1.3.3 BGP 汇总 ·········· 105
  - ➢ 原理 ·········· 105
  - ➢ 任务拓扑 ·········· 106
  - ➢ 实施步骤 ·········· 106
  - ➢ 任务验证 ·········· 109
  - ➢ 问题与思考 ·········· 112

第 2 章 IS-IS 路由协议 ·········· 113

2.1 IS-IS 协议基础 ·········· 113
- 2.1.1 IS-IS 协议邻居建立 ·········· 113
  - ➢ 原理 ·········· 113
  - ➢ 任务拓扑 ·········· 114
  - ➢ 实施步骤 ·········· 114
  - ➢ 任务验证 ·········· 116
  - ➢ 问题与思考 ·········· 117
- 2.1.2 IS-IS 协议路由渗透 ·········· 117
  - ➢ 原理 ·········· 117
  - ➢ 任务拓扑 ·········· 118
  - ➢ 实施步骤 ·········· 118
  - ➢ 任务验证 ·········· 123
  - ➢ 问题与思考 ·········· 126

2.2 IS-IS 特性 ·········· 127
- 2.2.1 IS-IS 协议认证 ·········· 127
  - ➢ 原理 ·········· 127
  - ➢ 任务拓扑 ·········· 127
  - ➢ 实施步骤 ·········· 127
  - ➢ 任务验证 ·········· 130
  - ➢ 问题与思考 ·········· 130
- 2.2.2 IS-IS 协议汇总 ·········· 130
  - ➢ 原理 ·········· 130
  - ➢ 任务拓扑 ·········· 131
  - ➢ 实施步骤 ·········· 131

    ➢ 任务验证 ··· 133
    ➢ 问题与思考 ··· 136

第 3 章　MPLS 与 VPN 应用 ·· 137

3.1　MPLS 协议 ··· 137
3.1.1　静态 MPLS 协议 ·· 137
    ➢ 原理 ·· 137
    ➢ 任务拓扑 ·· 138
    ➢ 实施步骤 ·· 138
    ➢ 任务验证 ·· 141
    ➢ 问题与思考 ··· 144
3.1.2　动态 MPLS LDP 协议 ··· 144
    ➢ 原理 ·· 144
    ➢ 任务拓扑 ·· 145
    ➢ 实施步骤 ·· 145
    ➢ 任务验证 ·· 148
    ➢ 问题与思考 ··· 151
3.2　MPLS VPN 协议 ··· 151
3.2.1　单域 VPN 实例 ··· 151
    ➢ 原理 ·· 151
    ➢ 任务拓扑 ·· 153
    ➢ 实施步骤 ·· 154
    ➢ 任务验证 ·· 159
    ➢ 问题与思考 ··· 165
3.2.2　MPLS VPN-Hub Spoke ··· 165
    ➢ 原理 ·· 165
    ➢ 任务拓扑 ·· 165
    ➢ 实施步骤 ·· 165
    ➢ 任务验证 ·· 170
    ➢ 问题与思考 ··· 176
3.3　跨域 MPLS VPN ·· 176
3.3.1　跨域 MPLS VPN-OptionA ·· 176
    ➢ 原理 ·· 176
    ➢ 任务拓扑 ·· 176
    ➢ 实施步骤 ·· 177
    ➢ 任务验证 ·· 182
    ➢ 问题与思考 ··· 191
3.3.2　跨域 MPLS VPN-OptionB ·· 191
    ➢ 原理 ·· 191

- 任务拓扑 ········ 192
- 实施步骤 ········ 192
- 任务验证 ········ 196
- 问题与思考 ········ 205

3.3.3 跨域 MPLS VPN-OptionC（1） ········ 205
- 原理 ········ 205
- 任务拓扑 ········ 206
- 实施步骤 ········ 206
- 任务验证 ········ 210
- 问题与思考 ········ 224

3.3.4 跨域 MPLS VPN-OptionC（2） ········ 224
- 原理 ········ 224
- 任务拓扑 ········ 224
- 实施步骤 ········ 224
- 任务验证 ········ 229
- 问题与思考 ········ 239

第 4 章 GRE Over IPSec VPN 协议 ········ 241

4.1 GRE 隧道 ········ 241
- 原理 ········ 241
- 任务拓扑 ········ 241
- 实施步骤 ········ 242
- 任务验证 ········ 243
- 问题与思考 ········ 245

4.2 GRE Over IPSec VPN 隧道 ········ 245
- 原理 ········ 245
- 任务拓扑 ········ 247
- 实施步骤 ········ 247
- 任务验证 ········ 249
- 问题与思考 ········ 252

第 5 章 IPv6 协议 ········ 253

5.1 IPv6 路由协议 ········ 253

5.1.1 IPv6 静态路由协议 ········ 253
- 原理 ········ 253
- 任务拓扑 ········ 253
- 实施步骤 ········ 254
- 任务验证 ········ 256
- 问题与思考 ········ 259

5.1.2 OSPFv3 路由协议 ··· 260
   ➢ 原理 ··· 260
   ➢ 任务拓扑 ··· 260
   ➢ 实施步骤 ··· 260
   ➢ 任务验证 ··· 262
   ➢ 问题与思考 ·· 265
5.1.3 BGP4+路由协议 ··· 265
   ➢ 原理 ··· 265
   ➢ 任务拓扑 ··· 265
   ➢ 实施步骤 ··· 265
   ➢ 任务验证 ··· 268
   ➢ 问题与思考 ·· 270

5.2 IPv6 过渡技术 ·· 270
 5.2.1 IPv6 手工隧道（GRE）·· 270
   ➢ 原理 ··· 270
   ➢ 任务拓扑 ··· 271
   ➢ 实施步骤 ··· 271
   ➢ 任务验证 ··· 273
   ➢ 问题与思考 ·· 275
 5.2.2 IPv6 自动隧道（6to4）·· 275
   ➢ 原理 ··· 275
   ➢ 任务拓扑 ··· 276
   ➢ 实施步骤 ··· 276
   ➢ 任务验证 ··· 278
   ➢ 问题与思考 ·· 280
 5.2.3 IPv6 自动隧道（ISATAP）··· 280
   ➢ 原理 ··· 280
   ➢ 任务拓扑 ··· 281
   ➢ 实施步骤 ··· 281
   ➢ 任务验证 ··· 284
   ➢ 问题与思考 ·· 287

第 1 章　BGP 路由协议

1.1　BGP 路径属性

1.1.1　Origin 属性

➢ 原理

Origin 属性被用于标识路由的来源，在路由产生时会自动设置，如果去往同一个网络存在多条不同 Origin 属性的 BGP 路由时，在其他条件相同的情况下，路由的优选原则是 IGP > EGP > Incomplete，通过 IGP 通告的路由优先进入路由表。

➢ 任务拓扑

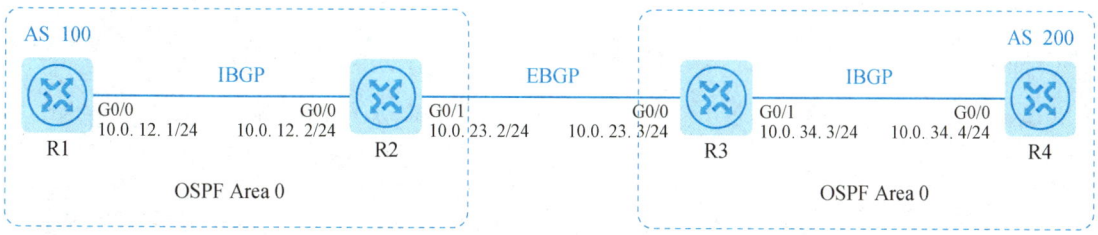

➢ 实施步骤

1. 根据任务拓扑配置各设备接口 IP 地址，各设备配置 loopback 0 接口（R1:10.1.1.1/32）。

```
Ruijie>enable
Password:ruijie
Ruijie# configure terminal
Ruijie(config)#hostname    R1
R1(config)#interface    gigabitEthernet 0/0
R1(config-if-GigabitEthernet 0/0)#no    switchport
R1(config-if-GigabitEthernet 0/0)#ip address    10.0.12.1 24
R1(config-if-GigabitEthernet 0/0)#exit
R1(config)#interface    loopback 0
R1(config-if-Loopback 0)#ip address    10.1.1.1 32
R1(config-if-Loopback 0)#exit
R1(config)#

Ruijie>enable
Password:ruijie
Ruijie# configure terminal
Ruijie(config)#hostname    R2
```

```
R2(config)#interface   gigabitEthernet 0/0
R2(config-if-GigabitEthernet 0/0)#no   switchport
R2(config-if-GigabitEthernet 0/0)#ip address    10.0.12.2 24
R2(config-if-GigabitEthernet 0/0)#exit
R2(config)#interface   gigabitEthernet 0/1
R2(config-if-GigabitEthernet 0/1)#no   switchport
R2(config-if-GigabitEthernet 0/1)#ip address    10.0.23.2 24
R2(config-if-GigabitEthernet 0/1)#exit
R2(config)#interface   loopback 0
R2(config-if-Loopback 0)#ip address    10.2.2.2 32
R2(config-if-Loopback 0)#exit
R2(config)#

Ruijie>enable
Password:ruijie
Ruijie# configure terminal
Ruijie(config)#hostname   R3
R3(config)#interface   gigabitEthernet 0/0
R3(config-if-GigabitEthernet 0/0)#no   switchport
R3(config-if-GigabitEthernet 0/0)#ip address    10.0.23.3 24
R3(config-if-GigabitEthernet 0/0)#exit
R3(config)#interface   gigabitEthernet 0/1
R3(config-if-GigabitEthernet 0/1)#no   switchport
R3(config-if-GigabitEthernet 0/1)#ip address    10.0.34.3 24
R3(config-if-GigabitEthernet 0/1)#exit
R3(config)#interface   loopback 0
R3(config-if-Loopback 0)#ip address    10.3.3.3 32
R3(config-if-Loopback 0)#exit
R3(config)#

Ruijie>enable
Password:ruijie
Ruijie#configure   terminal
Ruijie(config)#hostname   R4
R4(config)#interface   gigabitEthernet 0/0
R4(config-if-GigabitEthernet 0/0)#no   switchport
R4(config-if-GigabitEthernet 0/0)#ip address    10.0.34.4 24
R4(config-if-GigabitEthernet 0/0)#exit
R4(config)#interface   loopback 0
R4(config-if-Loopback 0)#ip address    10.4.4.4 32
R4(config-if-Loopback 0)#exit
R4(config)#
```

2. 在 AS 100 内部和 AS 200 内部配置 OSPF 协议，实现互通。

```
R1(config)#router   ospf   1
R1(config-router)#router-id   1.1.1.1
Change router-id and update OSPF process! [yes/no]:y
R1(config-router)#network   10.1.1.1 0.0.0.0 area   0
R1(config-router)#network   10.0.12.0 0.0.0.255 area   0
R1(config-router)#exit
R1(config)#


R2(config)#router   ospf   1
R2(config-router)#router-id   2.2.2.2
Change router-id and update OSPF process! [yes/no]:y
```

```
R2(config-router)#network    10.0.12.0 0.0.0.255 area    0
R2(config-router)#network    10.2.2.2 0.0.0.0 area    0
R2(config-router)#exit
R2(config)#


R3(config)#router   ospf   1
R3(config-router)#router-id    3.3.3.3
Change router-id and update OSPF process! [yes/no]:y
R3(config-router)#network    10.0.34.0 0.0.0.255 area    0
R3(config-router)#network    10.3.3.3 0.0.0.0 area    0
R3(config-router)#exit
R3(config)#


R4(config)#router   ospf   1
R4(config-router)#router-id    4.4.4.4
Change router-id and update OSPF process! [yes/no]:y
R4(config-router)#network    10.0.34.0 0.0.0.255 area    0
R4(config-router)#network    10.4.4.4 0.0.0.0 area    0
R4(config-router)#exit
R4(config)#
```

3．在 AS 100 内部和 AS 200 内部建立 IBGP 邻居关系，在 AS 100 和 AS 200 之间建立 EBGP 邻居关系，在 AS 内部通过 loopback 0 接口建立 IBGP 邻居关系，更新源地址配置为 loopback 0 接口。

```
R1(config)#router   bgp   100
R1(config-router)#bgp   router-id   1.1.1.1
R1(config-router)#neighbor    10.2.2.2 remote-as    100
R1(config-router)#neighbor    10.2.2.2 update-source    loopback 0
R1(config-router)#exit
R1(config)#

R2(config)#router   bgp   100
R2(config-router)#bgp   router-id   2.2.2.2
R2(config-router)#neighbor    10.1.1.1 remote-as    100
R2(config-router)#neighbor    10.1.1.1 update-source    loopback 0
R2(config-router)#neighbor    10.1.1.1 next-hop-self     //修改下一跳地址，使 R1 能正常接收路由
R2(config-router)#neighbor    10.0.23.3 remote-as    200
R2(config-router)#exit
R2(config)#
R3(config)#router   bgp   200
R3(config-router)#bgp   router-id   3.3.3.3
R3(config-router)#neighbor    10.0.23.2 remote-as    100
R3(config-router)#neighbor    10.4.4.4 remote-as    200
R3(config-router)#neighbor    10.4.4.4 update-source    loopback 0
R3(config-router)#neighbor    10.4.4.4 next-hop-self
R3(config-router)#exit
R3(config)#

R4(config)#router   bgp   200
R4(config-router)#bgp   router-id   4.4.4.4
R4(config-router)#neighbor    10.3.3.3 remote-as    200
R4(config-router)#neighbor    10.3.3.3 update-source    loopback 0
R4(config-router)#exit
R4(config)#
```

4. 在 R1 设备创建 loopback 1 接口，并将该接口通过 network 命令通告到 BGP 路由表中。

```
R1(config)#interface   loopback 1
R1(config-if-Loopback 1)#ip address    100.1.1.1 32
R1(config-if-Loopback 1)#exit

R1(config)#router   bgp    100
R1(config-router)#network   100.1.1.1 mask   255.255.255.255
R1(config-router)#exit
R1(config)#
```

5. 在 R4 设备创建 loopback 1 接口，并将该接口通过 redistribute 方式通告到 BGP 路由表中。

```
R4(config)#interface   loopback 1
R4(config-if-Loopback 1)#ip address    200.1.1.1 32
R4(config-if-Loopback 1)#exit

R4(config)#access-list   1 permit   200.1.1.1 0.0.0.0
R4(config)#route-map   Lo1 permit   10
R4(config-route-map)#match   ip address   1
R4(config-route-map)#exit
R4(config)#
R4(config)#router   bgp   200
R4(config-router)#redistribute   connected   route-map   Lo1
R4(config-router)#exit
R4(config)#
```

➢ 任务验证

1. 检查 OSPF 协议邻居建立情况，状态【State】为 Full，表示邻居关系建立完成。

```
R2#show   ip ospf   neighbor

OSPF process 1, 1 Neighbors, 1 is Full:
Neighbor ID    Pri    State    BFD State    Dead Time    Address       Interface
1.1.1.1        1      Full/BDR    -         00:00:38     10.0.12.1     GigabitEthernet 0/0
R2#
```

2. 检查 BGP 路由协议邻居关系建立情况，【Up/Down】显示为时间参数，则表示 BGP 邻居关系建立完成。

```
R2#show   ip bgp   summary
For address family: IPv4 Unicast
BGP router identifier 2.2.2.2, local AS number 100
BGP table version is 3
2 BGP AS-PATH entries
0 BGP Community entries
2 BGP Prefix entries (Maximum-prefix:4294967295)

Neighbor      V    AS     MsgRcvd  MsgSent   TblVer   InQ OutQ Up/Down    State/PfxRcd
10.0.23.3     4    200    9        7         3        0   0    00:05:18   0
10.1.1.1      4    100    10       11        3        0   0    00:05:47   0

Total number of neighbors 2, established neighbors 2

R2#
```

3. 检查 BGP 路由表，其 Origin 属性显示在【Path】字段右侧，R1 设备通过 network 命令通告的 BGP 路由显示为【i】，R4 设备通过 redistribute 方式通告的 BGP 路由显示为【?】。

```
R1#show    ip bgp
BGP table version is 3, local router ID is 1.1.1.1
Status codes: s suppressed, d damped, h history, * valid, > best, i - internal,
              S Stale, b - backup entry, m - multipath, f Filter, a additional-path
Origin codes: i - IGP, e - EGP, ? - incomplete

    Network          Next Hop              Metric      LocPrf      Weight Path
*>  100.1.1.1/32     0.0.0.0               0                       32768    i
*>i 200.1.1.1/32    10.2.2.2               0           100         0 200 ?

Total number of prefixes 2
R1#
```

➢ 问题与思考

1. Origin 属性属于 BGP 路径属性中的哪一类属性？
2. Origin 属性除包含 IGP 和 Incomplete 属性外，还包含第三种属性，如何实现这种属性？

1.1.2　AS-Path 属性

➢ 原理

AS-Path 属性是公认必遵属性。AS-Path 属性描述到达目标网络所要经过的 AS 号序列，主要是为了避免 AS 之间产生路由环路。

如果去往同一个网络存在多条不同 AS-Path 属性的 BGP 路由时，在其他条件相同的情况下，AS-Path 列表最短的路由优先进入路由表。

➢ 任务拓扑

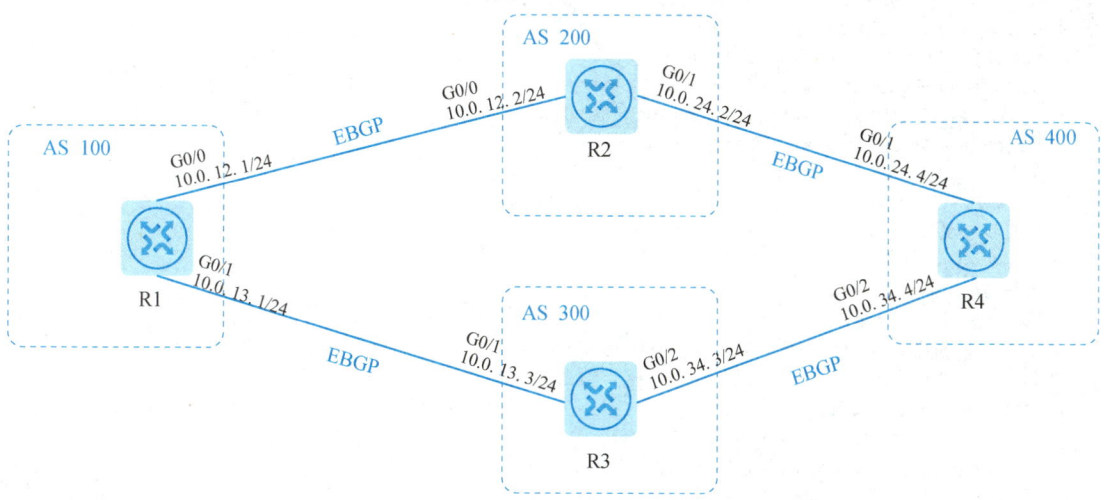

➢ 实施步骤

1. 根据任务拓扑配置各设备接口 IP 地址。

```
Ruijie>enable
Password:ruijie
Ruijie#configure terminal
Ruijie(config)#hostname    R1
R1(config)#interface    gigabitEthernet 0/0
R1(config-if-GigabitEthernet 0/0)#no    switchport
R1(config-if-GigabitEthernet 0/0)#ip address    10.0.12.1 24
R1(config-if-GigabitEthernet 0/0)#exit
R1(config)#interface    gigabitEthernet 0/1
R1(config-if-GigabitEthernet 0/1)#no    switchport
R1(config-if-GigabitEthernet 0/1)#ip address    10.0.13.1 24
R1(config-if-GigabitEthernet 0/1)#exit
R1(config)#

Ruijie>enable
Password:ruijie
Ruijie#configure terminal
Ruijie(config)#hostname    R2
R2(config)#interface    gigabitEthernet 0/0
R2(config-if-GigabitEthernet 0/0)#no    switchport
R2(config-if-GigabitEthernet 0/0)#ip address    10.0.12.2 24
R2(config-if-GigabitEthernet 0/0)#exit
R2(config)#interface    gigabitEthernet 0/1
R2(config-if-GigabitEthernet 0/1)#no    switchport
R2(config-if-GigabitEthernet 0/1)#ip address    10.0.24.2 24
R2(config-if-GigabitEthernet 0/1)#exit
R2(config)#

Ruijie>enable
Password:ruijie
Ruijie#configure terminal
Ruijie(config)#hostname    R3
R3(config)#interface    gigabitEthernet 0/1
R3(config-if-GigabitEthernet 0/1)#no    switchport
R3(config-if-GigabitEthernet 0/1)#ip address    10.0.13.3 24
R3(config-if-GigabitEthernet 0/1)#exit
R3(config)#interface    gigabitEthernet 0/2
R3(config-if-GigabitEthernet 0/2)#no    switchport
R3(config-if-GigabitEthernet 0/2)#ip address    10.0.34.3 24
R3(config-if-GigabitEthernet 0/2)#exit
R3(config)#

Ruijie>enable
Password:ruijie
Ruijie#configure terminal
R4(config)#interface    gigabitEthernet 0/1
R4(config-if-GigabitEthernet 0/1)#no    switchport
R4(config-if-GigabitEthernet 0/1)#ip address    10.0.24.4 24
R4(config-if-GigabitEthernet 0/1)#ex
R4(config)#interface    gigabitEthernet 0/2
R4(config-if-GigabitEthernet 0/2)#no    switchport
R4(config-if-GigabitEthernet 0/2)#ip address    10.0.34.4 24
R4(config-if-GigabitEthernet 0/2)#ex
R4(config)#
```

2. 根据任务拓扑配置 BGP 协议，在所有设备之间建立 EBGP 邻居关系。

```
R1(config)#router    bgp    100
R1(config-router)#bgp    router-id    1.1.1.1
R1(config-router)#neighbor    10.0.12.2 remote-as    200
R1(config-router)#neighbor    10.0.13.3 remote-as    300
R1(config-router)#exit

R2(config)#router    bgp    200
R2(config-router)#bgp    router-id    2.2.2.2
R2(config-router)#neighbor    10.0.12.1 remote-as    100
R2(config-router)#neighbor    10.0.24.4 remote-as    400
R2(config-router)#exit
R2(config)#

R3(config)#router    bgp    300
R3(config-router)#neighbor    10.0.13.1 remote-as    100
R3(config-router)#neighbor    10.0.34.4 remote-as    400
R3(config-router)#exit
R3(config)#

R4(config)#router    bgp    400
R4(config-router)#bgp    router-id    4.4.4.4
R4(config-router)#neighbor    10.0.24.2    remote-as    200
R4(config-router)#neighbor    10.0.34.3    remote-as    300
R4(config-router)# exit
R4(config)#
```

3. 在 R4 设备创建一个 loopback 0 接口，并将该接口通告到 BGP 路由表中。

```
R4(config)#interface    loopback 0
R4(config-if-Loopback 0)#ip address    40.1.1.1 32
R4(config-if-Loopback 0)#exit
R4(config)#router    bgp    400
R4(config-router)#network    40.1.1.1 mask    255.255.255.255
R4(config-router)#exit
R4(config)#
```

4. 在 R2 设备创建一个 loopback 0 接口，配置的 IP 地址与 R4 设备的 loopback 0 接口相同，并将该接口通告到 BGP 路由表中。

```
R2(config)#interface    loopback 0
R2(config-if-Loopback 0)#ip address    40.1.1.1 32
R2(config-if-Loopback 0)#exit
R2(config)#router    bgp    200
R2(config-router)#network    40.1.1.1    mask 255.255.255.255
R2(config-router)#exit
R2(config)#
```

➢ 任务验证

1. 检查 BGP 邻居关系建立情况。

```
R1#show ip bgp    summary
For address family: IPv4 Unicast
BGP router identifier 1.1.1.1, local AS number 100
BGP table version is 6
0 BGP AS-PATH entries
0 BGP Community entries
```

```
0 BGP Prefix entries (Maximum-prefix:4294967295)

Neighbor      V      AS  MsgRcvd MsgSent    TblVer   InQ OutQ Up/Down   State/PfxRcd
10.0.12.2     4     200      4       3         6      0    0  00:01:19       0
10.0.13.3     4     300      4       2         6      0    0  00:01:10       0

Total number of neighbors 2, established neighbors 2

R1#
```

2. 检查 BGP 路由通告的路由信息，40.1.1.1/32 路由优选下一跳地址 10.0.12.2 传递的路由。

```
R1#show ip bgp
BGP table version is 7, local router ID is 1.1.1.1
Status codes: s suppressed, d damped, h history, * valid, > best, i - internal,
              S Stale, b - backup entry, m - multipath, f Filter, a additional-path
Origin codes: i - IGP, e - EGP, ? - incomplete

    Network          Next Hop         Metric     LocPrf        Weight Path
*b  40.1.1.1/32      10.0.13.3        0                        0 300 400 i
*>                   10.0.12.2        0                        0 200 400 i

Total number of prefixes 1
R1#
```

3. 在 R2 设备创建 loopback 0 接口通告相同网段路由，优先选择 AS-Path 列表短的路由。

```
R1#show    ip bgp
BGP table version is 8, local router ID is 1.1.1.1
Status codes: s suppressed, d damped, h history, * valid, > best, i - internal,
              S Stale, b - backup entry, m - multipath, f Filter, a additional-path
Origin codes: i - IGP, e - EGP, ? - incomplete

    Network          Next Hop         Metric     LocPrf        Weight Path
*b  40.1.1.1/32      10.0.13.3        0                        0 300 400 i
*>                   10.0.12.2        0                        0 200 i

Total number of prefixes 1
R1#
```

> 问题与思考

AS-Path 属性除能够记录 BGP 路由经过 AS 路径及执行路径优先外，还有什么作用？

1.1.3 Next-Hop 属性

> 原理

Next-Hop 属性是公认必遵属性，用于指定到达目标网络的下一跳地址。

当 BGP 路由器收到 EBGP 邻居传来的路由后，将检查 Next-Hop 属性，如果 Next-Hop 属性值（IP 地址）不可达，则显示此路由不可用（不会出现 >）；当 BGP 路由器从 EBGP 邻居接收路由再传递给 IBGP 邻居时，需要通过 next-hop-self 命令修改下一跳地址，否则会出现不可达问题。

BGP 路由器向 EBGP 邻居传递路由时，会将该路由的 Next-Hop 设置为自己的更新源地址。

BGP 路由器收到 EBGP 邻居传来的路由后，再传递给自己的 IBGP 邻居时，会保持原来的 Next-Hop 属性值。

如果路由器收到某条 BGP 路由，该路由的 Next-Hop 与将要发送过去的 EBGP 邻居同属一个网段，那么该路由的 Next-Hop 将保持不变。

➢ 任务拓扑

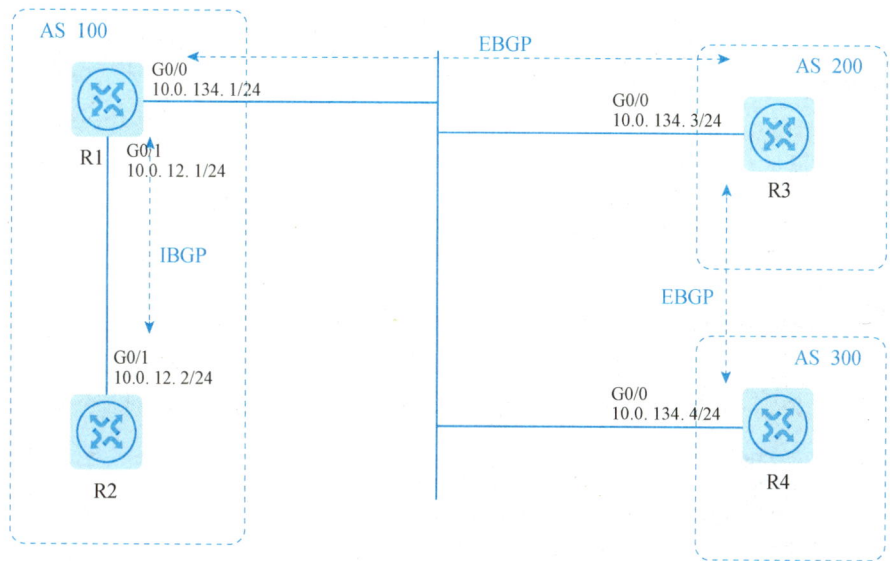

➢ 实施步骤

1. 根据任务拓扑配置各设备接口 IP 地址。

```
Ruijie>enable
Password:ruijie
Ruijie#configure terminal
Ruijie(config)#hostname   R1
R1(config)#interface   GigabitEthernet 0/0
R1(config-if-GigabitEthernet 0/0)#no switchport
R1(config-if-GigabitEthernet 0/0)#ip address   10.0.134.1 24
R1(config-if-GigabitEthernet 0/0)#exit
R1(config)#interface   gigabitEthernet 0/1
R1(config-if-GigabitEthernet 0/1)#no   switchport
R1(config-if-GigabitEthernet 0/1)#ip address   10.0.12.1 24
R1(config-if-GigabitEthernet 0/1)#exit
R1(config)#

Ruijie>enable
Password:ruijie
Ruijie#configure terminal
Ruijie(config)#hostname   R2
R2(config)#interface   gigabitEthernet 0/1
R2(config-if-GigabitEthernet 0/1)#no   switchport
R2(config-if-GigabitEthernet 0/1)#ip address   10.0.12.2 24
R2(config-if-GigabitEthernet 0/1)#exit
R2(config)#
```

```
Ruijie>enable
Password:ruijie
Ruijie#configure terminal
Ruijie(config)#hostname   R3
R3(config)#interface   gigabitEthernet 0/0
R3(config-if-GigabitEthernet 0/0)#no   switchport
R3(config-if-GigabitEthernet 0/0)#ip address    10.0.134.3 24
R3(config-if-GigabitEthernet 0/0)#exit
R3(config)#

Ruijie>enable
Password:ruijie
Ruijie#configure terminal
Ruijie(config)#hostname   R4
R4(config)#interface   gigabitEthernet 0/0
R4(config-if-GigabitEthernet 0/0)#no   switchport
R4(config-if-GigabitEthernet 0/0)#ip address    10.0.134.4   24
R4(config-if-GigabitEthernet 0/0)#exit
R4(config)#
```

2．BGP 路由协议邻居关系建立，从 EBGP 邻居接收路由并发送给 IBGP 邻居时，需要修改下一跳地址，否则 BGP 路由无效，在指定 IBGP 邻居时添加修改下一跳地址的命令。

```
R1(config)#router bgp   100
R1(config-router)#bgp    router-id   1.1.1.1
R1(config-router)#neighbor   10.0.12.2 remote-as   100
R1(config-router)#neighbor   10.0.12.2 next-hop-self
R1(config-router)#neighbor   10.0.134.3 remote-as   200
R1(config-router)#exit
R1(config)#

R2(config)#router   bgp   100
R2(config-router)#bgp    router 2.2.2.2
R2(config-router)#neighbor   10.0.12.2 remote
R2(config-router)#neighbor   10.0.12.1 remote-as   100
R2(config-router)#exit
R2(config)#

R3(config)#router bgp   200
R3(config-router)#bgp    router-id   3.3.3.3
R3(config-router)#neighbor   10.0.134.1 remote-as   100
R3(config-router)#neighbor   10.0.134.4 remote-as   300
R3(config-router)#exit
R3(config)#

R4(config)#router   bgp   300
R4(config-router)#bgp    router-id 4.4.4.4
R4(config-router)#neighbor   10.0.134.3 remote-as   200
R4(config-router)#exit
```

3．在 R2 设备和 R4 设备分别创建 loopback 0 接口，并将两个接口的网段通告到 BGP 路由表中。

```
R2(config)#interface   loopback 0
R2(config-if-Loopback 0)#ip address   20.0.1.1 32
R2(config-if-Loopback 0)#exit
```

```
R2(config)#router bgp   100
R2(config-router)#network   20.0.1.1 mask 255.255.255.255
R2(config-router)#exit
R2(config)#


R4(config)#interface   loopback 0
R4(config-if-Loopback 0)#ip address   40.0.1.1 32
R4(config-if-Loopback 0)#exit
R4(config)#


R4(config)#router bgp   300
R4(config-router)#network   40.0.1.1 mask   255.255.255.255
R4(config-router)#exit
R4(config)
```

➢ 任务验证

1. 在 R1 设备通过 show   ip bgp   summary 命令检查 BGP 邻居关系建立情况。

```
R1#show   ip bgp   summary
For address family: IPv4 Unicast
BGP router identifier 1.1.1.1, local AS number 100
BGP table version is 3
2 BGP AS-PATH entries
0 BGP Community entries
2 BGP Prefix entries (Maximum-prefix:4294967295)

Neighbor       V     AS MsgRcvd MsgSent    TblVer   InQ OutQ Up/Down     State/PfxRcd
10.0.12.2      4     100    37      37         3     0    0 00:31:29          1
10.0.134.3     4     200    35      34         3     0    0 00:31:10          1

Total number of neighbors 2, established neighbors 2

R1#
```

2. 检查 R2 设备通告的 BGP 路由 20.0.1.1/32 的下一跳地址信息，R1 设备收到 R2 设备（IBGP 邻居）通告的 BGP 路由下一跳地址为 R2 设备的更新源地址，R3 设备收到 R1 设备（EBGP 邻居）通告的 BGP 路由下一跳地址为 R1 设备的更新源地址。

```
R1#show   ip bgp
BGP table version is 4, local router ID is 1.1.1.1
Status codes: s suppressed, d damped, h history, * valid, > best, i - internal,
              S Stale, b - backup entry, m - multipath, f Filter, a additional-path
Origin codes: i - IGP, e - EGP, ? - incomplete

     Network          Next Hop           Metric      LocPrf      Weight Path
*>i 20.0.1.1/32       10.0.12.2               0         100           0 i

Total number of prefixes 1
R1#


R3#show   ip bgp
BGP table version is 4, local router ID is 3.3.3.3
Status codes: s suppressed, d damped, h history, * valid, > best, i - internal,
```

```
            S Stale, b - backup entry, m - multipath, f Filter, a additional-path
Origin codes: i - IGP, e - EGP, ? - incomplete

     Network          Next Hop          Metric      LocPrf      Weight Path
*>   20.0.1.1/32      10.0.134.1        0                       0 100 i

Total number of prefixes 1
R3#
```

3．R4 设备收到 BGP 路由的下一跳地址，由于下一跳地址与 R1 设备的更新源地址处于相同网段，为避免次优路径的产生，R4 设备收到的 BGP 路由下一跳地址仍然是 R1 设备的更新源地址。

```
R4#show   ip bgp
BGP table version is 4, local router ID is 4.4.4.4
Status codes: s suppressed, d damped, h history, * valid, > best, i - internal,
            S Stale, b - backup entry, m - multipath, f Filter, a additional-path
Origin codes: i - IGP, e - EGP, ? - incomplete

     Network          Next Hop          Metric      LocPrf      Weight Path
*>   20.0.1.1/32      10.0.134.1        0                       0 200 100 i

Total number of prefixes 1
R4#

R4#show   ip bgp   20.0.1.1
BGP routing table entry for 20.0.1.1/32(#0x7f414f321bd8)
Paths: (1 available, best #1, table Default-IP-Routing-Table)
  Advertised to update-groups:
    2

  200 100
    10.0.134.1 from 10.0.134.3 (3.3.3.3)
      Origin IGP, metric 0, localpref 100, valid, external, best
      Last update: Mon Jul 17 07:03:26 2023
      RX ID: 0,TX ID: 0
R4#
```

4．检查 R4 设备通告的 BGP 路由 40.0.1.1/32 的下一跳地址信息，R3 设备接收 R4 设备（EBGP 邻居）传递的 BGP 路由，下一跳地址为 R4 设备的更新源地址，R1 设备与 R3 设备建立 EBGP 邻居关系，未与 R4 设备建立 EBGP 邻居关系，即 BGP 路由从 R3 设备传递给 R1 设备，为避免产生次优路径，R1 设备收到的 BGP 路由下一跳地址为 R4 设备的更新源地址。

```
R3#show   ip bgp
BGP table version is 6, local router ID is 3.3.3.3
Status codes: s suppressed, d damped, h history, * valid, > best, i - internal,
            S Stale, b - backup entry, m - multipath, f Filter, a additional-path
Origin codes: i - IGP, e - EGP, ? - incomplete

     Network          Next Hop          Metric      LocPrf      Weight Path
*>   40.0.1.1/32      10.0.134.4        0                       0 300 i

Total number of prefixes 1
R3#
```

```
R1#show    ip bgp
BGP table version is 6, local router ID is 1.1.1.1
Status codes: s suppressed, d damped, h history, * valid, > best, i - internal,
              S Stale, b - backup entry, m - multipath, f Filter, a additional-path
Origin codes: i - IGP, e - EGP, ? - incomplete

     Network          Next Hop            Metric       LocPrf       Weight Path
*>   40.0.1.1/32      10.0.134.4          0                         0 200 300 i

Total number of prefixes 1
R1#
```

5. R2 设备接收 R1 设备（IBGP 邻居）传递的 BGP 路由，此时默认从 EBGP 邻居接收到的路由传递给 IBGP 邻居时不改变下一跳地址，仍然是 R4 设备的更新源地址，这样会导致 R2 设备收到的 BGP 路由下一跳地址不可达，从而成为无效路由，需在 R1 设备通过 next-hop-self 命令修改下一跳地址。

```
R2#show    ip bgp
BGP table version is 4, local router ID is 2.2.2.2
Status codes: s suppressed, d damped, h history, * valid, > best, i - internal,
              S Stale, b - backup entry, m - multipath, f Filter, a additional-path
Origin codes: i - IGP, e - EGP, ? - incomplete

     Network          Next Hop            Metric       LocPrf       Weight Path
*>i  40.0.1.1/32      10.0.12.1           0            100          0 200 300 i

Total number of prefixes 1
R2#


R2#show    ip bgp    40.0.1.1
BGP routing table entry for 40.0.1.1/32(#0x7fe0c4321bd8)
Paths: (1 available, best #1, table Default-IP-Routing-Table)
  Not advertised to any peer

  200 300
     10.0.12.1 from 10.0.12.1 (1.1.1.1)
        Origin IGP, metric 0, localpref 100, valid, internal, best
        Last update: Mon Jul 17 07:16:22 2023
        RX ID: 0,TX ID: 0
R2#
```

➢ 问题与思考

BGP 路由从 EBGP 邻居接收并传递给 IBGP 邻居时，默认不修改下一跳地址的原因是什么？

1.1.4　Local Preference 属性

➢ 原理

Local Preference 属性（本地优先级）是公认自决属性，用于告知 AS 内部的 BGP 路由器，哪一条路径是离开本 AS 的最佳路径；Local Preference 属性通常只在 AS 内部的 IBGP 邻居之间传递，不会传递给 EBGP 邻居。

> 任务拓扑

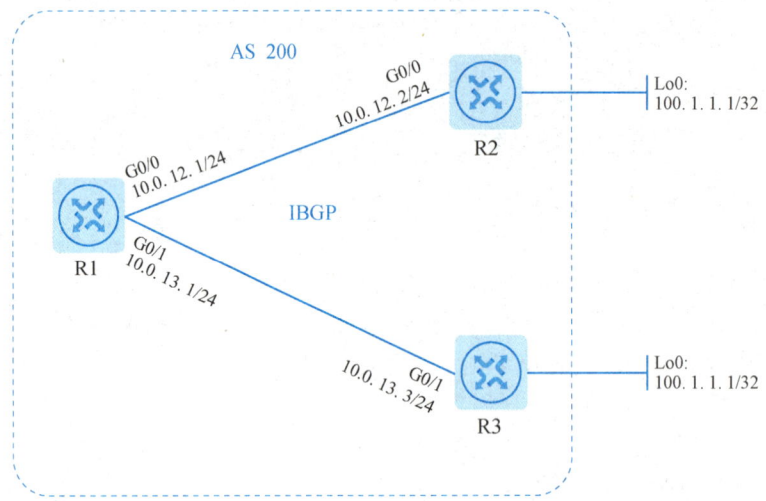

> 实施步骤

1. 根据任务拓扑配置各设备接口 IP 地址。

```
Ruijie>enable
Password:ruijie
Ruijie#configure terminal
Ruijie(config)#hostname    R1
R1(config)#interface    gigabitEthernet 0/0
R1(config-if-GigabitEthernet 0/0)#no switchport
R1(config-if-GigabitEthernet 0/0)#ip address    10.0.12.1 24
R1(config-if-GigabitEthernet 0/0)#exit
R1(config)#interface    gigabitEthernet 0/1
R1(config-if-GigabitEthernet 0/1)#no    switchport
R1(config-if-GigabitEthernet 0/1)#ip address    10.0.13.1 24
R1(config-if-GigabitEthernet 0/1)#exit
R1(config)#

Ruijie>enable
Password:ruijie
Ruijie#configure terminal
Ruijie(config)#hostname    R2
R2(config)#interface    gigabitEthernet 0/0
R2(config-if-GigabitEthernet 0/0)# no switchport
R2(config-if-GigabitEthernet 0/0)#ip address    10.0.12.2 24
R2(config-if-GigabitEthernet 0/0)#exit
R2(config)#

Ruijie>enable
Password:ruijie
Ruijie#configure terminal
Ruijie(config)#hostname    R3
R3(config)#interface    gigabitEthernet 0/1
R3(config-if-GigabitEthernet 0/1)# no switchport
```

```
R3(config-if-GigabitEthernet 0/1)#ip address    10.0.13.3 24
R3(config-if-GigabitEthernet 0/1)#exit
R3(config)#
```

2. 在 AS 内部配置 BGP 协议，通过物理接口建立 IBGP 邻居关系。

```
R1(config)#router   bgp   100
R1(config-router)#bgp    router-id   1.1.1.1
R1(config-router)#neighbor   10.0.12.2 remote-as    100
R1(config-router)#neighbor   10.0.13.3 remote-as    100
R1(config-router)#exit
R1(config)#

R2(config)#router   bgp   100
R2(config-router)#bgp    router-id   2.2.2.2
R2(config-router)#neighbor   10.0.12.1 remote-as    100
R2(config-router)#exit
R2(config)#

R3(config)#router   bgp   100
R3(config-router)#bgp    router-id   3.3.3.3
R3(config-router)#neighbor   10.0.13.1 remote-as    100
R3(config-router)#exit
R3(config)#
```

3. 在 R2 设备和 R3 设备分别创建 loopback 0 接口，两个接口的 IP 地址处于相同网段，并将两个接口通告到 BGP 路由表中。

```
R2(config-router)#interface Loopback 0
R2(config-if-Loopback 0)#ip address    100.1.1.1 32
R2(config-if-Loopback 0)#exit
R2(config)#router   bgp   100
R2(config-router)#network   100.1.1.1 mask    255.255.255.255
R2(config-router)#exit
R2(config)#

R3(config)#interface   loopback 0
R3(config-if-Loopback 0)#ip address    100.1.1.1 32
R3(config-if-Loopback 0)#exit
R3(config)#router   bgp   100
R3(config-router)#network   100.1.1.1 mask    255.255.255.255
R3(config-router)#exit
R3(config)#
```

4. 在 R1 设备通过 route-map 命令设置 local-preference。

```
R1(config)#access-list   1 permit    100.1.1.1 0.0.0.0
R1(config)#route-map Loc permit    10
R1(config-route-map)#match    ip address    1
R1(config-route-map)#set   local-preference    300
R1(config-route-map)#exit
R1(config)#route-map   Loc permit 20
R1(config-route-map)#exit
R1(config)#router   bgp   100
R1(config-router)#neighbor   10.0.13.3 route-map Loc in
R1(config-router)#exit
R1(config)#
```

➢ 任务验证

1. 检查 IP 地址配置情况。

```
R1#show    ip interface    brief
Interface          IP-Address(Pri)       IP-Address(Sec)      Status              Protocol
GigabitEthernet 0/0    10.0.12.1/24      no address           up                  up
GigabitEthernet 0/1    10.0.13.1/24      no address           up                  up
VLAN 1                 no address        no address           up                  down
R1#

R2#show    ip interface    brief
Interface          IP-Address(Pri)       IP-Address(Sec)      Status              Protocol
GigabitEthernet 0/0    10.0.12.2/24      no address           up                  up
Loopback 0             100.1.1.1/32      no address           up                  up
VLAN 1                 no address        no address           up                  down
R2#

R3#show    ip interface    brief
Interface          IP-Address(Pri)       IP-Address(Sec)      Status              Protocol
GigabitEthernet 0/1    10.0.13.3/24      no address           up                  up
Loopback 0             100.1.1.1/32      no address           up                  up
VLAN 1                 no address        no address           up                  down
R3#
```

2. 检查 BGP 邻居关系建立情况。

```
R1#show    ip bgp    summary
For address family: IPv4 Unicast
BGP router identifier 1.1.1.1, local AS number 100
BGP table version is 3
0 BGP AS-PATH entries
0 BGP Community entries
0 BGP Prefix entries (Maximum-prefix:4294967295)

Neighbor         V        AS  MsgRcvd  MsgSent    TblVer   InQ OutQ Up/Down      State/PfxRcd
10.0.12.2        4       100       79       77         2     0    0 01:14:53     0
10.0.13.3        4       100       81       78         2     0    0 01:14:41     0

Total number of neighbors 2, established neighbors 2

R1#
```

3. 在 R1 设备查看 R2 设备和 R3 设备通告的路由信息。

```
R1#show    ip bgp
BGP table version is 3, local router ID is 1.1.1.1
Status codes: s suppressed, d damped, h history, * valid, > best, i - internal,
              S Stale, b - backup entry, m - multipath, f Filter, a additional-path
Origin codes: i - IGP, e - EGP, ? - incomplete

     Network            Next Hop          Metric       LocPrf       Weight Path
*bi 100.1.1.1/32        10.0.13.3         0            100          0      i
*>i                     10.0.12.2         0            100          0      i

Total number of prefixes 1
R1#
```

4．根据 BGP 路由的 Local Preference 参数选择最优路径。

```
R1#show    ip bgp
BGP table version is 3, local router ID is 1.1.1.1
Status codes: s suppressed, d damped, h history, * valid, > best, i - internal,
              S Stale, b - backup entry, m - multipath, f Filter, a additional-path
Origin codes: i - IGP, e - EGP, ? - incomplete

     Network          Next Hop            Metric      LocPrf      Weight    Path
*>i 100.1.1.1/32      10.0.13.3             0           300          0       i
*bi                   10.0.12.2             0           100          0       i

Total number of prefixes 1
R1#
```

➢ 问题与思考

关于 Local Preference 属性实现路径优选的配置方法有两种，除文中提到的 route-map 配置外，还能配置什么命令实现路径优选？两种配置方法有什么区别？

1.1.5 MED 属性

➢ 原理

MED（Multi-Exit Discriminator）属性是可选不可传递属性，是一种度量值，用于向外部对等体指出进入 AS 的首选路径。MED 属性仅在相邻的两个 AS 之间传递，收到此属性的 AS 不会再将其通告给任何其他第三方 AS。MED 属性可以手动配置，若 BGP 路由传递中没有配置 MED 属性，会将该路由的 MED 值按默认值来处理。

➢ 任务拓扑

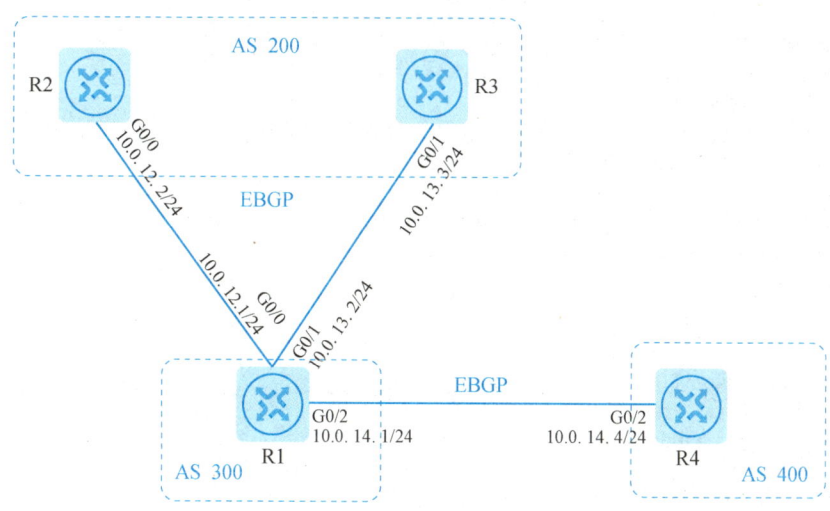

➢ 实施步骤

1．根据任务拓扑配置各设备接口 IP 地址。

Ruijie>enable

```
Password:ruijie
Ruijie(config)#hostname    R1
R1(config)#interface    gigabitEthernet 0/0
R1(config-if-GigabitEthernet 0/0)#no    switchport
R1(config-if-GigabitEthernet 0/0)#ip address    10.0.12.1 24
R1(config-if-GigabitEthernet 0/0)#exit
R1(config)#interface    gigabitEthernet 0/1
R1(config-if-GigabitEthernet 0/1)#no    switchport
R1(config-if-GigabitEthernet 0/1)#ip address    10.0.13.1 24
R1(config-if-GigabitEthernet 0/1)#exit
R1(config)#interface    gigabitEthernet 0/2
R1(config-if-GigabitEthernet 0/2)#no    switchport
R1(config-if-GigabitEthernet 0/2)#ip address    10.0.14.1 24
R1(config-if-GigabitEthernet 0/2)#exit
R1(config)#

Ruijie>enable
Password:ruijie
Ruijie#configure terminal
Ruijie(config)#hostname    R2
R2(config)#interface    gigabitEthernet 0/0
R2(config-if-GigabitEthernet 0/0)#ip address    10.0.12.2 24
R2(config-if-GigabitEthernet 0/0)#exit
R2(config)#

Ruijie>enable
Password:ruijie
Ruijie#configure terminal
Ruijie(config)#hostname    R3
R3(config)#interface    gigabitEthernet 0/1
R3(config-if-GigabitEthernet 0/1)#ip address    10.0.13.3 24
R3(config-if-GigabitEthernet 0/1)#exit
R3(config)#

Ruijie>enable
Password:ruijie
Ruijie#configure terminal
Ruijie(config)#hostname    R4
R4(config)#interface    gigabitEthernet 0/2
R4(config-if-GigabitEthernet 0/2)#no switchport
R4(config-if-GigabitEthernet 0/2)#ip address    10.0.14.4 24
R4(config-if-GigabitEthernet 0/2)#exit
R4(config)#
```

2. 在各设备之间建立 EBGP 邻居关系。

```
R1(config)#router bgp    300
R1(config-router)#bgp    router-id    1.1.1.1
R1(config-router)#neighbor    10.0.12.2 remote-as    200
R1(config-router)#neighbor    10.0.13.3 remote-as    200
R1(config-router)#neighbor    10.0.14.4 remote-as    400
R1(config-router)#exit
R1(config)#

R2(config)#router    bgp    200
R2(config-router)#bgp    router-id    2.2.2.2
```

```
R2(config-router)#neighbor    10.0.12.1 remote-as    300
R2(config-router)#exit
R2(config)#


R3(config)#router    bgp    200
R3(config-router)#bgp    router-id    3.3.3.3
R3(config-router)#neighbor    10.0.13.1 remote-as    300
R3(config-router)#exit
R3(config)#


R4(config)#router    bgp    400
R4(config-router)#bgp    router-id    4.4.4.4
R4(config-router)#neighbor    10.0.14.1 remote-as    300
R4(config-router)#exit
R4(config)#
```

3．在 R2 设备和 R3 设备分别创建 loopback 0 接口，并将两个接口通告到 BGP 路由表中。

```
R2(config)#interface    loopback 0
R2(config-if-Loopback 0)#ip address    192.168.10.1 24
R2(config-if-Loopback 0)#exit
R2(config)#router    bgp    200
R2(config-router)#network    192.168.10.0 mask    255.255.255.0
R2(config-router)#exit
R2(config)#


R3(config)#interface    loopback 0
R3(config-if-Loopback 0)#ip address    192.168.10.1 24
R3(config-if-Loopback 0)#exit
R3(config)#router bgp    200
R3(config-router)#network 192.168.10.0 mask    255.255.255.0
R3(config-router)#exit
R3(config)#
```

4．通过 route-map 命令修改 MED 值，选择最优路径。

```
R1(config)#access-list 1 permit    192.168.10.0 0.0.0.255
R1(config)#route-map MED permit    10
R1(config-route-map)#match    ip address    1
R1(config-route-map)#set    metric 100
R1(config-route-map)#exit
R1(config)#route-map MED permit    20
R1(config-route-map)#exit
R1(config)#router    bgp    300
R1(config-router)#neighbor    10.0.12.2 route-map MED in
R1(config-router)#exit
R1(config)#
```

➢ 任务验证

1．检查 IP 地址配置情况。

```
R1#show    ip interface    brief
Interface            IP-Address(Pri)      IP-Address(Sec)    Status        Protocol
GigabitEthernet 0/0      10.0.12.1/24         no address         up              up
GigabitEthernet 0/1      10.0.13.1/24         no address         up              up
GigabitEthernet 0/2      10.0.14.1/24         no address         up              up
```

```
VLAN 1              no address      no address      up          down
R1#


R2#show    ip interface    brief
Interface           IP-Address(Pri)    IP-Address(Sec)    Status          Protocol
GigabitEthernet 0/0     10.0.12.2/24       no address         up              up
Loopback 0              192.168.10.1/24    no address         up              up
VLAN 1                  no address         no address         up              down
R2#


R3#show    ip interface    brief
Interface           IP-Address(Pri)    IP-Address(Sec)    Status          Protocol
GigabitEthernet 0/1     10.0.13.3/24       no address         up              up
Loopback 0              192.168.10.1/24    no address         up              up
VLAN 1                  no address         no address         up              down
R3#


R4#show    ip interface    brief
Interface           IP-Address(Pri)    IP-Address(Sec)    Status          Protocol
GigabitEthernet 0/2     10.0.14.4/24       no address         up              up
VLAN 1                  no address         no address         up              down
R4#
```

2．检查 BGP 邻居关系建立情况。

```
R1#show    ip bgp    summary
For address family: IPv4 Unicast
BGP router identifier 1.1.1.1, local AS number 300
BGP table version is 6
0 BGP AS-PATH entries
0 BGP Community entries
0 BGP Prefix entries (Maximum-prefix:4294967295)

Neighbor       V     AS MsgRcvd MsgSent    TblVer    InQ OutQ Up/Down     State/PfxRcd
10.0.12.2      4    200    74      75         6      0    0 01:05:55           0
10.0.13.3      4    200    71      73         6      0    0 01:05:43           0
10.0.14.4      4    400    69      71         6      0    0 01:05:26           0

Total number of neighbors 3, established neighbors 3

R1#
```

3．检查 BGP 路由传递情况，此时 BGP 路由优选下一跳地址 10.0.12.2。

```
R1#show    ip bgp
BGP table version is 7, local router ID is 1.1.1.1
Status codes: s suppressed, d damped, h history, * valid, > best, i - internal,
              S Stale, b - backup entry, m - multipath, f Filter, a additional-path
Origin codes: i - IGP, e - EGP, ? - incomplete

     Network         Next Hop         Metric      LocPrf      Weight Path
*b   192.168.10.0    10.0.13.3        0                       0 200 i
*>                   10.0.12.2        0                       0 200 i
```

```
Total number of prefixes 1
R1#
```

4．设置 MED 值后，选择 MED 值小的路径，MED 值越小，路径越优先。

```
R1#show    ip bgp
BGP table version is 8, local router ID is 1.1.1.1
Status codes: s suppressed, d damped, h history, * valid, > best, i - internal,
              S Stale, b - backup entry, m - multipath, f Filter, a additional-path
Origin codes: i - IGP, e - EGP, ? - incomplete

     Network          Next Hop          Metric      LocPrf      Weight Path
*>   192.168.10.0     10.0.13.3         0                       0 200 i
*b                    10.0.12.2         100                     0 200 i

Total number of prefixes 1
R1#
```

5．MED 值只在相邻的两个 AS 间传递，不再传递给下一个 AS，在 R4 设备上查询的 MED 值为 0。

```
R4#show    ip bgp
BGP table version is 4, local router ID is 4.4.4.4
Status codes: s suppressed, d damped, h history, * valid, > best, i - internal,
              S Stale, b - backup entry, m - multipath, f Filter, a additional-path
Origin codes: i - IGP, e - EGP, ? - incomplete

     Network          Next Hop          Metric      LocPrf      Weight Path
*>   192.168.10.0     10.0.14.1         0                       0 300 200 i

Total number of prefixes 1
R4#
```

> 问题与思考

在 BGP 路由上通过 MED 值执行路径选择时，如何比较去往同一个目的地但来自不同 AS 的路由？

1.1.6 Community 属性

> 原理

Community 属性（团体属性）是可选可传递属性，用于标识一些有相同性质的路由前缀，相当于给路由打标记，以便简化路由策略的执行。

Community 属性不限于一个网络或一个自治系统，没有物理边界，一条路由可以携带多个团体属性；而 Community 属性在 BGP 邻居之间默认不传递，需要针对邻居使用 neighbor x.x.x.x send-community 命令才能将 Community 属性发送出去。

Community 属性分为公认团体属性和私有团体属性，公认团体属性为事先约定好的一些团体属性，可以直接使用，而私有团体属性是根据需要而自定义的团体属性。

公认团体属性采用名称格式表示团体属性值。

公认团体属性	说　　明
internet	收到携带此属性的 BGP 路由，可以向任何 BGP 邻居发送此路由，默认情况下所有路由都属于此属性
local-AS	收到携带此属性的 BGP 路由，将不向 AS 外发送此路由，也不向 AS 内其他子 AS（BGP 联邦）发送此路由
no-advertise	收到携带此属性的 BGP 路由，将不向任何 BGP 邻居发送此路由
no-export	收到携带此属性的 BGP 路由，将不向任何 EBGP 邻居发送此路由（联邦 EBGP 仍会传递）

私有团体属性采用十进制整数格式，即 AA:NN 格式，其中 AA 表示 AS 号，NN 是自定义的编号，这种形式更易于记忆。

➢ **任务拓扑**

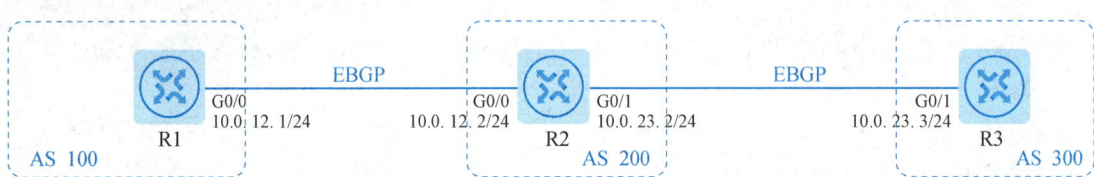

➢ **实施步骤**

1. 根据任务拓扑配置各设备接口 IP 地址。

```
Ruijie>enable
Password:ruijie
Ruijie#configure terminal
Ruijie(config)#hostname   R1
R1(config)#interface   gigabitEthernet 0/0
R1(config-if-GigabitEthernet 0/0)#no   switchport
R1(config-if-GigabitEthernet 0/0)#ip address    10.0.12.1 24
R1(config-if-GigabitEthernet 0/0)#exit
R1(config)#

Ruijie>enable
Password:ruijie
Ruijie#configure terminal
Ruijie(config)#hostname   R2
R2(config)#interface   gigabitEthernet 0/0
R2(config-if-GigabitEthernet 0/0)#no   switchport
R2(config-if-GigabitEthernet 0/0)#ip address    10.0.12.2 24
R2(config-if-GigabitEthernet 0/0)#exit
R2(config)#interface   gigabitEthernet 0/1
R2(config-if-GigabitEthernet 0/1)#no   switchport
R2(config-if-GigabitEthernet 0/1)#ip address    10.0.23.2 24
R2(config-if-GigabitEthernet 0/1)#exit
R2(config)#

Ruijie>enable
Password:ruijie
Ruijie#configure terminal
Ruijie(config)#hostname   R3
R3(config)#interface   gigabitEthernet 0/1
```

```
R3(config-if-GigabitEthernet 0/1)#no    switchport
R3(config-if-GigabitEthernet 0/1)#ip address    10.0.23.3 24
R3(config-if-GigabitEthernet 0/1)#exit
R3(config)#
```

2．在各设备之间建立 EBGP 邻居关系，手动配置 BGP 协议的 Router ID。

```
R1(config)#router   bgp   100
R1(config-router)#bgp   router-id   1.1.1.1
R1(config-router)#neighbor   10.0.12.2 remote-as   200
R1(config-router)#exit
R1(config)#

R2(config)#router bgp   200
R2(config-router)#bgp   router-id   2.2.2.2
R2(config-router)#neighbor   10.0.12.1 remote-as   100
R2(config-router)#neighbor   10.0.23.3 remote-as   300
R2(config-router)#exit
R2(config)#

R3(config)#router bgp   300
R3(config-router)#bgp   router-id   3.3.3.3
R3(config-router)#neighbor   10.0.23.2 remote-as   200
R3(config-router)#exit
R3(config)#
```

3．在 R1 设备创建 loopback 0 接口，并将该接口通告到 BGP 路由表中。

```
R1(config)#interface   loopback 0
R1(config-if-Loopback 0)#ip address   100.1.1.1 32
R1(config-if-Loopback 0)#exit
R1(config)#router   bgp   100
R1(config-router)#network   100.1.1.1 mask   255.255.255.255
R1(config-router)#exit
R1(config)#
```

4．通过 route-map 命令设置 Community 属性值时，需要指定邻居开启发送 Community 属性值的功能，否则无法发送 Community 属性值。不同公认团体属性值传递的范围不同，internet 属性可传递给所有 BGP 邻居。

```
R1(config)#access-list 1 permit   100.1.1.1 0.0.0.0
R1(config)#route-map Com permit   10
R1(config-route-map)#match   ip address 1
R1(config-route-map)#set   community   internet
R1(config-route-map)#exit
R1(config)#route-map Com permit   20
R1(config-route-map)#exit
R1(config)#router   bgp   100
R1(config-router)#neighbor   10.0.12.2 route-map   Com out
R1(config-router)#neighbor   10.0.12.2 send-community         //团体属性必须添加配置命令
R1(config-router)#exit
R1(config)#

R2(config)#router   bgp   200
R2(config-router)#neighbor   10.0.23.3 send-community
R2(config-router)#exit
R2(config)#
```

5. 将 Community 属性值设置为 local-AS，若收到的 BGP 路由携带此属性，则不再向 AS 外发送此路由。

```
R1(config)#access-list 1 permit    100.1.1.1 0.0.0.0
R1(config)#route-map Com permit    10
R1(config-route-map)#match    ip address 1
R1(config-route-map)#set    community    local-AS
R1(config-route-map)#exit
R1(config)#route-map Com permit    20
R1(config-route-map)#exit
R1(config)#
R1(config)#router    bgp    100
R1(config-router)#neighbor    10.0.12.2 route-map    Com out
R1(config-router)#exit
R1(config)#
```

➢ 任务验证

1. 检查 IP 地址配置情况。

```
R1#show    ip interface    brief
Interface          IP-Address(Pri)      IP-Address(Sec)    Status           Protocol
GigabitEthernet 0/0    10.0.12.1/24        no address        up               up
VLAN 1                  no address         no address        up               down
R1#

R2#show    ip interface    brief
Interface          IP-Address(Pri)      IP-Address(Sec)    Status           Protocol
GigabitEthernet 0/0    10.0.12.2/24        no address        up               up
GigabitEthernet 0/1    10.0.23.2/24        no address        up               up
VLAN 1                  no address         no address        up               down
R2#

R3#show    ip interface    brief
Interface          IP-Address(Pri)      IP-Address(Sec)    Status           Protocol
GigabitEthernet 0/1    10.0.23.3/24        no address        up               up
VLAN 1                  no address         no address        up               down
R3#
```

2. 检查 BGP 邻居关系建立情况。

```
R2#show    ip bgp    summary
For address family: IPv4 Unicast
BGP router identifier 2.2.2.2, local AS number 200
BGP table version is 1
0 BGP AS-PATH entries
0 BGP Community entries
0 BGP Prefix entries (Maximum-prefix:4294967295)

Neighbor       V      AS MsgRcvd MsgSent    TblVer    InQ OutQ Up/Down    State/PfxRcd
10.0.12.1      4      100      8        7         1      0    0 00:05:19          0
10.0.23.3      4      300      7        5         1      0    0 00:05:00          0

Total number of neighbors 2, established neighbors 2

R2#
```

3．检查 BGP 路由通告的路由信息。

```
R2#show    ip bgp
BGP table version is 5, local router ID is 2.2.2.2
Status codes: s suppressed, d damped, h history, * valid, > best, i - internal,
              S Stale, b - backup entry, m - multipath, f Filter, a additional-path
Origin codes: i - IGP, e - EGP, ? - incomplete

     Network           Next Hop            Metric       LocPrf     Weight Path
*>   100.1.1.1/32      10.0.12.1           0                       0 100 i

Total number of prefixes 1
R2#
```

4．检查 BGP 路由携带的 Community 属性，当 BGP 邻居开启发送 Community 属性值的功能后，BGP 路由携带 Community 属性，否则无法将 Community 属性传递给 BGP 邻居，internet 属性能通过 BGP 路由传递给所有的 BGP 邻居。

```
R2#show    ip bgp 100.1.1.1
BGP routing table entry for 100.1.1.1/32(#0x7f23d03e9cf0)
Paths: (1 available, best #1, table Default-IP-Routing-Table)
  Advertised to update-groups:
  1

  100
    10.0.12.1 from 10.0.12.1 (1.1.1.1)
       Origin IGP, metric 0, localpref 100, valid, external, best
    Community: internet
    Last update: Tue Jun 20 07:02:30 2023
    RX ID: 0,TX ID: 0
R2#

R3#show    ip bgp    100.1.1.1
BGP routing table entry for 100.1.1.1/32(#0x7fa168f04b30)
Paths: (1 available, best #1, table Default-IP-Routing-Table)
  Advertised to update-groups:
  1

  200 100
    10.0.23.2 from 10.0.23.2 (2.2.2.2)
       Origin IGP, metric 0, localpref 100, valid, external, best
    Community: internet
    Last update: Tue Jun 20 07:20:31 2023
    RX ID: 0,TX ID: 0
R3#
```

5．检查配置 local-AS 属性值后的结果，保证此路由不再传递给 AS 300 路由器。

```
R2#show    ip bgp    100.0.1.1
BGP routing table entry for 100.0.1.1/32(#0x7f3c247eecb8)
Paths: (1 available, best #1, table Default-IP-Routing-Table, not advertised outside local AS)
  Not advertised to any peer    //不通告给任何 BGP 邻居

  100
    10.0.12.1 from 10.0.12.1 (1.1.1.1)
       Origin IGP, metric 0, localpref 100, valid, external, best
```

```
                    Community: local-AS
                    Last update: Tue Jul 18 07:56:49 2023
                    RX ID: 0,TX ID: 0
R2#

R3#show ip bgp    100.0.1.1
% Network not in table
R3#
```

➢ 问题与思考

通过实操验证公认团体属性值 no-export 与 local-AS 两者的区别。

1.2 BGP 选路原则

1.2.1 选路原则 1——优选 Weight 值大

➢ 原理

Weight 属性只在本地有效，无法传递给其他 BGP 邻居。某路由携带的 Weight 值越大，则该路由越优先。本地始发路由的默认值是 32768，从其他 BGP 邻居过来的路由的默认值是 0。Weight 属性为锐捷厂商的私有属性。

➢ 任务拓扑

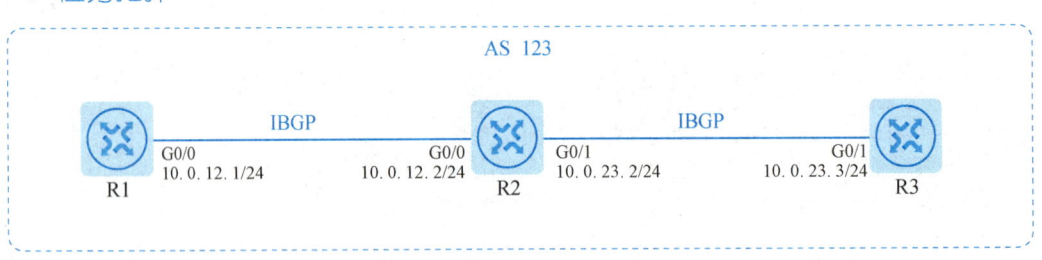

➢ 实施步骤

1. 根据任务拓扑配置各设备接口 IP 地址。

```
Ruijie>enable
Password:ruijie
Ruijie#configure terminal
Ruijie(config)#hostname   R1
R1(config)#interface    gigabitEthernet 0/0
R1(config-if-GigabitEthernet 0/0)#no    switchport
R1(config-if-GigabitEthernet 0/0)#ip address    10.0.12.1 24
R1(config-if-GigabitEthernet 0/0)#exit
R1(config)#interface    loopback 0
R1(config-if-Loopback 0)#ip address    10.1.1.1 32
R1(config-if-Loopback 0)#exit
```

```
Ruijie>enable
Password:ruijie
Ruijie#configure terminal
Ruijie(config)#hostname    R2
R2(config)#interface    gigabitEthernet 0/0
R2(config-if-GigabitEthernet 0/0)#no    switchport
R2(config-if-GigabitEthernet 0/0)#ip address    10.0.12.2 24
R2(config-if-GigabitEthernet 0/0)#exit
R2(config)#interface    gigabitEthernet 0/1
R2(config-if-GigabitEthernet 0/1)#no    switchport
R2(config-if-GigabitEthernet 0/1)#ip address    10.0.23.2 24
R2(config-if-GigabitEthernet 0/1)#exit
R2(config)#interface    loopback 0
R2(config-if-Loopback 0)#ip address    10.1.2.2 32
R2(config-if-Loopback 0)#exit
R2(config)#

Ruijie>enable
Password:ruijie
Ruijie#configure terminal
Ruijie(config)#hostname    R3
R3(config)#interface    gigabitEthernet 0/1
R3(config-if-GigabitEthernet 0/1)#no    switchport
R3(config-if-GigabitEthernet 0/1)#ip address    10.0.23.3 24
R3(config-if-GigabitEthernet 0/1)#exit
R3(config)#interface    loopback 0
R3(config-if-Loopback 0)#ip address    10.1.3.3 32
R3(config-if-Loopback 0)#exit
R3(config)#
```

2. 在 AS 内部通过逻辑接口建立 BGP 邻居关系，在 AS 内部配置 OSPF 协议，使 AS 内部逻辑接口能够互通。

```
R1(config)#router    ospf    10
R1(config-router)#router-id    1.1.1.1
Change router-id and update OSPF process! [yes/no]:y
R1(config-router)#network    10.0.12.0 0.0.0.255 area    0
R1(config-router)#network    10.1.1.1 0.0.0.0 area    0
R1(config-router)#exit
R1(config)#

R2(config)#router    ospf    10
R2(config-router)#router-id    2.2.2.2
Change router-id and update OSPF process! [yes/no]:y
R2(config-router)#network    10.0.12.0 0.0.0.255 area    0
R2(config-router)#network    10.0.23.0 0.0.0.255 area    0
R2(config-router)#network    10.1.2.2 0.0.0.0 area    0
R2(config-router)#exit
R2(config)#

R3(config)#router    ospf    10
R3(config-router)#router-id    3.3.3.3
Change router-id and update OSPF process! [yes/no]:y
R3(config-router)#network    10.0.23.0 0.0.0.255 area    0
R3(config-router)#network    10.1.3.3 0.0.0.0 area    0
R3(config-router)#exit
R3(config)#
```

3. 在 AS 内部配置 BGP 协议，通过 loopback 0 接口建立 IBGP 邻居关系。

```
R1(config)#router   bgp   123
R1(config-router)#bgp   router-id   1.1.1.1
R1(config-router)#neighbor   10.1.2.2 remote-as   100
R1(config-router)#neighbor   10.1.2.2 update-source   loopback 0
R1(config-router)#neighbor   10.1.2.2 remote-as   123
R1(config-router)#neighbor   10.1.2.2 update-source   loopback 0
R1(config-router)#exit
R1(config)#

R2(config)#router   bgp   123
R2(config-router)#bgp   router 2.2.2.2
R2(conflg-router)#neighbor   10.1.1.1 remote-as   123
R2(config-router)#neighbor   10.1.1.1 update-source loopback 0
R2(config-router)#neighbor   10.1.3.3 remote-as   123
R2(config-router)#neighbor   10.1.3.3 update-source   loopback 0
R2(config-router)#exit
R2(config)#

R3(config)#router   bgp   123
R3(config-router)#bgp   router-id   3.3.3.3
R3(config-router)#neighbor   10.1.2.2 remote-as   123
R3(config-router)#neighbor   10.1.2.2 update-source loopback 0
R3(config-router)#exit
R3(config)#
```

4. 在 R1 设备和 R3 设备分别创建 loopback 1 接口，并将两个接口通告到 BGP 路由表中。

```
R1(config)#interface   loopback 1
R1(config-if-Loopback 1)#ip address   200.1.1.1 32
R1(config-if-Loopback 1)#exit
R1(config)#
R1(config)#router   bgp   123
R1(config-router)#network   200.1.1.1 mask   255.255.255.255
R1(config-router)#exit
R1(config)#

R3(config)#interface   loopback 1
R3(config-if-Loopback 1)#ip address   200.1.1.1 32
R3(config-if-Loopback 1)#exit
R3(config)#router bgp   123
R3(config-router)#network   200.1.1.1 mask   255.255.255.255
R3(config-router)#exit
R3(config)#
```

5. 在 R2 设备通过 route-map 命令设置 Weight 值，使路由器根据 Weight 值选择最优路径。

```
R2(config)#access-list 1 permit   200.1.1.1 0.0.0.0
R2(config)#route-map Weight permit   10
R2(config-route-map)#match   ip address   1
R2(config-route-map)#set   weight   1000
R2(config-route-map)#exit
R2(config)#route-map   Weight permit   20
R2(config-route-map)#exit
```

```
R2(config)#router  bgp   123
R2(config-router)#neighbor   10.1.3.3 route-map Weight in
R2(config-router)#exit
R2(config)#
```

> 任务验证

1. 检查 IP 地址配置情况。

```
R1#show   ip interface   brief
Interface            IP-Address(Pri)      IP-Address(Sec)     Status          Protocol
GigabitEthernet 0/0  10.0.12.1/24         no address          up              up
Loopback 0           10.1.1.1/32          no address          up              up
Loopback 1           200.1.1.1/32         no address          up              up
VLAN 1               no address           no address          up              down
R1#


Interface            IP-Address(Pri)      IP-Address(Sec)     Status          Protocol
GigabitEthernet 0/0  10.0.12.2/24         no address          up              up
GigabitEthernet 0/1  10.0.23.2/24         no address          up              up
Loopback 0           10.1.2.2/32          no address          up              up
VLAN 1               no address           no address          up              down
R2#


R3#show   ip interface   brief
Interface            IP-Address(Pri)      IP-Address(Sec)     Status          Protocol
GigabitEthernet 0/1  10.0.23.3/24         no address          up              up
Loopback 0           10.1.3.3/32          no address          up              up
Loopback 1           200.1.1.1/32         no address          up              up
VLAN 1               no address           no address          up              down
R3#
```

2. 检查 OSPF 邻居关系建立情况及路由传递情况。

```
R2#show    ip ospf    neighbor

OSPF process 10, 2 Neighbors, 2 is Full:
Neighbor ID     Pri     State       BFD State    Dead Time   Address      Interface
1.1.1.1         1       Full/BDR    -            00:00:38    10.0.12.1    GigabitEthernet 0/0
3.3.3.3         1       Full/DR     -            00:00:36    10.0.23.3    GigabitEthernet 0/1
R2#


R1#show    ip route

Codes:   C - Connected, L - Local, S - Static
         R - RIP, O - OSPF, B - BGP, I - IS-IS, V - Overflow route
         N1 - OSPF NSSA external type 1, N2 - OSPF NSSA external type 2
         E1 - OSPF external type 1, E2 - OSPF external type 2
         SU - IS-IS summary, L1 - IS-IS level-1, L2 - IS-IS level-2
         IA - Inter area, EV - BGP EVPN, A - Arp to host
         LA - Local aggregate route
         * - candidate default

Gateway of last resort is no set
C        10.0.12.0/24 is directly connected, GigabitEthernet 0/0
C        10.0.12.1/32 is local host.
```

```
O      10.0.23.0/24 [110/2] via 10.0.12.2, 00:52:20, GigabitEthernet 0/0
C      10.1.1.1/32 is local host.
O      10.1.2.2/32 [110/1] via 10.0.12.2, 00:52:20, GigabitEthernet 0/0
O      10.1.3.3/32 [110/2] via 10.0.12.2, 00:51:53, GigabitEthernet 0/0
R1#
```

3．检查 BGP 邻居关系建立情况。

```
R2#show    ip bgp    summary
For address family: IPv4 Unicast
BGP router identifier 2.2.2.2, local AS number 123
BGP table version is 1
0 BGP AS-PATH entries
0 BGP Community entries
0 BGP Prefix entries (Maximum-prefix:4294967295)

Neighbor      V      AS     MsgRcvd MsgSent   TblVer  InQ OutQ Up/Down   State/PfxRcd
10.1.1.1      4      123    27      27        1       0   0    00:24:36  0
10.1.3.3      4      123    27      26        1       0   0    00:24:18  0

Total number of neighbors 2, established neighbors 2

R2#
```

4．检查 BGP 路由通告的路由信息。

```
R2#show ip bgp
BGP table version is 3, local router ID is 2.2.2.2
Status codes: s suppressed, d damped, h history, * valid, > best, i - internal,
              S Stale, b - backup entry, m - multipath, f Filter, a additional-path
Origin codes: i - IGP, e - EGP, ? - incomplete

     Network         Next Hop        Metric     LocPrf     Weight Path
*>i 200.1.1.1/32    10.1.1.1         0          100        0      i
*bi                 10.1.3.3         0          100        0      i

Total number of prefixes 1
R2#
```

5．检查 BGP 路由的最优路径，看其是否为 Weight 值最大的路径。

```
R2#show    ip bgp
BGP table version is 4, local router ID is 2.2.2.2
Status codes: s suppressed, d damped, h history, * valid, > best, i - internal,
              S Stale, b - backup entry, m - multipath, f Filter, a additional-path
Origin codes: i - IGP, e - EGP, ? - incomplete

     Network         Next Hop        Metric     LocPrf     Weight Path
*bi 200.1.1.1/32    10.1.1.1         0          100        0      i
*>i                 10.1.3.3         0          100        1000   i

Total number of prefixes 1
R2#
```

> 问题与思考

根据 Weight 值选择最优路径时，在 BGP 协议中可以通过哪几种方法进行设置？

1.2.2　选路原则 2——优选 Local Preference 值大

➤ 原理

Local Preference 属性只能在 IBGP 邻居之间传递，不能在 EBGP 邻居之间传递，如果在 EBGP 邻居之间收到的路由的路径中携带了 Local Preference 属性，则会触发 Notifacation 报文，造成会话中断。

默认情况下，本地始发路由的 Local Preference 值为 100，从 IBGP 邻居接收到路由的 Local Preference 值为 100，Local Preference 值越大，路径越优先。

Local Preference 属性可以通过以下三种方法进行设置：

（1）通过 IGP 路由引入 BGP 时关联 route-map 进行设置；

（2）针对 IBGP 邻居应用 IN/OUT 方向的 route-map，对从邻居接收到的或者通告给邻居的所有或部分路由进行设置；

（3）针对 EBGP 邻居应用 IN 方向的 route-map，对从邻居接收到的所有或部分路由进行设置。

➤ 任务拓扑

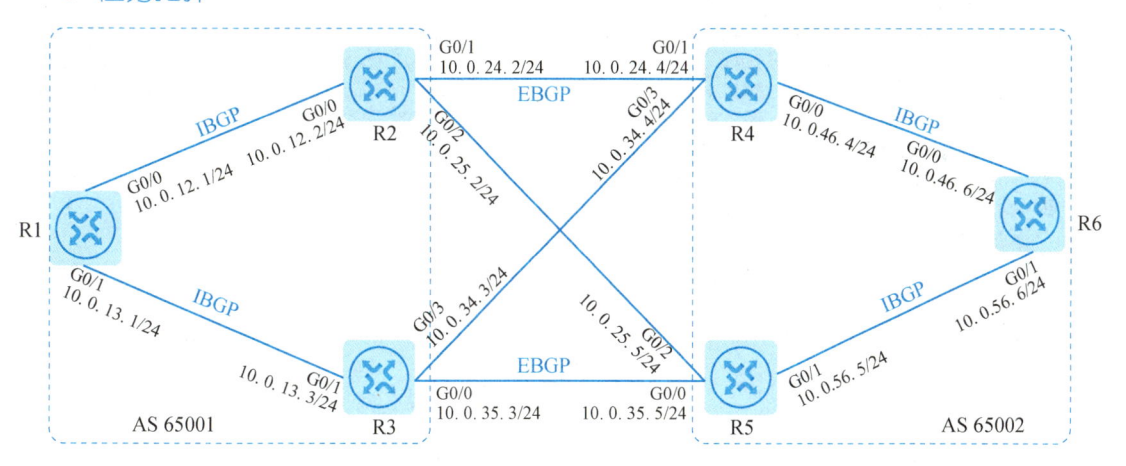

➤ 实施步骤

1. 根据任务拓扑配置各设备接口 IP 地址。

```
Ruijie>enable
Password:ruijie
Ruijie#configure terminal
Ruijie(config)#hostname    R1
R1(config)#interface    gigabitEthernet 0/0
R1(config-if-GigabitEthernet 0/0)#no    switchport
R1(config-if-GigabitEthernet 0/0)#ip address    10.0.12.1 24
R1(config-if-GigabitEthernet 0/0)#exit
R1(config)#interface    gigabitEthernet 0/1
R1(config-if-GigabitEthernet 0/1)#no    switchport
R1(config-if-GigabitEthernet 0/1)#ip address    10.0.13.1 24
R1(config-if-GigabitEthernet 0/1)#exit
```

```
Ruijie>enable
Password:ruijie
Ruijie#configure terminal
Ruijie(config)#hostname   R2
R2(config)#interface    gigabitEthernet 0/0
R2(config-if-GigabitEthernet 0/0)#no    switchport
R2(config-if-GigabitEthernet 0/0)#ip address    10.0.12.2 24
R2(config-if-GigabitEthernet 0/0)#exit
R2(config)#interface    gigabitEthernet 0/1
R2(config-if-GigabitEthernet 0/1)#no    switchport
R2(config-if-GigabitEthernet 0/1)#ip address    10.0.24.2 24
R2(config-if-GigabitEthernet 0/1)#exit
R2(config)#interface    gigabitEthernet 0/2
R2(config-if-GigabitEthernet 0/2)#no    switchport
R2(config-if-GigabitEthernet 0/2)#ip address    10.0.25.2 24
R2(config-if-GigabitEthernet 0/2)#exit
R2(config)#

Ruijie>enable
Password:ruijie
Ruijie#configure terminal
Ruijie(config)#hostname   R3
R3(config)#interface    gigabitEthernet 0/1
R3(config-if-GigabitEthernet 0/1)#no    switchport
R3(config-if-GigabitEthernet 0/1)#ip address    10.0.13.3 24
R3(config-if-GigabitEthernet 0/1)#exit
R3(config)#interface    gigabitEthernet 0/0
R3(config-if-GigabitEthernet 0/0)#no    switchport
R3(config-if-GigabitEthernet 0/0)#ip address    10.0.35.3 24
R3(config-if-GigabitEthernet 0/0)#exit
R3(config)#interface    gigabitEthernet 0/3
R3(config-if-GigabitEthernet 0/3)#no    switchport
R3(config-if-GigabitEthernet 0/3)#ip address    10.0.34.3 24
R3(config-if-GigabitEthernet 0/3)#exit
R3(config)#

Ruijie>enable
Password:ruijie
Ruijie#configure terminal
Ruijie(config)#hostname   R4
R4(config)#interface    gigabitEthernet 0/1
R4(config-if-GigabitEthernet 0/1)#no    switchport
R4(config-if-GigabitEthernet 0/1)#ip address    10.0.24.4 24
R4(config-if-GigabitEthernet 0/1)#exit
R4(config)#interface    gigabitEthernet 0/0
R4(config-if-GigabitEthernet 0/0)#no    switchport
R4(config-if-GigabitEthernet 0/0)#ip address    10.0.46.4 24
R4(config-if-GigabitEthernet 0/0)#exit
R4(config)#interface    gigabitEthernet 0/3
R4(config-if-GigabitEthernet 0/3)#no    switchport
R4(config-if-GigabitEthernet 0/3)#ip address    10.0.34.4 24
R4(config-if-GigabitEthernet 0/3)#exit
R4(config)#

Ruijie>enable
```

```
Password:ruijie
Ruijie#configure terminal
Ruijie(config)#hostname    R5
R5(config)#interface    gigabitEthernet 0/0
R5(config-if-GigabitEthernet 0/0)#no    switchport
R5(config-if-GigabitEthernet 0/0)#ip address    10.0.35.5 24
R5(config-if-GigabitEthernet 0/0)#exit
R5(config)#interface    gigabitEthernet 0/1
R5(config-if-GigabitEthernet 0/1)#no    switchport
R5(config-if-GigabitEthernet 0/1)#ip address    10.0.56.5 24
R5(config-if-GigabitEthernet 0/1)#exit
R5(config)#interface    gigabitEthernet 0/2
R5(config-if-GigabitEthernet 0/2)#no    switchport
R5(config-if-GigabitEthernet 0/2)#ip address    10.0.25.5 24
R5(config-if-GigabitEthernet 0/2)#exit
R5(config)#

Ruijie>enable
Password:ruijie
Ruijie#configure terminal
Ruijie(config)#hostname    R6
R6(config)#interface    gigabitEthernet 0/0
R6(config-if-GigabitEthernet 0/0)#no    switchport
R6(config-if-GigabitEthernet 0/0)#ip address    10.0.46.6 24
R6(config-if-GigabitEthernet 0/0)#exit
R6(config)#interface    gigabitEthernet 0/1
R6(config-if-GigabitEthernet 0/1)#no    switchport
R6(config-if-GigabitEthernet 0/1)#ip address    10.0.56.6 24
R6(config-if-GigabitEthernet 0/1)#exit
R6(config)#
```

2. BGP 协议配置，在 AS 65001 内部 R1、R2、R3 三个设备之间建立 IBGP 邻居关系，在 AS 65002 内部 R4、R5、R6 三个设备之间建立 IBGP 邻居关系，在 R2 设备与 R4、R5 两个设备之间建立 EBGP 邻居关系，在 R3 设备与 R4、R5 两个设备之间建立 EBGP 邻居关系。

```
R1(config)#router    bgp    65001
R1(config-router)#bgp    router-id    1.1.1.1
R1(config-router)#neighbor    10.0.12.2 remote-as    65001
R1(config-router)#neighbor    10.0.13.3 remote-as    65001
R1(config-router)#exit
R1(config)#

R2(config)#router    bgp    65001
R2(config-router)#bgp    router-id    2.2.2.2
R2(config-router)#neighbor    10.0.12.1 remote-as    65001
R2(config-router)#neighbor    10.0.12.1 next-hop-self
R2(config-router)#neighbor    10.0.24.4 remote-as    65002
R2(config-router)#neighbor    10.0.25.5 remote-as    65002
R2(config-router)#exit
R2(config)#

R3(config)#router    bgp    65001
R3(config-router)#bgp    router-id    3.3.3.3
R3(config-router)#neighbor    10.0.13.1 remote-as    65001
R3(config-router)#neighbor    10.0.13.1 next-hop-self
```

```
R3(config-router)#neighbor    10.0.34.4 remote-as      65002
R3(config-router)#neighbor    10.0.35.5 remote-as      65002
R3(config-router)#exit
R3(config)#

R4(config)#router bgp    65002
R4(config-router)#bgp    router-id    4.4.4.4
R4(config-router)#neighbor    10.0.24.2 remote-as      65001
R4(config-router)#neighbor    10.0.34.3 remote-as      65001
R4(config-router)#neighbor    10.0.46.6 remote-as      65002
R4(config-router)#neighbor    10.0.46.6 next-hop-self
R4(config-router)#exit
R4(config)#

R5(config)#router    bgp    65002
R5(config-router)#bgp    router-id    5.5.5.5
R5(config-router)#neighbor    10.0.35.3 remote-as      65001
R5(config-router)#neighbor    10.0.25.2 remote-as      65001
R5(config-router)#neighbor    10.0.56.6 remote-as      65002
R5(config-router)#neighbor    10.0.56.6 next-hop-self
R5(config-router)# exit
R5(config)#

R6(config)#router    bgp    65002
R6(config-router)#bgp    router-id    6.6.6.6
R6(config-router)#neighbor    10.0.46.4 remote-as      65002
R6(config-router)#neighbor    10.0.56.5 remote-as      65002
R6(config-router)#exit
R6(config)#
```

3．在 R1 设备和 R6 设备分别创建两个 loopback 接口，并分别将两个 loopback 接口的路由通告到 BGP 路由表中。

```
R1(config)#interface    loopback 0
R1(config-if-Loopback 0)#ip address    20.0.1.1 32
R1(config-if-Loopback 0)#exit
R1(config)#interface    loopback 1
R1(config-if-Loopback 1)#ip address    20.0.1.2 32
R1(config-if-Loopback 1)#exit
R1(config)#router    bgp    65001
R1(config-router)#network    20.0.1.1 mask    255.255.255.255
R1(config-router)#network    20.0.1.2 mask    255.255.255.255
R1(config-router)#exit
R1(config)#

R6(config)#interface    loopback 0
R6(config-if-Loopback 0)#ip address    30.0.1.1 32
R6(config-if-Loopback 0)#exit
R6(config)#interface    loopback 1
R6(config-if-Loopback 1)#ip address    30.0.1.2 32
R6(config-if-Loopback 1)#exit
R6(config)#router    bgp    65002
R6(config-router)#network    30.0.1.1    mask    255.255.255.255
R6(config-router)#network    30.0.1.2    mask    255.255.255.255
R6(config-router)#exit
R6(config)#
```

4. 在 R2 设备和 R3 设备分别对 R4 设备和 R5 设备应用 IN 方向的 route-map，用 R1 设备访问 30.0.1.1，选择 R1-R2-R4-R6 路径，接着访问 30.0.1.2，选择 R1-R3-R5-R6 路径。

```
R2(config)#access-list 1 permit   30.0.1.1 0.0.0.0
R2(config)#route-map From_R4 permit   10
R2(config-route-map)#match   ip address   1
R2(config-route-map)#set   local-preference 500
R2(config-route-map)#exit
R2(config)#route-map From_R4 permit   20
R2(config-route-map)#exit
R2(config)#router   bgp   65001
R2(config-router)#neighbor   10.0.24.4 route-map   From_R4 in
R2(config-router)#exit
R2(config)#

R3(config)#access-list 2 permit   30.0.1.2 0.0.0.0
R3(config)#route-map From_R5 permit   10
R3(config-route-map)#match   ip address   2
R3(config-route-map)#set   local-preference 500
R3(config-route-map)#exit
R3(config)#route-map From_R5 permit   20
R3(config-route-map)#exit
R3(config)#router bgp   65001
R3(config-router)#neighbor   10.0.35.5 route-map From_R5 in
R3(config-router)#exit
R3(config)#
```

5. 在 R4 设备和 R5 设备分别对 R2 设备和 R3 设备应用 IN 方向的 route-map，用 R6 设备访问 20.0.1.1，选择 R6-R4-R2-R1 路径，接着访问 20.0.1.2，选择 R6-R5-R3-R1 路径。

```
R4(config)#access-list 1 permit 20.0.1.1 0.0.0.0
R4(config)#route-map From_R2 permit   10
R4(config-route-map)#match   ip address   1
R4(config-route-map)#set   local-preference 400
R4(config-route-map)#exit
R4(config)#route-map   From_R2 permit   20
R4(config-route-map)#exit
R4(config)#router   bgp   65002
R4(config-router)#neighbor   10.0.24.2 route-map From_R2 in
R4(config-router)#exit
R4(config)#

R5(config)#access-list   1   permit   20.0.1.2 0.0.0.0
R5(config)#route-map From_R3 permit   10
R5(config-route-map)#match   ip address   1
R5(config-route-map)#set   local-preference   400
R5(config-route-map)#exit
R5(config)#route-map From_R3 permit   20
R5(config-route-map)#exit
R5(config)#router   bgp   65002
R5(config-router)#neighbor   10.0.35.3 route-map From_R3 in
R5(config-router)#exit
R5(config)#
```

> 任务验证

1. 检查 IP 地址配置情况。

```
R1#show    ip interface    brief
Interface              IP-Address(Pri)     IP-Address(Sec)     Status          Protocol
GigabitEthernet 0/0    10.0.12.1/24        no address          up              up
GigabitEthernet 0/1    10.0.13.1/24        no address          up              up
Loopback 0             20.0.1.1/32         no address          up              up
Loopback 1             20.0.1.2/32         no address          up              up
VLAN 1                 no address          no address          up              down
R1#

R2#show    ip interface    brief
Interface              IP-Address(Pri)     IP-Address(Sec)     Status          Protocol
GigabitEthernet 0/0    10.0.12.2/24        no address          up              up
GigabitEthernet 0/1    10.0.24.2/24        no address          up              up
GigabitEthernet 0/2    10.0.25.2/24        no address          up              up
VLAN 1                 no address          no address          up              down
R2#

R3#show    ip interface    brief
Interface              IP-Address(Pri)     IP-Address(Sec)     Status          Protocol
GigabitEthernet 0/0    10.0.35.3/24        no address          up              up
GigabitEthernet 0/1    10.0.13.3/24        no address          up              up
GigabitEthernet 0/3    10.0.34.3/24        no address          up              up
VLAN 1                 no address          no address          up              down
R3#

R4#show    ip interface    brief
Interface              IP-Address(Pri)     IP-Address(Sec)     Status          Protocol
GigabitEthernet 0/0    10.0.46.4/24        no address          up              up
GigabitEthernet 0/1    10.0.24.4/24        no address          up              up
GigabitEthernet 0/3    10.0.34.4/24        no address          up              up
VLAN 1                 no address          no address          up              down
R4#

R5#show    ip interface brief
Interface              IP-Address(Pri)     IP-Address(Sec)     Status          Protocol
GigabitEthernet 0/0    10.0.35.5/24        no address          up              up
GigabitEthernet 0/1    10.0.56.5/24        no address          up              up
GigabitEthernet 0/2    10.0.25.5/24        no address          up              up
VLAN 1                 no address          no address          up              down
R5#

R6#show    ip interface    brief
Interface              IP-Address(Pri)     IP-Address(Sec)     Status          Protocol
GigabitEthernet 0/0    10.0.46.6/24        no address          up              up
GigabitEthernet 0/1    10.0.56.6/24        no address          up              up
Loopback 0             30.0.1.1/32         no address          up              up
Loopback 1             30.0.1.2/32         no address          up              up
VLAN 1                 no address          no address          up              down
R6#
```

2. 检查 BGP 邻居关系建立情况。

```
R1#show    ip bgp    summary
For address family: IPv4 Unicast
BGP router identifier 1.1.1.1, local AS number 65001
BGP table version is 6
0 BGP AS-PATH entries
0 BGP Community entries
0 BGP Prefix entries (Maximum-prefix:4294967295)

Neighbor      V      AS  MsgRcvd MsgSent    TblVer   InQ OutQ Up/Down    State/PfxRcd
10.0.12.2     4    65001    22      19         5     0    0 00:13:43        0
10.0.13.3     4    65001    23      18         5     0    0 00:12:57        0

Total number of neighbors 2, established neighbors 2

R1#

R3#show    ip bgp    summary
For address family: IPv4 Unicast
BGP router identifier 3.3.3.3, local AS number 65001
BGP table version is 11
0 BGP AS-PATH entries
0 BGP Community entries
0 BGP Prefix entries (Maximum-prefix:4294967295)

Neighbor      V      AS  MsgRcvd MsgSent    TblVer   InQ OutQ Up/Down    State/PfxRcd
10.0.13.1     4    65001    21      23        11     0    0 00:14:17        0
10.0.34.4     4    65002    22      21        11     0    0 00:13:02        0
10.0.35.5     4    65002    21      19        11     0    0 00:12:14        0

Total number of neighbors 3, established neighbors 3

R3#

R4#show    ip bgp    summary
For address family: IPv4 Unicast
BGP router identifier 4.4.4.4, local AS number 65002
BGP table version is 11
0 BGP AS-PATH entries
0 BGP Community entries
0 BGP Prefix entries (Maximum-prefix:4294967295)

Neighbor      V      AS  MsgRcvd MsgSent    TblVer   InQ OutQ Up/Down    State/PfxRcd
10.0.24.2     4    65001    21      22        11     0    0 00:14:05        0
10.0.34.3     4    65001    23      21        11     0    0 00:13:22        0
10.0.46.6     4    65002    19      20         9     0    0 00:12:04        0

Total number of neighbors 3, established neighbors 3

R4#


R6#show    ip bgp    summary
For address family: IPv4 Unicast
BGP router identifier 6.6.6.6, local AS number 65002
BGP table version is 5
```

0 BGP AS-PATH entries
0 BGP Community entries
0 BGP Prefix entries (Maximum-prefix:4294967295)

Neighbor	V	AS	MsgRcvd	MsgSent	TblVer	InQ	OutQ	Up/Down	State/PfxRcd
10.0.46.4	4	65002	21	17	5	0	0	00:11:32	0
10.0.56.5	4	65002	21	16	5	0	0	00:11:26	0

Total number of neighbors 2, established neighbors 2

R6#

3. 在 R1 设备检查 BGP 路由关于 30.0.1.1/32 和 30.0.1.2/32 的通告情况。此时，在 R1 设备访问 30.0.1.1/32 和 30.0.1.2/32 网段，优选 R1-R2-R4-R6 路径。

```
R1#show    ip bgp
BGP table version is 8, local router ID is 1.1.1.1
Status codes: s suppressed, d damped, h history, * valid, > best, i - internal,
              S Stale, b - backup entry, m - multipath, f Filter, a additional-path
Origin codes: i - IGP, e - EGP, ? - incomplete

     Network          Next Hop         Metric      LocPrf     Weight Path
*>   20.0.1.1/32      0.0.0.0          0                      32768    i
*>   20.0.1.2/32      0.0.0.0          0                      32768    i
*bi  30.0.1.1/32      10.0.13.3        0           100        0 65002 i
*>i                   10.0.12.2        0           100        0 65002 i
*bi  30.0.1.2/32      10.0.13.3        0           100        0 65002 i
*>i                   10.0.12.2        0           100        0 65002 i

Total number of prefixes 4
R1#

R2#show    ip bgp
BGP table version is 15, local router ID is 2.2.2.2
Status codes: s suppressed, d damped, h history, * valid, > best, i - internal,
              S Stale, b - backup entry, m - multipath, f Filter, a additional-path
Origin codes: i - IGP, e - EGP, ? - incomplete

     Network          Next Hop         Metric      LocPrf     Weight Path
*>i  20.0.1.1/32      10.0.12.1        0           100        0        i
*>i  20.0.1.2/32      10.0.12.1        0           100        0        i
*b   30.0.1.1/32      10.0.25.5        0                      0 65002 i
*>                    10.0.24.4        0                      0 65002 i
*>   30.0.1.2/32      10.0.24.4        0                      0 65002 i
*b                    10.0.25.5        0                      0 65002 i

Total number of prefixes 4
R2#
```

4. 在 R6 设备检查 BGP 路由关于 20.0.1.1/32 和 20.0.1.2/32 的通告情况。此时，在 R6 设备访问 20.0.1.1/32 与 20.0.1.2/32 网段，优选 R6-R4-R2-R1 路径。

```
R6#show    ip bgp
BGP table version is 8, local router ID is 6.6.6.6
Status codes: s suppressed, d damped, h history, * valid, > best, i - internal,
              S Stale, b - backup entry, m - multipath, f Filter, a additional-path
```

Origin codes: i - IGP, e - EGP, ? - incomplete

	Network	Next Hop	Metric	LocPrf	Weight Path
*>i	20.0.1.1/32	10.0.46.4	0	100	0 65001 i
*bi		10.0.56.5	0	100	0 65001 i
*>i	20.0.1.2/32	10.0.46.4	0	100	0 65001 i
*bi		10.0.56.5	0	100	0 65001 i
*>	30.0.1.1/32	0.0.0.0	0		32768 i
*>	30.0.1.2/32	0.0.0.0	0		32768 i

Total number of prefixes 4
R6#

R4#show   ip bgp
BGP table version is 15, local router ID is 4.4.4.4
Status codes: s suppressed, d damped, h history, * valid, > best, i - internal,
              S Stale, b - backup entry, m - multipath, f Filter, a additional-path
Origin codes: i - IGP, e - EGP, ? - incomplete

	Network	Next Hop	Metric	LocPrf	Weight Path
*>	20.0.1.1/32	10.0.24.2	0		0 65001 i
*b		10.0.34.3	0		0 65001 i
*b	20.0.1.2/32	10.0.34.3	0		0 65001 i
*>		10.0.24.2	0		0 65001 i
*>i	30.0.1.1/32	10.0.46.6	0	100	0 i
*>i	30.0.1.2/32	10.0.46.6	0	100	0 i

Total number of prefixes 4
R4#

5. 选择 Local Preference 值大的路径访问。访问 30.0.1.1/32，优选 R1-R2-R4-R6 路径，访问 30.0.1.2/32，优选 R1-R3-R5-R6 路径。

R1#show   ip bgp
BGP table version is 8, local router ID is 1.1.1.1

Status codes: s suppressed, d damped, h history, * valid, > best, i - internal,
              S Stale, b - backup entry, m - multipath, f Filter, a additional-path
Origin codes: i - IGP, e - EGP, ? - incomplete

	Network	Next Hop	Metric	LocPrf	Weight Path
*>	20.0.1.1/32	0.0.0.0	0		32768 i
*>	20.0.1.2/32	0.0.0.0	0		32768 i
*bi	30.0.1.1/32	10.0.13.3	0	100	0 65002 i
*>i		10.0.12.2	0	500	0 65002 i
*>i	30.0.1.2/32	10.0.13.3	0	500	0 65002 i
*bi		10.0.12.2	0	100	0 65002 i

Total number of prefixes 4
R1#

R2#show ip bgp
BGP table version is 8, local router ID is 2.2.2.2
Status codes: s suppressed, d damped, h history, * valid, > best, i - internal,
              S Stale, b - backup entry, m - multipath, f Filter, a additional-path
Origin codes: i - IGP, e - EGP, ? - incomplete

```
    Network              Next Hop          Metric        LocPrf        Weight Path
*>i 20.0.1.1/32          10.0.12.1         0             100           0      i
*>i 20.0.1.2/32          10.0.12.1         0             100           0      i
*>  30.0.1.1/32          10.0.24.4         0             500                  0 65002 i
*b                       10.0.25.5         0                                  0 65002 i
*>  30.0.1.2/32          10.0.24.4         0                                  0 65002 i
*b                       10.0.25.5         0                                  0 65002 i

Total number of prefixes 4
R2#

R3#show ip bgp
BGP table version is 8, local router ID is 3.3.3.3
Status codes: s suppressed, d damped, h history, * valid, > best, i - internal,
              S Stale, b - backup entry, m - multipath, f Filter, a additional-path
Origin codes: i - IGP, e - EGP, ? - incomplete

    Network              Next Hop          Metric        LocPrf        Weight Path
*>i 20.0.1.1/32          10.0.13.1         0             100           0      i
*>i 20.0.1.2/32          10.0.13.1         0             100           0      i
*>  30.0.1.1/32          10.0.34.4         0                                  0 65002 i
*b                       10.0.35.5         0                                  0 65002 i
*b  30.0.1.2/32          10.0.34.4         0                                  0 65002 i
*>                       10.0.35.5         0             500                  0 65002 i

Total number of prefixes 4
R3#

R1#traceroute   30.0.1.1 source   20.0.1.1
   < press Ctrl+C to break >
Tracing the route to 30.0.1.1

  1         10.0.12.2       30 msec     19 msec    41 msec
  2         10.0.24.4       42 msec     32 msec    42 msec
  3         30.0.1.1        109 msec    46 msec    69 msec

R1#traceroute   30.0.1.2 source 20.0.1.2
   < press Ctrl+C to break >
Tracing the route to 30.0.1.2

  1         10.0.13.3       79 msec     38 msec    35 msec
  2         10.0.35.5       49 msec     48 msec    55 msec
  3         30.0.1.2        62 msec     58 msec    43 msec
R1#
```

6. 根据 Local Preference 值，在 R6 设备访问 20.0.1.1/32，选择 R6-R4-R2-R1 路径，接着访问 20.0.1.2/32，选择 R6-R5-R3-R1 路径，报文的来回路径一致。

```
R4#show ip bgp
BGP table version is 7, local router ID is 4.4.4.4
Status codes: s suppressed, d damped, h history, * valid, > best, i - internal,
              S Stale, b - backup entry, m - multipath, f Filter, a additional-path
Origin codes: i - IGP, e - EGP, ? - incomplete

    Network              Next Hop          Metric        LocPrf        Weight Path
```

```
*b   20.0.1.1/32     10.0.34.3            0                    0 65001 i
*>                   10.0.24.2            0        400         0 65001 i
*>   20.0.1.2/32     10.0.24.2            0                    0 65001 i
*b                   10.0.34.3            0                    0 65001 i
*>i  30.0.1.1/32     10.0.46.6            0        100       0    i
*>i  30.0.1.2/32     10.0.46.6            0        100       0    i
```

Total number of prefixes 4
R4#

R5#show ip bgp
BGP table version is 7, local router ID is 5.5.5.5
Status codes: s suppressed, d damped, h history, * valid, > best, i - internal,
              S Stale, b - backup entry, m - multipath, f Filter, a additional-path
Origin codes: i - IGP, e - EGP, ? - incomplete

```
     Network         Next Hop         Metric    LocPrf    Weight Path
*b   20.0.1.1/32     10.0.35.3            0                    0 65001 i
*>                   10.0.25.2            0                    0 65001 i
*b   20.0.1.2/32     10.0.25.2            0                    0 65001 i
*>                   10.0.35.3            0        400         0 65001 i
*>i  30.0.1.1/32     10.0.56.6            0        100       0    i
*>i  30.0.1.2/32     10.0.56.6            0        100       0    i
```

Total number of prefixes 4
R5#

R6#show   ip bgp
BGP table version is 4, local router ID is 6.6.6.6
Status codes: s suppressed, d damped, h history, * valid, > best, i - internal,
              S Stale, b - backup entry, m - multipath, f Filter, a additional-path
Origin codes: i - IGP, e - EGP, ? - incomplete

```
     Network         Next Hop         Metric    LocPrf    Weight Path
*>i  20.0.1.1/32     10.0.46.4            0        400         0 65001 i
*bi                  10.0.56.5            0        100         0 65001 i
*bi  20.0.1.2/32     10.0.46.4            0        100         0 65001 i
*>i                  10.0.56.5            0        400         0 65001 i
*>   30.0.1.1/32     0.0.0.0              0                32768    i
*>   30.0.1.2/32     0.0.0.0              0                32768    i
```

Total number of prefixes 4
R6#

R6#traceroute   20.0.1.1 source   30.0.1.1
  < press Ctrl+C to break >
Tracing the route to 20.0.1.1

```
1       10.0.46.4     20 msec    <1 msec    41 msec
2       10.0.24.2     55 msec    32 msec    60 msec
3       20.0.1.1      33 msec    55 msec    51 msec
```

R6#traceroute   20.0.1.2 source   30.0.1.2
  < press Ctrl+C to break >

```
Tracing the route to 20.0.1.2

1          10.0.56.5       34 msec      32 msec      9 msec
2          10.0.35.3       18 msec      56 msec      32 msec
3          20.0.1.2        36 msec      40 msec      46 msec
R6#
```

> 问题与思考

根据 BGP 选路原则 2 调整任务拓扑的路径访问，针对 20.0.1.2/32 的 BGP 路由设置互访路径（R1-R2-R5-R6），路径来回一致。

1.2.3 选路原则 3——优选 AS-Path 长度短

> 原理

当向 EBGP 邻居通告路由时，将自己的 AS 号加在 AS-Path 列表的最左端，该操作对 IBGP 邻居无法生效，这句话意味着路由更新在 AS 内部传递时，AS-Path 属性将会保持不变。

根据 AS-Path 属性的路径优选，将优选 AS-Path 路径长度最短的路由。

AS-Path 路径优选注意事项：

（1）可以通过对 EBGP 邻居使用 IN/OUT 方向的 route-map，使用 set as-path prepend 命令来添加 AS 号；

（2）route-map in 将 AS 号附加在原始 AS 号的左侧；

（3）route-map out 将 AS 号附加在原始 AS 号的右侧。

> 任务拓扑

> 实施步骤

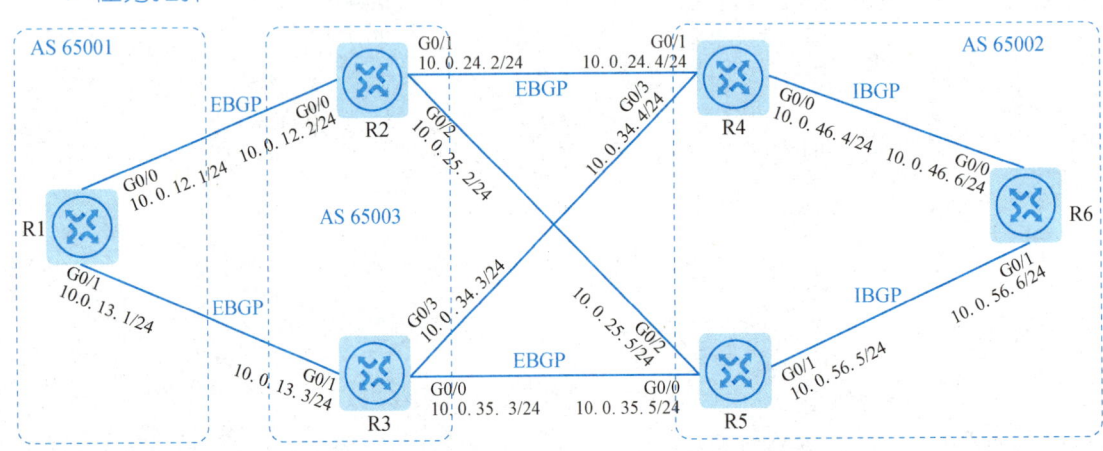

1. 根据任务拓扑配置各设备接口 IP 地址。

```
Ruijie>enable
Password:ruijie
Ruijie#configure terminal
Ruijie(config)#hostname    R1
R1(config)#interface    gigabitEthernet 0/0
```

```
R1(config-if-GigabitEthernet 0/0)#no    switchport
R1(config-if-GigabitEthernet 0/0)#ip address    10.0.12.1 24
R1(config-if-GigabitEthernet 0/0)#exit
R1(config)#interface    gigabitEthernet 0/1
R1(config-if-GigabitEthernet 0/1)#no    switchport
R1(config-if-GigabitEthernet 0/1)#ip address    10.0.13.1 24
R1(config-if-GigabitEthernet 0/1)#exit
R1(config)#

Ruijie>enable
Password:ruijie
Ruijie#configure terminal
Ruijie(config)#hostname    R2
R2(config)#interface    gigabitEthernet 0/0
R2(config-if-GigabitEthernet 0/0)#no    switchport
R2(config-if-GigabitEthernet 0/0)#ip address    10.0.12.2 24
R2(config-if-GigabitEthernet 0/0)#exit
R2(config)#interface    gigabitEthernet 0/1
R2(config-if-GigabitEthernet 0/1)#no    switchport
R2(config-if-GigabitEthernet 0/1)#ip address    10.0.24.2 24
R2(config-if-GigabitEthernet 0/1)#exit
R2(config)#interface    gigabitEthernet 0/2
R2(config-if-GigabitEthernet 0/2)#no    switchport
R2(config-if-GigabitEthernet 0/2)#ip address    10.0.25.2 24
R2(config-if-GigabitEthernet 0/2)#exit
R2(config)#

Ruijie>enable
Password:ruijie
Ruijie#configure terminal
Ruijie(config)#hostname    R3
R3(config)#interface    gigabitEthernet 0/1
R3(config-if-GigabitEthernet 0/1)#no    switchport
R3(config-if-GigabitEthernet 0/1)#ip address    10.0.13.3 24
R3(config-if-GigabitEthernet 0/1)#exit
R3(config)#interface    gigabitEthernet 0/0
R3(config-if-GigabitEthernet 0/0)#no    switchport
R3(config-if-GigabitEthernet 0/0)#ip address    10.0.35.3 24
R3(config-if-GigabitEthernet 0/0)#exit
R3(config)#interface    gigabitEthernet 0/3
R3(config-if-GigabitEthernet 0/3)#no    switchport
R3(config-if-GigabitEthernet 0/3)#ip address    10.0.34.3 24
R3(config-if-GigabitEthernet 0/3)#exit
R3(config)#

Ruijie>enable
Password:ruijie
Ruijie#configure terminal
Ruijie(config)#hostname    R4
R4(config)#interface    gigabitEthernet 0/1
R4(config-if-GigabitEthernet 0/1)#no    switchport
R4(config-if-GigabitEthernet 0/1)#ip address    10.0.24.4 24
R4(config-if-GigabitEthernet 0/1)#exit
R4(config)#interface    gigabitEthernet 0/0
R4(config-if-GigabitEthernet 0/0)#no    switchport
```

```
R4(config-if-GigabitEthernet 0/0)#ip address    10.0.46.4 24
R4(config-if-GigabitEthernet 0/0)#exit
R4(config)#interface    gigabitEthernet 0/3
R4(config-if-GigabitEthernet 0/3)#no    switchport
R4(config-if-GigabitEthernet 0/3)#ip address    10.0.34.4 24
R4(config-if-GigabitEthernet 0/3)#exit
R4(config)#


Ruijie>enable
Password:ruijie
Ruijie#configure terminal
Ruijie(config)#hostname    R5
R5(config)#interface    gigabitEthernet 0/0
R5(config-if-GigabitEthernet 0/0)#no    switchport
R5(config-if-GigabitEthernet 0/0)#ip address    10.0.25.5 24
R5(config-if-GigabitEthernet 0/0)#exit
R5(config)#interface    gigabitEthernet 0/0
R5(config-if-GigabitEthernet 0/0)#ip address    10.0.35.5 24
R5(config-if-GigabitEthernet 0/0)#exit
R5(config)#interface    gigabitEthernet 0/1
R5(config-if-GigabitEthernet 0/1)#no    switchport
R5(config-if-GigabitEthernet 0/1)#ip address    10.0.56.5 24
R5(config-if-GigabitEthernet 0/1)#exit
R5(config)#interface    gigabitEthernet 0/2
R5(config-if-GigabitEthernet 0/2)#no    switchport
R5(config-if-GigabitEthernet 0/2)#ip address    10.0.25.5 24
R5(config-if-GigabitEthernet 0/2)#exit
R5(config)#


Ruijie>enable
Password:ruijie
Ruijie#configure terminal
Ruijie(config)#hostname    R6
R6(config)#interface    gigabitEthernet 0/0
R6(config-if-GigabitEthernet 0/0)#no    switchport
R6(config-if-GigabitEthernet 0/0)#ip address    10.0.46.6 24
R6(config-if-GigabitEthernet 0/0)#exit
R6(config)#interface    gigabitEthernet 0/1
R6(config-if-GigabitEthernet 0/1)#no    switchport
R6(config-if-GigabitEthernet 0/1)#ip address    10.0.56.6 24
R6(config-if-GigabitEthernet 0/1)#exit
R6(config)#
```

2. 在各设备之间建立 BGP 邻居关系，在 AS 65002 内部通过物理接口建立 IBGP 邻居关系。

```
R1(config)#router    bgp    65001
R1(config-router)#bgp    router-id    1.1.1.1
R1(config-router)#neighbor    10.0.12.2 remote-as    65003
R1(config-router)#neighbor    10.0.13.3 remote-as    65003
R1(config-router)#exit
R1(config)#

R2(config)#router    bgp    65003
R2(config-router)#bgp    router-id    2.2.2.2
R2(config-router)#neighbor    10.0.12.1 remote-as    65001
```

```
R2(config-router)#neighbor    10.0.24.4 remote-as    65002
R2(config-router)#neighbor    10.0.25.5 remote-as    65002
R2(config-router)#exit
R2(config)#

R3(config)#router    bgp    65003
R3(config-router)#bgp router-id 3.3.3.3
R3(config-router)#neighbor    10.0.13.1 remote-as    65001
R3(config-router)#neighbor    10.0.35.5 remote-as    65002
R3(config-router)#neighbor    10.0.34.4 remote-as    65002
R3(config-router)#exit
R3(config)#

R4(config)#router    bgp    65002
R4(config-router)#bgp    router-id    4.4.4.4
R4(config-router)#neighbor    10.0.24.2 remote-as    65003
R4(config-router)#neighbor    10.0.34.3 remote-as    65003
R4(config-router)#neighbor    10.0.46.6 remote-as    65002
R4(config-router)#neighbor    10.0.46.6 next-hop-self
R4(config-router)#exit
R4(config)#

R5(config)#router    bgp    65002
R5(config-router)#bgp    router-id    5.5.5.5
R5(config-router)#neighbor    10.0.25.2 remote-as    65003
R5(config-router)#neighbor    10.0.35.3 remote-as    65003
R5(config-router)#neighbor    10.0.56.6 remote-as    65002
R5(config-router)#neighbor    10.0.56.6 next-hop-self
R5(config-router)#exit
R5(config)#

R6(config)#router    bgp    65002
R6(config-router)#bgp    router-id 6.6.6.6
R6(config-router)#neighbor    10.0.46.4 remote-as    65002
R6(config-router)#neighbor    10.0.56.5 remote-as    65002
R6(config-router)#exit
R6(config)#
```

3. 在 R1 设备和 R6 设备分别创建两个 loopback 接口，并将这些 loopback 接口通告到 BGP 路由表中。

```
R1(config)#interface    loopback 0
R1(config-if-Loopback 0)#ip address    10.0.1.1 32
R1(config-if-Loopback 0)#exit
R1(config)#interface    loopback 1
R1(config-if-Loopback 1)#ip address    11.0.1.1 32
R1(config-if-Loopback 1)#exit
R1(config)#router    bgp    65001
R1(config-router)#network    10.0.1.1 mask    255.255.255.255
R1(config-router)#network    11.0.1.1 mask    255.255.255.255
R1(config-router)#exit
R1(config)#


R6(config)#interface    loopback 0
R6(config-if-Loopback 0)#ip address    20.0.1.1 32
R6(config-if-Loopback 0)#exit
```

```
R6(config)#interface    loopback 1
R6(config-if-Loopback 1)#ip address    21.0.1.1 32
R6(config-if-Loopback 1)#exit
R6(config)#router   bgp    65002
R6(config-router)#network    20.0.1.1 mask    255.255.255.255
R6(config-router)#network    21.0.1.1 mask    255.255.255.255
R6(config-router)#exit
R6(config)#
```

4. 在 R2 设备和 R3 设备的 OUT 方向调用 route-map 修改 AS-Path 长度，根据选路原则优选 AS-Path 长度短的路径。

```
R2(config)#access-list 1 permit    21.0.1.1 0.0.0.0
R2(config)#route-map AS permit    10
R2(config-route-map)#match    ip address    1
R2(config-route-map)#set    as-path    prepend 100 200 300       //设置 AS Path 长度
R2(config-route-map)#exit
R2(config)#route-map    AS permit    20
R2(config-route-map)#exit
R2(config)#router    bgp    65003
R2(config-router)#neighbor 10.0.12.1 route-map AS out
R2(config-router)#exit
R2(config)#

R3(config)#access-list 2 permit    20.0.1.1 0.0.0.0
R3(config)#route-map AS2 permit    10
R3(config-route-map)#match    ip address    2
R3(config-route-map)#set    as-path prepend    200 300 400
R3(config-route-map)#exit
R3(config)#route-map AS2 permit    20
R3(config-route-map)#exit
R3(config)#router   bgp    65003
R3(config-router)#neighbor    10.0.13.1 route-map AS2 out
R3(config-router)#exit
R3(config)#
```

5. 在 R4 设备针对 21.0.1.1/32 路由设置 AS-Path，将 AS-Path 路径加长，使访问路径选择 R5 设备，即访问 21.0.1.1/32 的 R1-R3-R5-R6 路径设备。

```
R4(config)#access-list    1 permit    21.0.1.1 0.0.0.0
R4(config)#route-map    AS permit    10
R4(config-route-map)#match    ip address    1
R4(config-route-map)#set    as-path prepend    500 600
R4(config-route-map)#exit
R4(config)#route-map    AS permit    20
R4(config-route-map)#exit
R4(config)#router bgp    65002
R4(config-router)#neighbor    10.0.24.2 route-map AS out
R4(config-router)#exit
R4(config)#router bgp    65002
R4(config-router)#neighbor    10.0.34.3 route-map AS out
R4(config-router)#exit
R4(config)#

R5(config)#access-list    1 permit    21.0.1.1 0.0.0.0
R5(config)#route-map AS permit    10
```

```
R5(config-route-map)#match    ip address    1
R5(config-route-map)#set    as-path prepend    400 500
R5(config-route-map)#exit
R5(config)#route-map AS permit    20
R5(config-route-map)#exit
R5(config)#router    bgp    65002
R5(config-router)#neighbor    10.0.25.2 route-map    AS out
R5(config-router)#exit
R5(config)#
```

6. 在 R6 设备调用 route-map 的 IN 方向设置 BGP 的 AS-Path 属性，选择 AS-Path 长度短的路径访问目标。

```
R6(config)#access-list    1 permit    10.0.1.1 0.0.0.0
R6(config)#route-map AS1 permit    10
R6(config-route-map)#match    ip address    1
R6(config-route-map)#set    as-path prepend 30 40 50
R6(config-route-map)#exit
R6(config)#route-map    AS1    permit    20
R6(config-route-map)#exit
R6(config)#router bgp    65002
R6(config-router)#neighbor    10.0.46.4 route-map AS1 in
R6(config-router)#exit

R6(config)#access-list 2 permit    11.0.1.1 0.0.0.0
R6(config)#route-map AS2 permit    10
R6(config-route-map)#match    ip address    2
R6(config-route-map)#set    as-path prepend    10 20 30 40
R6(config-route-map)#exit
R6(config)#route-map AS permit    20
R6(config-route-map)#exit
R6(config)#router bgp    65002
R6(config-router)#neighbor    10.0.56.5 route-map    AS2 in
R6(config-router)#exit
R6(config)#
```

> 任务验证

1. 检查 IP 地址配置情况。

```
R1#show    ip interface    brief
Interface            IP-Address(Pri)       IP-Address(Sec)    Status         Protocol
GigabitEthernet 0/0   10.0.12.1/24         no address         up             up
GigabitEthernet 0/1   10.0.13.1/24         no address         up             up
Loopback 0            10.0.1.1/32          no address         up             up
Loopback 1            11.0.1.1/32          no address         up             up
VLAN 1                no address           no address         up             down
R1#

R2#show    ip interface    brief
Interface            IP-Address(Pri)       IP-Address(Sec)    Status         Protocol
GigabitEthernet 0/0   10.0.12.2/24         no address         up             up
GigabitEthernet 0/1   10.0.24.2/24         no address         up             up
GigabitEthernet 0/2   10.0.25.2/24         no address         up             up
VLAN 1                no address           no address         up             down
R2#
```

```
R3#show    ip interface   brief
Interface            IP-Address(Pri)     IP-Address(Sec)     Status          Protocol
GigabitEthernet 0/0  10.0.35.3/24        no address          up              up
GigabitEthernet 0/1  10.0.13.3/24        no address          up              up
GigabitEthernet 0/3  10.0.34.3/24        no address          up              up
VLAN 1               no address          no address          up              down
R3#


R4#show    ip interface   brief
Interface            IP-Address(Pri)     IP-Address(Sec)     Status          Protocol
GigabitEthernet 0/0  10.0.46.4/24        no address          up              up
GigabitEthernet 0/1  10.0.24.4/24        no address          up              up
GigabitEthernet 0/3  10.0.34.4/24        no address          up              up
VLAN 1               no address          no address          up              down
R4#


R5#show   ip interface   brief
Interface            IP-Address(Pri)     IP-Address(Sec)     Status          Protocol
GigabitEthernet 0/0  10.0.35.5/24        no address          up              up
GigabitEthernet 0/1  10.0.56.5/24        no address          up              up
GigabitEthernet 0/2  10.0.25.5/24        no address          up              up
VLAN 1               no address          no address          up              down
R5#


R6#show   ip interface   brief
Interface            IP-Address(Pri)     IP-Address(Sec)     Status          Protocol
GigabitEthernet 0/0  10.0.46.6/24        no address          up              up
GigabitEthernet 0/1  10.0.56.6/24        no address          up              up
Loopback 0           20.0.1.1/32         no address          up              up
Loopback 1           21.0.1.1/32         no address          up              up
VLAN 1               no address          no address          up              down
R6#
```

2. 检查 BGP 邻居关系建立情况。

```
R1#show    ip bgp   summary
For address family: IPv4 Unicast
BGP router identifier 1.1.1.1, local AS number 65001
BGP table version is 5
0 BGP AS-PATH entries
0 BGP Community entries
0 BGP Prefix entries (Maximum-prefix:4294967295)

Neighbor       V    AS  MsgRcvd MsgSent   TblVer  InQ OutQ Up/Down   State/PfxRcd
10.0.12.2      4   65003    231    228       5    0    0  03:43:50       0
10.0.13.3      4   65003    231    228       5    0    0  03:43:20       0

Total number of neighbors 2, established neighbors 2

R1#



R3#show    ip bgp   summary
For address family: IPv4 Unicast
```

BGP router identifier 3.3.3.3, local AS number 65003
BGP table version is 5
0 BGP AS-PATH entries
0 BGP Community entries
0 BGP Prefix entries (Maximum-prefix:4294967295)

Neighbor	V	AS	MsgRcvd	MsgSent	TblVer	InQ	OutQ	Up/Down	State/PfxRcd
10.0.13.1	4	65001	229	230	5	0	0	03:43:29	0
10.0.34.4	4	65002	228	228	5	0	0	03:42:49	0
10.0.35.5	4	65002	228	227	5	0	0	03:42:12	0

Total number of neighbors 3, established neighbors 3

R3#

R4#show ip bgp summary
For address family: IPv4 Unicast
BGP router identifier 4.4.4.4, local AS number 65002
BGP table version is 7
0 BGP AS-PATH entries
0 BGP Community entries
0 BGP Prefix entries (Maximum-prefix:4294967295)

Neighbor	V	AS	MsgRcvd	MsgSent	TblVer	InQ	OutQ	Up/Down	State/PfxRcd
10.0.24.2	4	65003	229	227	7	0	0	03:43:24	0
10.0.34.3	4	65003	229	226	7	0	0	03:43:12	0
10.0.46.6	4	65002	225	227	7	0	0	03:42:08	0

Total number of neighbors 3, established neighbors 3

R4#

R6#show ip bgp summary
For address family: IPv4 Unicast
BGP router identifier 6.6.6.6, local AS number 65002
BGP table version is 2
0 BGP AS-PATH entries
0 BGP Community entries
0 BGP Prefix entries (Maximum-prefix:4294967295)

Neighbor	V	AS	MsgRcvd	MsgSent	TblVer	InQ	OutQ	Up/Down	State/PfxRcd
10.0.46.4	4	65002	229	224	1	0	0	03:41:22	0
10.0.56.5	4	65002	233	223	1	0	0	03:41:20	0

Total number of neighbors 2, established neighbors 2

R6#

3. 检查 BGP 路由通告的路由信息，此时按照默认选路规则优选路径。

R1#show ip bgp
BGP table version is 12, local router ID is 1.1.1.1
Status codes: s suppressed, d damped, h history, * valid, > best, i - internal,
              S Stale, b - backup entry, m - multipath, f Filter, a additional-path

Origin codes: i - IGP, e - EGP, ? - incomplete

	Network	Next Hop	Metric	LocPrf	Weight Path
*>	10.0.1.1/32	0.0.0.0	0		32768 i
*>	11.0.1.1/32	0.0.0.0	0		32768 i
*b	20.0.1.1/32	10.0.13.3	0		0 65003 65002 i
*>		10.0.12.2	0		0 65003 65002 i
*b	21.0.1.1/32	10.0.13.3	0		0 65003 65002 i
*>		10.0.12.2	0		0 65003 65002 i

Total number of prefixes 4
R1#

R6#show   ip bgp
BGP table version is 4, local router ID is 6.6.6.6
Status codes: s suppressed, d damped, h history, * valid, > best, i - internal,
              S Stale, b - backup entry, m - multipath, f Filter, a additional-path
Origin codes: i - IGP, e - EGP, ? - incomplete

	Network	Next Hop	Metric	LocPrf	Weight Path
*>i	10.0.1.1/32	10.0.46.4	0	100	0 65003 65001 i
*bi		10.0.56.5	0	100	0 65003 65001 i
*>i	11.0.1.1/32	10.0.46.4	0	100	0 65003 65001 i
*bi		10.0.56.5	0	100	0 65003 65001 i
*>	20.0.1.1/32	0.0.0.0	0		32768 i
*>	21.0.1.1/32	0.0.0.0	0		32768 i

Total number of prefixes 4
R6#

4. 在 R1 设备检查 20.0.1.1/32 的 BGP 路由，其路径选择 AS-Path 长度短的路径。实验发现，在设备 OUT 方向调用 route-map，其 AS 号添加在 AS-Path 列表右侧，且因 AS-Path 列表增长转为备份路径。

R1#show   ip bgp
BGP table version is 13, local router ID is 1.1.1.1
Status codes: s suppressed, d damped, h history, * valid, > best, i - internal,
              S Stale, b - backup entry, m - multipath, f Filter, a additional-path
Origin codes: i - IGP, e - EGP, ? - incomplete

	Network	Next Hop	Metric	LocPrf	Weight Path
*>	10.0.1.1/32	0.0.0.0	0		32768 i
*>	11.0.1.1/32	0.0.0.0	0		32768 i
*b	20.0.1.1/32	10.0.13.3	0		0 65003 200 300 400 65002 i
*>		10.0.12.2	0		0 65003 65002 i
*>	21.0.1.1/32	10.0.13.3	0		0 65003 65002 i
*b		10.0.12.2	0		0 65003 100 200 300 65002 500 600 i

Total number of prefixes 4
R1#

R1#traceroute   21.0.1.1 source   11.0.1.1
 < press Ctrl+C to break >

```
Tracing the route to 21.0.1.1

 1       10.0.13.3       1 msec     <1 msec    2 msec
 2       10.0.13.3       <1 msec    <1 msec
         10.0.35.5       21 msec
 3       21.0.1.1        2 msec     3 msec     2 msec
R1#
```

5. 在 R6 设备向 R4 设备与 R5 设备的 IN 方向调用 route-map 设置 AS-Path，选择 AS-Path 长度短的路径访问。在设备 IN 方向调用 route-map，其 AS 号添加在 AS-Path 列表左侧。

```
R6#show    ip bgp
BGP table version is 4, local router ID is 6.6.6.6
Status codes: s suppressed, d damped, h history, * valid, > best, i - internal,
              S Stale, b - backup entry, m - multipath, f Filter, a additional-path
Origin codes: i - IGP, e - EGP, ? - incomplete

    Network            Next Hop         Metric      LocPrf     Weight Path
*bi 10.0.1.1/32        10.0.56.5        0           100        0 30 40 50 65003 65001 i
*>i                    10.0.46.4        0           100        0 65003 65001 i
*bi 11.0.1.1/32        10.0.46.4        0           100        0 10 20 30 40 65003 65001 i
*>i                    10.0.56.5        0           100        0 65003 65001 i
*>  20.0.1.1/32        0.0.0.0          0                      32768     i
*>  21.0.1.1/32        0.0.0.0          0                      32768     i

Total number of prefixes 4
R6#
```

> 问题与思考

如果在 BGP 进程下使用 bgp bestpath as-path ignore 命令跳过 AS-Path 的长度比较，那么该命令对 AS-Path 属性的环路检测是否生效？

1.2.4　选路原则 4——Origin 优选 i>e>?

> 原理

根据 Origin 属性选择最优路径，三种不同 Origin 属性优先顺序：IGP>EGP>incomplete。Origin 属性在 BGP 路由表中一直携带。

Origin 属性配置方式：

（1）将 IGP 路由重分布到 BGP 时关联 route-map 进行设置；

（2）默认情况下，由 network 方式产生的 BGP 路由的 Origin 属性为 IGP，而由 redistribute 方式产生的 BGP 路由的 Origin 属性为【?】；

（3）对 BGP 邻居应用 IN/OUT 方向的 route-map 进行设置。

➢ 任务拓扑

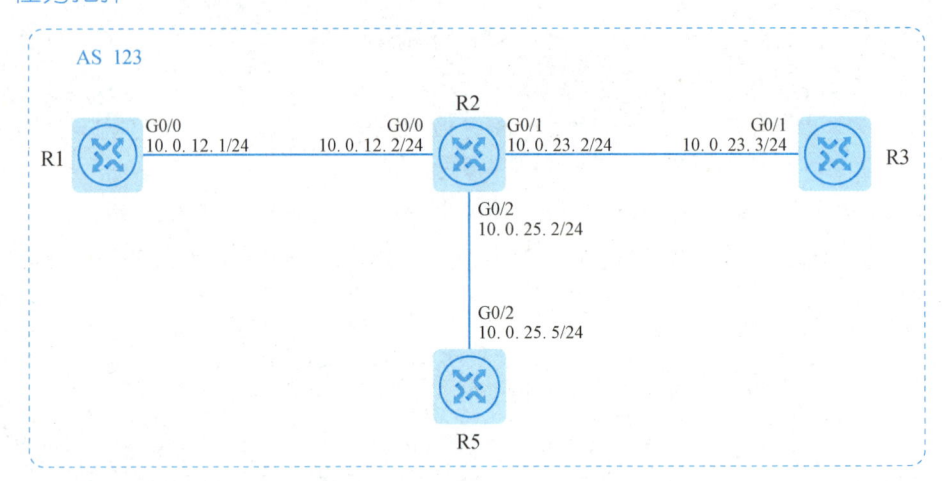

```
AS 123
                                      R2
           G0/0              G0/0    G0/1              G0/1
   R1      10.0.12.1/24      10.0.12.2/24  10.0.23.2/24      10.0.23.3/24    R3
                                     G0/2
                                     10.0.25.2/24

                                     G0/2
                                     10.0.25.5/24
                                      R5
```

➢ 实施步骤

1. 根据任务拓扑配置各设备接口 IP 地址，并分别在 R1 设备、R3 设备、R5 设备创建 loopback 0 接口，地址一致配置为 20.1.1.1/32。

```
Ruijie>enable
Password:ruijie
Ruijie#configure terminal
Ruijie(config)#hostname   R1
R1(config)#interface    gigabitEthernet 0/0
R1(config-if-GigabitEthernet 0/0)#no    switchport
R1(config-if-GigabitEthernet 0/0)#ip address    10.0.12.1 24
R1(config-if-GigabitEthernet 0/0)#exit
R1(config)#interface    loopback 0
R1(config-if-Loopback 0)#ip address 20.1.1.1 32
R1(config-if-Loopback 0)#exit
R1(config)#

Ruijie>enable
Password:ruijie
Ruijie#configure terminal
Ruijie(config)#hostname   R2
R2(config)#interface   gigabitEthernet 0/0
R2(config-if-GigabitEthernet 0/0)#no    switchport
R2(config-if-GigabitEthernet 0/0)#ip address    10.0.12.2 24
R2(config-if-GigabitEthernet 0/0)#exit
R2(config)#interface    gigabitEthernet 0/1
R2(config-if-GigabitEthernet 0/2)#no    switchport
R2(config-if-GigabitEthernet 0/2)#ip address    10.0.25.2 24
R2(config-if-GigabitEthernet 0/2)#exit
R2(config)#

Ruijie>enable
Password:ruijie
Ruijie#configure terminal
Ruijie(config)# hostname R3
R3(config)#interface    gigabitEthernet 0/1
```

```
R3(config-if-GigabitEthernet 0/1)#no    switchport
R3(config-if-GigabitEthernet 0/1)#ip address    10.0.23.3 24
R3(config-if-GigabitEthernet 0/1)#exit
R3(config)#interface    loopback 0
R3(config-if-Loopback 0)#ip address    20.1.1.1 32
R3(config-if-Loopback 0)#exit
R3(config)#

Ruijie>enable
Password:ruijie
Ruijie#configure terminal
Ruijie(config)#hostname    R5
R5(config)#interface    gigabitEthernet 0/2
R5(config-if-GigabitEthernet 0/2)#no switchport
R5(config-if-GigabitEthernet 0/2)#ip address    10.0.25.5 24
R5(config-if-GigabitEthernet 0/2)#exit
R5(config)#interface    loopback 0
R5(config-if-Loopback 0)#ip address    20.1.1.1 32
R5(config-if-Loopback 0)#exit
R5(config)#
```

2. 在 AS 123 内部通过物理接口建立 IBGP 邻居关系。

```
R1(config)#router    bgp    123
R1(config-router)#bgp    router-id    1.1.1.1
R1(config-router)#neighbor    10.0.12.2 remote-as    123
R1(config)#exit

R2(config)#router    bgp    123
R2(config-router)#bgp    router-id    2.2.2.2
R2(config-router)#neighbor 10.0.12.1 remote-as    123
R2(config-router)#neighbor    10.0.23.3 remote-as    123
R2(config-router)#neighbor    10.0.25.5 remote-as    123
R2(config-router)#exit
R2(config)#

R3(config)#router    bgp    123
R3(config-router)#bgp    router-id    3.3.3.3
R3(config-router)#neighbor    10.0.23.2 remote-as    123
R3(config-router)#exit
R3(config)#

R5(config)#router    bgp    123
R5(config-router)#bgp    router-id 5.5.5.5
R5(config-router)#neighbor    10.0.25.2 remote-as    123
R5(config-router)#exit
R5(config)#
```

3. 在 R1 设备、R3 设备、R5 设备通过 network 命令通告 BGP 路由 20.1.1.1/32。

```
R1(config)#router    bgp    123
R1(config-router)#network    20.1.1.1 mask    255.255.255.255
R1(config-router)#exit
R1(config)#

R3(config)#router    bgp    123
R3(config-router)#network    20.1.1.1 mask    255.255.255.255
```

```
R3(config-router)#exit
R3(config)#

R5(config)#router    bgp    123
R5(config-router)#network    20.1.1.1 mask    255.255.255.255
R5(config-router)#exit
R5(config)#
```

4. 在 R1 设备、R3 设备对 R2 设备使用 OUT 方向的 route-map，修改 Origin 属性分别为【?，e】（R5 设备始发的路由默认为 i）。

```
R1(config)#route-map    ruijie permit    10
R1(config-route-map)#set origin incomplete
R1(config-route-map)#exit
R1(config)#route-map ruijie deny 20
R1(config-route-map)#exit
R1(config)#router    bgp    123
R1(config-router)#neighbor    10.0.12.2 route-map    ruijie out
R1(config-router)#exit
R1(config)#

R3(config)#route-map ruijie permit    10
R3(config-route-map)#set origin egp
R3(config-route-map)#exit
R3(config)#route-map ruijie deny 20
R3(config-route-map)#exit
R3(config)#router    bgp    123
R3(config-router)#neighbor    10.0.23.2 route-map ruijie out
R3(config-router)#exit
R3(config)#
```

➢ 任务验证

1. 检查 BGP 邻居关系建立情况。

```
R2#show    ip bgp    summary
For address family: IPv4 Unicast
BGP router identifier 2.2.2.2, local AS number 123
BGP table version is 1
0 BGP AS-PATH entries
0 BGP Community entries
0 BGP Prefix entries (Maximum-prefix:4294967295)

Neighbor      V      AS MsgRcvd MsgSent    TblVer    InQ OutQ Up/Down    State/PfxRcd
10.0.12.1     4      123      5         5         1    0    0 00:02:01        0
10.0.23.3     4      123      4         3         1    0    0 00:01:37        0
10.0.25.5     4      123      4         3         1    0    0 00:01:19        0

Total number of neighbors 3, established neighbors 3

R2#
```

2. 检查 BGP 路由通告的路由信息是否都通过 network 命令通告路由，Origin 属性为【i】。

```
R2#show    ip bgp
BGP table version is 2, local router ID is 2.2.2.2
Status codes: s suppressed, d damped, h history, * valid, > best, i - internal,
```

```
              S Stale, b - backup entry, m - multipath, f Filter, a additional-path
Origin codes: i - IGP, e - EGP, ? - incomplete

     Network          Next Hop          Metric      LocPrf      Weight Path
 * i 20.1.1.1/32      10.0.25.5           0          100          0      i
 *bi                  10.0.23.3           0          100          0      i
 *>i                  10.0.12.1           0          100          0      i

Total number of prefixes 1
R2#
```

3. 不同 Origin 属性之间的优先顺序选择是 i>e>?，R2 设备优先选择 R5 设备产生的 BGP 路由，此路由会被加入到 IP 路由表中。

```
R2#show   ip bgp
BGP table version is 2, local router ID is 2.2.2.2
Status codes: s suppressed, d damped, h history, * valid, > best, i - internal,
              S Stale, b - backup entry, m - multipath, f Filter, a additional-path
Origin codes: i - IGP, e - EGP, ? - incomplete

     Network          Next Hop          Metric      LocPrf      Weight Path
 *>i 20.1.1.1/32      10.0.25.5           0          100          0      i
 *bi                  10.0.23.3           0          100          0      e
 * i                  10.0.12.1           0          100          0      ?

Total number of prefixes 1
R2#
```

> ➢ 问题与思考

在 R1 设备新创建 loopback 1 接口地址为 100.0.1.1/32，将此地址通过重分布方式通告到 BGP 路由表中，重分布方式仅通告 loopback 1 接口，并在 R2 设备接收到的 Origin 属性为【e】。

1.2.5　选路原则 5——优选 MED 值小

> ➢ 原理

MED 值在 AS 之间交换，发送给 EBGP 邻居后，仅在 EBGP 邻居所属的 AS 内部传递 MED 值，不再传递给下一个 AS，默认情况下，当路径来自同一个 AS 中的不同 EBGP 邻居时，路由器才会比较它们的 MED 属性，MED 值为 BGP 选路策略常用的一种路径属性。

默认情况下，BGP 只能比较来自同一个 AS 路由的 MED 值，可以使用 bgp always-compare-med 修改，修改后可以比较不同 AS 路由的 MED 值。

MED 属性在比较时优先选择 MED 值较小的 BGP 路由。MED 在 IBGP 邻居和 EBGP 邻居传递时置值不同。

1. 将 BGP 路由通告给 EBGP 邻居时：

（1）如果此 BGP 路由是本地始发（network 或 redistribute）的，则携带相应的 MED 值发送给 EBGP 邻居；

（2）如果此 BGP 路由是其他 BGP 邻居传递的，则将此路由通告给 EBGP 邻居时，不携带

MED 属性（不对 EBGP 邻居使用 route-map 时，将空白的 MED 值以 0 填充）。

2．将 BGP 路由通告给 IBGP 邻居时：

（1）如果此 BGP 路由携带 MED 值，则携带 MED 值发送给 IBGP 邻居；

（2）如果此 BGP 路由不携带 MED 值，则将 MED 值设置为 0，通告给 IBGP 邻居。

➢ 任务拓扑

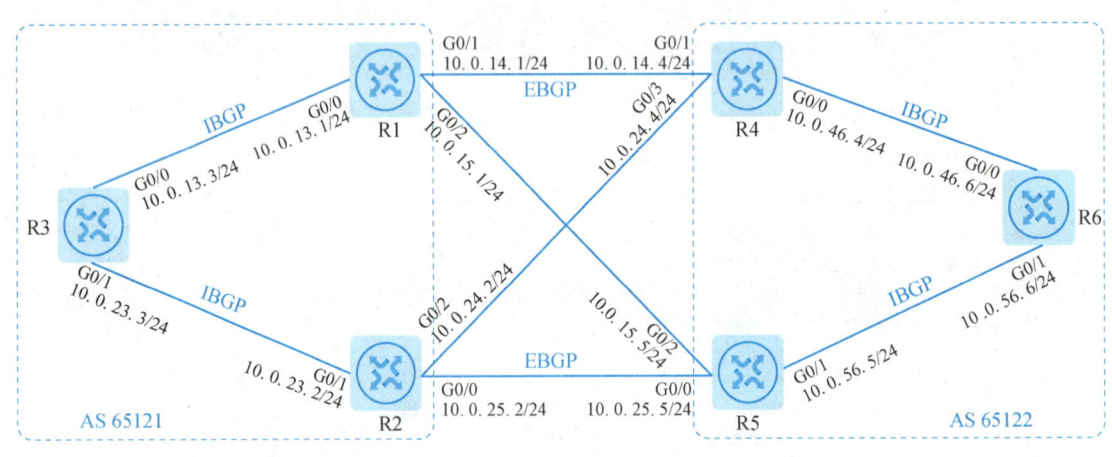

➢ 实施步骤

1．根据任务拓扑配置各设备接口 IP 地址，在 R6 设备创建 loopback 0 和 loopback 1 接口。

```
Ruijie>enable
Password:ruijie
Ruijie#configure terminal
Ruijie(config)#hostname   R1
R1(config)#interface   gigabitEthernet 0/0
R1(config-if-GigabitEthernet 0/0)#no   switchport
R1(config-if-GigabitEthernet 0/0)#ip address   10.0.13.1 24
R1(config-if-GigabitEthernet 0/0)#exit
R1(config)#interface   gigabitEthernet 0/1
R1(config-if-GigabitEthernet 0/1)#no   switchport
R1(config-if-GigabitEthernet 0/1)#ip address   10.0.14.1 24
R1(config-if-GigabitEthernet 0/1)#exit
R1(config)#interface   gigabitEthernet 0/2
R1(config-if-GigabitEthernet 0/2)#no   switchport
R1(config-if-GigabitEthernet 0/2)#ip address   10.0.15.1 24
R1(config-if-GigabitEthernet 0/2)#exit
R1(config)#

Ruijie>enable
Password:ruijie
Ruijie#configure terminal
Ruijie(config)#hostname   R2
R2(config)#interface   gigabitEthernet 0/1
R2(config-if-GigabitEthernet 0/1)#no switchport
R2(config-if-GigabitEthernet 0/1)#ip address   10.0.23.2 24
R2(config-if-GigabitEthernet 0/1)#exit
R2(config)#interface   gigabitEthernet 0/0
```

```
R2(config-if-GigabitEthernet 0/0)#no    switchport
R2(config-if-GigabitEthernet 0/0)#ip address    10.0.25.2 24
R2(config-if-GigabitEthernet 0/0)#exit
R2(config)#interface    gigabitEthernet 0/2
R2(config-if-GigabitEthernet 0/2)#no    switchport
R2(config-if-GigabitEthernet 0/2)#ip address    10.0.24.2 24
R2(config-if-GigabitEthernet 0/2)#exit
R2(config)#

Ruijie>enable
Password:ruijie
Ruijie#configure terminal
Ruijie(config)#hostname    R3
R3(config)#interface    gigabitEthernet 0/0
R3(config-if-GigabitEthernet 0/0)#no    switchport
R3(config-if-GigabitEthernet 0/0)#ip address    10.0.13.3 24
R3(config-if-GigabitEthernet 0/0)#exit
R3(config)#interface    gigabitEthernet 0/1
R3(config-if-GigabitEthernet 0/1)#no    switchport
R3(config-if-GigabitEthernet 0/1)#ip address    10.0.23.3 24
R3(config-if-GigabitEthernet 0/1)#exit
R3(config)#

Ruijie>enable
Password:ruijie
Ruijie#configure terminal
Ruijie(config)#hostname    R4
R4(config)#interface    gigabitEthernet 0/1
R4(config-if-GigabitEthernet 0/1)#no    switchport
R4(config-if-GigabitEthernet 0/1)#ip address    10.0.14.4 24
R4(config-if-GigabitEthernet 0/1)#exit
R4(config)#interface    gigabitEthernet 0/2
R4(config-if-GigabitEthernet 0/2)#no    switchport
R4(config-if-GigabitEthernet 0/2)#ip address    10.0.24.4 24
R4(config-if-GigabitEthernet 0/2)#exit
R4(config)#interface    gigabitEthernet 0/0
R4(config-if-GigabitEthernet 0/0)#no    switchport
R4(config-if-GigabitEthernet 0/0)#ip address    10.0.46.4 24
R4(config-if-GigabitEthernet 0/0)#exit
R4(config)#

Ruijie>enable
Password:ruijie
Ruijie#configure terminal
Ruijie(config)#hostname    R5
R5(config)#interface    gigabitEthernet 0/0
R5(config-if-GigabitEthernet 0/0)#no    switchport
R5(config-if-GigabitEthernet 0/0)#ip address    10.0.25.5 24
R5(config-if-GigabitEthernet 0/0)#exit
R5(config)#interface    gigabitEthernet 0/1
R5(config-if-GigabitEthernet 0/1)#no    switchport
R5(config-if-GigabitEthernet 0/1)#ip address    10.0.56.5 24
R5(config-if-GigabitEthernet 0/1)#exit
R5(config)#interface    gigabitEthernet 0/2
R5(config-if-GigabitEthernet 0/2)#no    switchport
```

```
R5(config-if-GigabitEthernet 0/2)#ip address    10.0.15.5 24
R5(config-if-GigabitEthernet 0/2)#exit
R5(config)#


Ruijie>enable
Password:ruijie
Ruijie#configure terminal
Ruijie(config)#hostname   R6
R6(config)#interface    gigabitEthernet 0/0
R6(config-if-GigabitEthernet 0/0)#no switchport
R6(config-if-GigabitEthernet 0/0)#ip address    10.0.46.6 24
R6(config-if-GigabitEthernet 0/0)#exit
R6(config)#interface    gigabitEthernet 0/1
R6(config-if-GigabitEthernet 0/1)#no   switchport
R6(config-if-GigabitEthernet 0/1)#ip address    10.0.56.6 24
R6(config-if-GigabitEthernet 0/1)#exit
R6(config)#interface    loopback 0
R6(config-if-Loopback 0)#ip address    20.0.1.1 32
R6(config-if-Loopback 0)#exit
R6(config)#interface    loopback 1
R6(config-if-Loopback 1)#ip address    20.0.2.1 32
R6(config-if-Loopback 1)#exit
R6(config)#
```

2．在各设备之间建立 BGP 邻居关系。

```
R1(config)#router    bgp   65121
R1(config-router)#bgp    router-id    1.1.1.1
R1(config-router)#neighbor    10.0.13.3 remote-as    65121
R1(config-router)#neighbor    10.0.15.5 remote-as    65122
R1(config-router)#neighbor    10.0.14.4 remote-as 65122
R1(config-router)#neighbor    10.0.13.3 next-hop-self
R1(config-router)#exit
R1(config)#


R2(config)#router    bgp   65121
R2(config-router)#bgp    router-id    2.2.2.2
R2(config-router)#neighbor    10.0.23.3 remote-as    65121
R2(config-router)#neighbor    10.0.23.3 next-hop-self
R2(config-router)#neighbor    10.0.24.4 remote-as 65122
R2(config-router)#neighbor    10.0.25.5 remote-as    65122
R2(config-router)#exit
R2(config)#


R3(config)#router    bgp   65121
R3(config-router)#bgp    router-id    3.3.3.3
R3(config-router)#neighbor    10.0.13.1 remote-as    65121
R3(config-router)#neighbor    10.0.23.2 remote-as    65121
R3(config-router)#exit
R3(config)#


R4(config)#router    bgp   65122
R4(config-router)#bgp    router-id    4.4.4.4
R4(config-router)#neighbor    10.0.14.1 remote-as    65121
R4(config-router)#neighbor    10.0.24.2 remote-as    65121
R4(config-router)#neighbor    10.0.46.6 remote-as    65122
```

```
R4(config-router)#neighbor    10.0.46.6 next-hop-self
R4(config-router)#exit
R4(config)#


R5(config)#router    bgp   65122
R5(config-router)#bgp    router-id    5.5.5.5
R5(config-router)#neighbor    10.0.15.1 remote-as    65121
R5(config-router)#neighbor    10.0.25.2 remote-as    65121
R5(config-router)#neighbor    10.0.56.6 remote-as 65122
R5(config-router)#neighbor    10.0.56.6 next-hop-self
R5(config-router)#exit
R5(config)#



R6(config)#router    bgp   65122
R6(config-router)#bgp    router-id    6.6.6.6
R6(config-router)#neighbor    10.0.46.4 remote-as    65122
R6(config-router)#neighbor    10.0.56.5 remote-as    65122
R6(config-router)#exit
R6(config)#
```

3. 在 R6 设备将 loopback 0 和 loopback 1 接口通告到 BGP 路由表中。

```
R6(config)#router    bgp   65122
R6(config-router)#network    20.0.1.1 mask    255.255.255.255
R6(config-router)#network    20.0.2.1 mask    255.255.255.255
R6(config-router)#exit
R6(config)#
```

4. 在 R4 设备和 R5 设备分别对 R1 设备、R2 设备应用 OUT 方向的 route-map 设置路由，根据 MED 值选路，MED 值越小，路径越优先。

```
R4(config)#access-list 1 permit    20.0.1.1 0.0.0.0
R4(config)#access-list 2 permit    20.0.2.1 0.0.0.0
R4(config)#route-map To_R2 permit    10
R4(config-route-map)#set metric 200
R4(config-route-map)#exit
R4(config)#route-map To_R1 permit    10
R4(config-route-map)#match    ip address    1
R4(config-route-map)#set    metric 100
R4(config-route-map)#exit
R4(config)#route-map To_R1 permit    20
R4(config-route-map)#match    ip address    2
R4(config-route-map)#set metric 300
R4(config-route-map)#exit
R4(config)#

R4(config)#router    bgp   65122
R4(config-router)#neighbor    10.0.14.1 route-map    To_R1 out
R4(config-router)#neighbor    10.0.24.2 route-map    To_R2 out
R4(config-router)#exit
R4(config)#

R5(config)#access-list    1 permit    20.0.1.1 0.0.0.0
R5(config)#access-list 2 permit    20.0.2.1 0.0.0.0
R5(config)#route-map To_R2 permit    10
```

```
R5(config-route-map)#match    ip address   1
R5(config-route-map)#set    metric 300
R5(config-route-map)#exit
R5(config)#route-map To_R2 permit   20
R5(config-route-map)#match    ip address   2
R5(config-route-map)#set    metric 100
R5(config-route-map)#exit
R5(config)#route-map    To_R1 permit    10
R5(config-route-map)#set    metric 200
R5(config-route-map)#exit
R5(config)#router    bgp   65122
R5(config-router)#neighbor    10.0.15.1 route-map To_R1 out
R5(config-router)#neighbor    10.0.25.2 route-map   To_R2 out
R5(config-router)#exit
R5(config)#
```

> 任务验证

1. 检查 BGP 邻居关系建立情况。

```
R1#show    ip bgp    summary
For address family: IPv4 Unicast
BGP router identifier 1.1.1.1, local AS number 65121
BGP table version is 3
1 BGP AS-PATH entries
0 BGP Community entries
1 BGP Prefix entries (Maximum-prefix:4294967295)

Neighbor       V      AS MsgRcvd MsgSent    TblVer    InQ OutQ Up/Down    State/PfxRcd
10.0.13.3      4      65121     44      45         3    0   0 00:41:35        0
10.0.14.4      4      65122     45      43         3    0   0 00:40:34        0
10.0.15.5      4      65122     44      43         3    0   0 00:40:05        1

Total number of neighbors 3, established neighbors 3

R1#

R3#show    ip bgp    summary
For address family: IPv4 Unicast
BGP router identifier 3.3.3.3, local AS number 65121
BGP table version is 2
0 BGP AS-PATH entries
0 BGP Community entries
0 BGP Prefix entries (Maximum-prefix:4294967295)

Neighbor       V      AS MsgRcvd MsgSent    TblVer    InQ OutQ Up/Down    State/PfxRcd
10.0.13.1      4      65121     47      42         1    0   0 00:42:04        0
10.0.23.2      4      65121     46      41         1    0   0 00:41:31        0

Total number of neighbors 2, established neighbors 2

R3#


R5#show    ip bgp    summary
For address family: IPv4 Unicast
```

```
BGP router identifier 5.5.5.5, local AS number 65122
BGP table version is 3
0 BGP AS-PATH entries
0 BGP Community entries
0 BGP Prefix entries (Maximum-prefix:4294967295)

Neighbor        V    AS    MsgRcvd  MsgSent   TblVer  InQ OutQ Up/Down    State/PfxRcd
10.0.15.1       4    65121    45      43       3      0   0    00:40:51        0
10.0.25.2       4    65121    44      42       3      0   0    00:40:43        0
10.0.56.6       4    65122    45      40       1      0   0    00:40:13        0

Total number of neighbors 3, established neighbors 3

R5#


R6#show   ip bgp   summary
For address family: IPv4 Unicast
BGP router identifier 6.6.6.6, local AS number 65122
BGP table version is 5
0 BGP AS-PATH entries
0 BGP Community entries
0 BGP Prefix entries (Maximum-prefix:4294967295)

Neighbor        V    AS    MsgRcvd  MsgSent   TblVer  InQ OutQ Up/Down    State/PfxRcd
10.0.46.4       4    65122    43      45       5      0   0    00:40:29        0
10.0.56.5       4    65122    43      44       5      0   0    00:40:22        0

Total number of neighbors 2, established neighbors 2

R6
```

2．检查 BGP 路由通告的路由信息，此时 20.0.1.1/32 的路由根据默认选路规则优选 10.0.13.1（R1 设备）传递路由。

```
R3#show   ip bgp
BGP table version is 2, local router ID is 3.3.3.3
Status codes: s suppressed, d damped, h history, * valid, > best, i - internal,
              S Stale, b - backup entry, m - multipath, f Filter, a additional-path
Origin codes: i - IGP, e - EGP, ? - incomplete

     Network           Next Hop           Metric     LocPrf      Weight Path
*bi 20.0.1.1/32       10.0.23.2            0          100         0 65122 i
*>i                   10.0.13.1            0          100         0 65122 i

Total number of prefixes 1
R3#
```

3．根据 route-map 命令设置 MED 值，此时 BGP 路由选路优选 MED 值小的路径访问目标地址。

```
R1#show   ip   bgp
BGP table version is 10, local router ID is 1.1.1.1
Status codes: s suppressed, d damped, h history, * valid, > best, i - internal,
              S Stale, b - backup entry, m - multipath, f Filter, a additional-path
Origin codes: i - IGP, e - EGP, ? - incomplete
```

	Network	Next Hop	Metric	LocPrf	Weight Path
*b	20.0.1.1/32	10.0.15.5	200		0 65122 i
*>		10.0.14.4	100		0 65122 i
*b	20.0.2.1/32	10.0.14.4	300		0 65122 i
*>		10.0.15.5	200		0 65122 i

Total number of prefixes 2
R1#

R2#show   ip bgp
BGP table version is 10, local router ID is 2.2.2.2
Status codes: s suppressed, d damped, h history, * valid, > best, i - internal,
              S Stale, b - backup entry, m - multipath, f Filter, a additional-path
Origin codes: i - IGP, e - EGP, ? - incomplete

	Network	Next Hop	Metric	LocPrf	Weight Path
*b	20.0.1.1/32	10.0.25.5	300		0 65122 i
*>		10.0.24.4	200		0 65122 i
*b	20.0.2.1/32	10.0.24.4	200		0 65122 i
*>		10.0.25.5	100		0 65122 i

Total number of prefixes 2
R2#

R3#show   ip bgp
BGP table version is 2, local router ID is 3.3.3.3
Status codes: s suppressed, d damped, h history, * valid, > best, i - internal,
              S Stale, b - backup entry, m - multipath, f Filter, a additional-path
Origin codes: i - IGP, e - EGP, ? - incomplete

	Network	Next Hop	Metric	LocPrf	Weight Path
*bi	20.0.1.1/32	10.0.23.2	200	100	0 65122 i
*>i		10.0.13.1	100	100	0 65122 i
*>i	20.0.2.1/32	10.0.23.2	100	100	0 65122 i
*bi		10.0.13.1	200	100	0 65122 i

Total number of prefixes 2
R3#

> 问题与思考

1. 将以下拓扑图完成配置。AS 1 内通过物理接口配置 IBGP 邻居，AS 1 和 AS 2 内通过物理接口配置 IBGP 邻居，AS 1 和 AS 2 配置 EBGP 邻居。

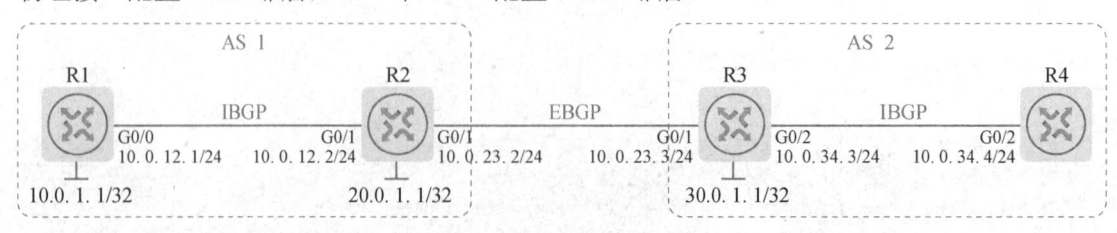

2. 将 R1 设备、R2 设备、R3 设备的 loopback 接口通过 network 命令通告到 BGP 路由表中。

3．通过 route-map 命令设置 R3 设备 loopback 接口的 BGP 路由的 MED 值为 301，观察 BGP 路由选择路径的情况。

4．将 R1 设备的 loopback 接口修改为 redistribute 方式通告到 BGP 路由表中，写明通过 network 方式和 redistribute 方式通告的 BGP 路由的区别。

1.2.6 选路原则 6——优选 EBGP 路由

➢ 原理

EBGP 路由指从 EBGP 邻居接收到的 BGP 路由，联盟 EBGP 路由指在 AS 内部建立联盟 EBGP 邻居传递的 BGP 路由，IBGP 路由指在 AS 内部建立 IBGP 邻居传递的路由。

在一些特殊的网络拓扑中无法通过以上 5 条选路原则选择最优路由时，会根据 EBGP 传递来的路由来进行优选，该原则用来执行路径优选策略较少。

➢ 任务拓扑

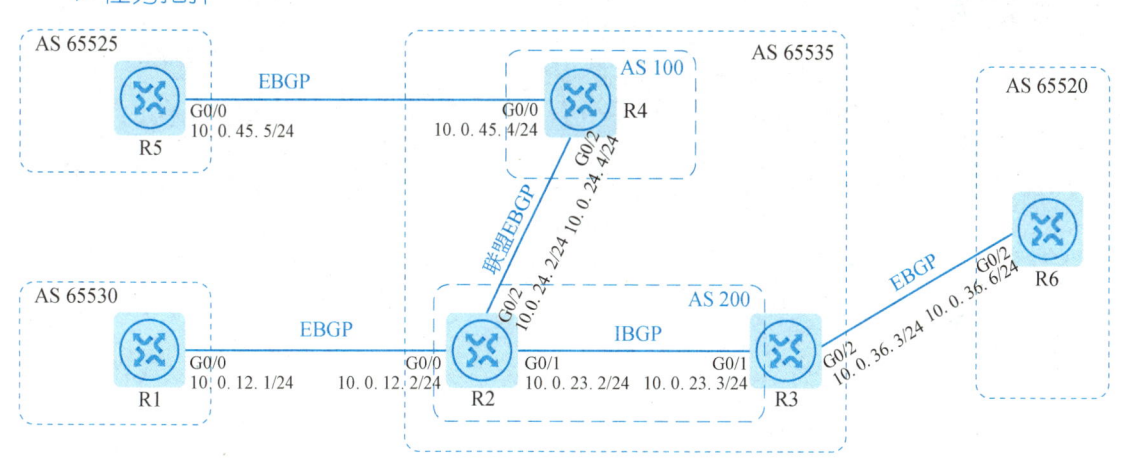

➢ 实施步骤

1．根据任务拓扑配置各设备接口 IP 地址，并分别在 R1 设备、R5 设备、R6 设备创建 loopback 0 接口，地址一致配置为 20.0.1.1/32。

```
Ruijie>enable
Password:ruijie
Ruijie#configure terminal
Ruijie(config)#hostname   R1
R1(config)#interface   gigabitEthernet 0/0
R1(config-if-GigabitEthernet 0/0)#no   switchport
R1(config-if-GigabitEthernet 0/0)#ip address   10.0.12.1 24
R1(config-if-GigabitEthernet 0/0)#exit
R1(config)#interface   loopback 0
R1(config-if-Loopback 0)#ip address   20.0.1.1 32
R1(config-if-Loopback 0)#exit
R1(config)#

Ruijie>enable
Password:ruijie
```

```
Ruijie#configure terminal
Ruijie(config)#hostname    R2
R2(config)#interface    gigabitEthernet 0/2
R2(config-if-GigabitEthernet 0/2)#no    switchport
R2(config-if-GigabitEthernet 0/2)#ip address    10.0.24.2 24
R2(config-if-GigabitEthernet 0/2)#exit
R2(config)#interface    gigabitEthernet 0/0
R2(config-if-GigabitEthernet 0/0)#no    switchport
R2(config-if-GigabitEthernet 0/0)#ip address    10.0.12.2 24
R2(config-if-GigabitEthernet 0/0)#exit
R2(config)#interface    gigabitEthernet 0/1
R2(config-if-GigabitEthernet 0/1)#no    switchport
R2(config-if-GigabitEthernet 0/1)#ip address    10.0.23.2 24
R2(config-if-GigabitEthernet 0/1)#exit
R2(config)#

Ruijie>enable
Password:ruijie
Ruijie#configure terminal
Ruijie(config)#hostname    R3
R3(config)#interface    gigabitEthernet 0/1
R3(config-if-GigabitEthernet 0/1)#no    switchport
R3(config-if-GigabitEthernet 0/1)#ip address    10.0.23.3 24
R3(config-if-GigabitEthernet 0/1)#exit
R3(config)#interface    gigabitEthernet 0/2
R3(config-if-GigabitEthernet 0/2)#no    switchport
R3(config-if-GigabitEthernet 0/2)#ip address    10.0.36.3 24
R3(config-if-GigabitEthernet 0/2)#exit
R3(config)#

Ruijie>enable
Password:ruijie
Ruijie#configure terminal
Ruijie(config)# hostname    R4
R4(config)#interface    gigabitEthernet 0/0
R4(config-if-GigabitEthernet 0/0)#no    switchport
R4(config-if-GigabitEthernet 0/0)#ip address    10.0.45.4 24
R4(config-if-GigabitEthernet 0/0)#exit
R4(config)#interface    gigabitEthernet 0/2
R4(config-if-GigabitEthernet 0/2)#no    switchport
R4(config-if-GigabitEthernet 0/2)#ip address    10.0.24.4 24
R4(config-if-GigabitEthernet 0/2)#exit
R4(config)#

Ruijie>enable
Password:ruijie
Ruijie#configure terminal
Ruijie(config)#hostname    R5
R5(config)#interface    gigabitEthernet 0/0
R5(config-if-GigabitEthernet 0/0)#no    switchport
R5(config-if-GigabitEthernet 0/0)#ip address    10.0.45.5 24
R5(config-if-GigabitEthernet 0/0)#exit
R5(config)#interface    loopback 0
R5(config-if-Loopback 0)#ip address    20.0.1.1 32
R5(config-if-Loopback 0)#exit
```

R5(config)#

Ruijie>enable
Password:ruijie
Ruijie#configure terminal
Ruijie(config)#hostname   R6
R6(config)#interface   gigabitEthernet 0/2
R6(config-if-GigabitEthernet 0/2)#no   switchport
R6(config-if-GigabitEthernet 0/2)#ip address   10.0.36.6 24
R6(config-if-GigabitEthernet 0/2)#exit
R6(config)#interface   loopback 0
R6(config-if-Loopback 0)#ip address   20.0.1.1 32
R6(config-if-Loopback 0)#exit
R6(config)#

2. 在各 AS 的设备之间建立 BGP 邻居关系，在 AS 65535 内部配置 BGP 联盟（关于 BGP 联盟在后续实验会详细介绍）。

```
R1(config)#router    bgp    65530
R1(config-router)#bgp    router-id    1.1.1.1
R1(config-router)#neighbor    10.0.12.2 remote-as 65535
R1(config-router)#exit
R1(config)#

R2(config)#router    bgp    200
R2(config-router)#bgp    router-id    2.2.2.2
R2(config-router)#bgp    confederation identifier    65535
R2(config-router)#neighbor    10.0.12.1 remote-as    65530
R2(config-router)#exit
R2(config)#router    bgp    200
R2(config-router)#neighbor    10.0.23.3 remote-as    200
R2(config-router)#exit
R2(config)#router    bgp    200
R2(config-router)#bgp    confederation    peers    100
R2(config-router)#neighbor    10.0.24.4 remote-as    100
R2(config-router)#exit
R2(config)#

R3(config)#router    bgp    200
R3(config-router)#bgp    confederation    identifier 65535
R3(config-router)#neighbor    10.0.23.2 remote-as    200
R3(config-router)#neighbor    10.0.36.6 remote-as    65520
R3(config-router)#exit
R3(config)#

R4(config)#router    bgp    100
R4(config-router)#bgp    router-id    4.4.4.4
R4(config-router)#bgp    confederation    identifier    65535
R4(config-router)#bgp    confederation    peers    200
R4(config-router)#neighbor    10.0.24.2 remote-as    200
R4(config-router)#neighbor    10.0.45.5 remote-as    65525
R4(config)#

R5(config)#router    bgp    65525
R5(config-router)#bgp    router-id    5.5.5.5
R5(config-router)#neighbor    10.0.45.5 remote-as    65535
```

```
R5(config-router)#neighbor    10.0.45.4 remote-as    65535
R5(config-router)#exit
R5(config)#

R6(config)#router    bgp    65520
R6(config-router)#bgp    router-id    6.6.6.6
R6(config-router)#neighbor    10.0.36.3 remote-as    65535
R6(config-router)#exit
R6(config)#
```

3. 在 R1 设备、R5 设备、R6 设备将 loopback 0 接口通告到 BGP 路由表中。

```
R1(config)#router    bgp    65530
R1(config-router)#network    20.0.1.1 mask    255.255.255.255
R1(config-router)#exit
R1(config)#

R5(config)#router    bgp    65525
R5(config-router)#network    20.0.1.1 mask    255.255.255.255
R5(config-router)#exit
R5(config)#

R6(config)#router    bgp    65520
R6(config-router)#network    20.0.1.1 mask    255.255.255.255
R6(config-router)#exit
R6(config)#
```

➢ 任务验证

1. 检查 BGP 邻居关系建立情况。

```
R2#show    ip bgp    summary
For address family: IPv4 Unicast
BGP router identifier 2.2.2.2, local AS number 200
BGP table version is 1
0 BGP AS-PATH entries
0 BGP Community entries
0 BGP Prefix entries (Maximum-prefix:4294967295)

Neighbor        V     AS MsgRcvd MsgSent    TblVer   InQ OutQ Up/Down    State/PfxRcd
10.0.12.1       4     65530    57       55         1    0    0 00:53:09       0
10.0.23.3       4     200      55       53         1    0    0 00:52:33       0
10.0.24.4       4     100      47       45         1    0    0 00:43:10       0

Total number of neighbors 3, established neighbors 3

R2#


R3#show    ip bgp    summary
For address family: IPv4 Unicast
BGP router identifier 10.0.36.3, local AS number 200
BGP table version is 1
0 BGP AS-PATH entries
0 BGP Community entries
0 BGP Prefix entries (Maximum-prefix:4294967295)
```

```
Neighbor           V     AS MsgRcvd MsgSent    TblVer   InQ OutQ Up/Down    State/PfxRcd
10.0.23.2          4     200    55       54        1    0    0 00:52:58    0
10.0.36.6          4     65520  19       17        1    0    0 00:15:59    0

Total number of neighbors 2, established neighbors 2

R3#


R4#show    ip bgp    summary
For address family: IPv4 Unicast
BGP router identifier 4.4.4.4, local AS number 100
BGP table version is 1
0 BGP AS-PATH entries
0 BGP Community entries
0 BGP Prefix entries (Maximum-prefix:4294967295)

Neighbor           V     AS MsgRcvd MsgSent    TblVer   InQ OutQ Up/Down    State/PfxRcd
10.0.24.2          4     200    46       45        1    0    0 00:43:23    0
10.0.45.5          4     65525  10       9         1    0    0 00:05:57    0

Total number of neighbors 2, established neighbors 2

R4#
```

2.检查 BGP 路由根据各设备通告的路由信息,在 R2 设备检查此路由的优选路径是否是从 EBGP 邻居(R1 设备)传递的路由。

```
R2#show    ip bgp
BGP table version is 2, local router ID is 2.2.2.2
Status codes: s suppressed, d damped, h history, * valid, > best, i - internal,
              S Stale, b - backup entry, m - multipath, f Filter, a additional-path
Origin codes: i - IGP, e - EGP, ? - incomplete

     Network           Next Hop          Metric      LocPrf      Weight Path
* i 20.0.1.1/32       10.0.36.6         0           100         0 65520 i
*                     10.0.45.5         0           100         0 (100) 65525 i
*>                    10.0.12.1         0                       0 65530 i

Total number of prefixes 1
R2#
```

> 问题与思考

在任务拓扑中 AS 65535 内部的联盟 EBGP 邻居传递的 EBGP 与 IBGP 传递的 BGP 路由是否同理?

1.2.7 选路原则 7——优选最近的 IGP 邻居通告

> 原理

BGP 选路原则的第 7 条,优选到距离 BGP 下一跳地址最近的 IGP 邻居,即比较邻居更新源地址在本地 IGP 路由表中的 Metric 值，Metric 值越小，路径越优先。

➤ 任务拓扑

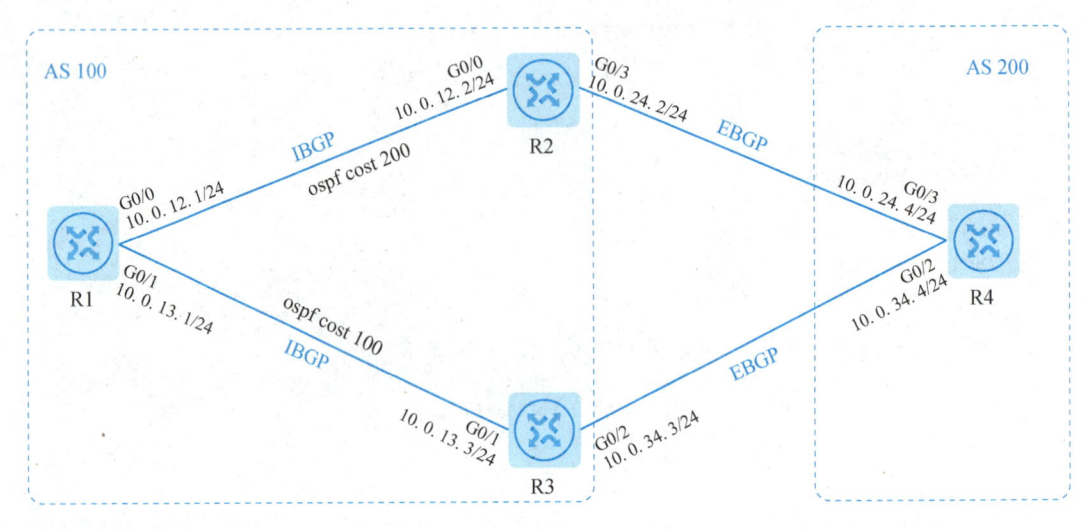

➤ 实施步骤

1. 根据任务拓扑配置各设备接口 IP 地址，在 R4 设备创建 loopback 0 接口地址 20.0.1.1/32。

```
Ruijie>enable
Password:ruijie
Ruijie#configure terminal
Ruijie(config)#hostname   R1
R1(config)#interface   loopback 0
R1(config-if-Loopback 0)#ip address 1.1.1.1 32
R1(config-if-Loopback 0)#exit
R1(config)#interface   gigabitEthernet 0/0
R1(config-if-GigabitEthernet 0/0)#no switchport
R1(config-if-GigabitEthernet 0/0)#ip address   10.0.12.1 24
R1(config-if-GigabitEthernet 0/0)#exit
R1(config)#interface   gigabitEthernet 0/1
R1(config-if-GigabitEthernet 0/1)#no   switchport
R1(config-if-GigabitEthernet 0/1)#ip address   10.0.13.1 24
R1(config-if-GigabitEthernet 0/1)#exit
R1(config)#

Ruijie>enable
Password:ruijie
Ruijie#configure terminal
Ruijie(config)#hostname   R2
R2(config)#interface   gigabitEthernet 0/0
R2(config-if-GigabitEthernet 0/0)#no   switchport
R2(config-if-GigabitEthernet 0/0)#ip address   10.0.12.2 24
R2(config-if-GigabitEthernet 0/0)#exit
R2(config)#interface   gigabitEthernet 0/3
R2(config-if-GigabitEthernet 0/3)#no   switchport
R2(config-if-GigabitEthernet 0/3)#ip address   10.0.24.2 24
R2(config-if-GigabitEthernet 0/3)#exit
```

```
R2(config)#

Ruijie>enable
Password:ruijie
Ruijie#configure terminal
Ruijie(config)#hostname   R3
R3(config)#interface   loopback 0
R3(config-if-Loopback 0)#ip address    3.3.3.3 32
R3(config-if-Loopback 0)#exit
R3(config)#interface   gigabitEthernet 0/1
R3(config-if-GigabitEthernet 0/1)#no   switchport
R3(config-if-GigabitEthernet 0/1)#ip address    10.0.13.3 24
R3(config-if-GigabitEthernet 0/1)#exit
R3(config)#interface   gigabitEthernet 0/2
R3(config-if-GigabitEthernet 0/2)#no   switchport
R3(config-if-GigabitEthernet 0/2)#ip address    10.0.34.3 24
R3(config-if-GigabitEthernet 0/2)#exit
R3(config)#

Ruijie>enable
Password:ruijie
Ruijie#configure terminal
Ruijie(config)#hostname   R4
R4(config)#interface   gigabitEthernet 0/3
R4(config-if-GigabitEthernet 0/3)#no   switchport
R4(config-if-GigabitEthernet 0/3)#ip address    10.0.24.4 24
R4(config-if-GigabitEthernet 0/3)#exit
R4(config)#interface   gigabitEthernet 0/2
R4(config-if-GigabitEthernet 0/2)#no   switchport
R4(config-if-GigabitEthernet 0/2)#ip address    10.0.34.4 24
R4(config-if-GigabitEthernet 0/2)#exit
R4(config)#interface   loopback 1
R4(config-if-Loopback 0)#ip address    20.0.1.1 32
R4(config-if-Loopback 0)#exit
R4(config)#
```

2. 在 AS 100 内配置 OSPF 协议，将各 loopback 0 接口通告到 OSPF 协议中。

```
R1(config)#router   ospf   10
R1(config-router)#router-id   1.1.1.1
Change router-id and update OSPF process! [yes/no]:y
R1(config-router)#network    10.0.12.0 0.0.0.255 area    0
R1(config-router)#network    10.0.13.0 0.0.0.255 area    0
R1(config-router)#network    1.1.1.1 0.0.0.0 are 0
R1(config-router)#exit
R1(config)#

R2(config)#router   ospf   10
R2(config-router)#router-id 2.2.2.2
R2(config-router)#network    2.2.2.2 0.0.0.0 area    0
R2(config-router)#network    10.0.12.0 0.0.0.255 are 0
R2(config-router)#exit

R3(config)#router   ospf   10
R3(config-router)#router-id    3.3.3.3
Change router-id and update OSPF process! [yes/no]:y
```

```
R3(config-router)#network    3.3.3.3 0.0.0.0 area    0
R3(config-router)#network    10.0.13.0 0.0.0.255 area    0
R3(config-router)#exit
R3(config)#
```

3. 在各设备之间建立 BGP 邻居关系,在 AS 100 内部通过 loopback 0 接口建立 IBGP 邻居关系。

```
R1(config)#router bgp    100
R1(config-router)#bgp    router-id    1.1.1.1
R1(config-router)#neighbor    2.2.2.2 remote-as    100
R1(config-router)#neighbor    2.2.2.2 update-source    loopback 0
R1(config-router)#neighbor    3.3.3.3 remote-as    100
R1(config-router)#neighbor    3.3.3.3 update-source    loopback 0
R1(config-router)#exit
R1(config)#


R2(config)#router    bgp    100
R2(config-router)#bgp    router-id    2.2.2.2
R2(config-router)#neighbor    1.1.1.1 remote-as    100
R2(config-router)#neighbor    1.1.1.1 update-source    loopback 0
R2(config-router)#neighbor    1.1.1.1 next-hop-self
R2(config-router)#neighbor    10.0.24.4 remote-as    200
R2(config-router)#exit
R2(config)#


R3(config)#router    bgp    100
R3(config-router)#bgp    router-id    3.3.3.3
R3(config-router)#neighbor    1.1.1.1 remote-as    100
R3(config-router)#neighbor    1.1.1.1 update-source loopback 0
R3(config-router)#neighbor    1.1.1.1 next-hop-self
R3(config-router)#neighbor    10.0.34.4 remote-as    200
R3(config-router)#exit
R3(config)#


R4(config)#router    bgp    200
R4(config-router)#bgp    router-id    4.4.4.4
R4(config-router)#neighbor    10.0.24.2 remote-as    100
R4(config-router)#neighbor    10.0.34.3 remote-as    100
R4(config-router)#exit
R4(config)#
```

4. 在 R4 设备将 loopback 1 接口通告到 BGP 路由表中。

```
R4(config)#router    bgp    200
R4(config-router)#network    20.0.1.1 mask    255.255.255.255
R4(config-router)#exit
R4(config)#
```

5. 将 AS 100 内的互联链路设置 OSPF 协议的 Cost 值,使其选择 IGP 的 Metric 值小的邻居传递的路由。

```
R1(config)#interface    gigabitEthernet 0/0
R1(config-if-GigabitEthernet 0/0)#ip ospf    cost 200
R1(config-if-GigabitEthernet 0/0)#exit
R1(config)#interface    gigabitEthernet 0/1
R1(config-if-GigabitEthernet 0/1)#ip ospf    cost 100
R1(config-if-GigabitEthernet 0/1)#exit
```

```
R1(config)#

R2(config)#interface    gigabitEthernet 0/0
R2(config-if-GigabitEthernet 0/0)#ip ospf    cost 200
R2(config-if-GigabitEthernet 0/0)#exit
R2(config)#

R3(config)#interface    gigabitEthernet 0/1
R3(config-if-GigabitEthernet 0/1)#ip ospf    cost 100
R3(config-if-GigabitEthernet 0/1)#exit
R3(config)#
```

➢ 任务验证

1. 检查全局 IP 路由表，此时全局 IP 路由表中的 Metric 值默认是 1。

```
R1#show    ip route

Codes:    C - Connected, L - Local, S - Static
          R - RIP, O - OSPF, B - BGP, I - IS-IS, V - Overflow route
          N1 - OSPF NSSA external type 1, N2 - OSPF NSSA external type 2
          E1 - OSPF external type 1, E2 - OSPF external type 2
          SU - IS-IS summary, L1 - IS-IS level-1, L2 - IS-IS level-2
          IA - Inter area, EV - BGP EVPN, A - Arp to host
          LA - Local aggregate route
          * - candidate default

Gateway of last resort is no set
C       1.1.1.1/32 is local host.
O       2.2.2.2/32 [110/1] via 10.0.12.2, 00:19:12, GigabitEthernet 0/0
O       3.3.3.3/32 [110/1] via 10.0.13.3, 00:04:53, GigabitEthernet 0/1
C       10.0.12.0/24 is directly connected, GigabitEthernet 0/0
C       10.0.12.1/32 is local host.
C       10.0.13.0/24 is directly connected, GigabitEthernet 0/1
C       10.0.13.1/32 is local host.
R1#
```

2. 检查 BGP 邻居关系建立情况。

```
R1#show    ip bgp    summary
For address family: IPv4 Unicast
BGP router identifier 1.1.1.1, local AS number 100
BGP table version is 1
0 BGP AS-PATH entries
0 BGP Community entries
0 BGP Prefix entries (Maximum-prefix:4294967295)

Neighbor           V     AS MsgRcvd MsgSent    TblVer   InQ OutQ Up/Down    State/PfxRcd
2.2.2.2            4    100      11      11         1     0    0 00:08:50         0
3.3.3.3            4    100       9       7         1     0    0 00:05:52         0

Total number of neighbors 2, established neighbors 2

R1#
```

```
R4#show    ip bgp    summary
For address family: IPv4 Unicast
BGP router identifier 4.4.4.4, local AS number 200
BGP table version is 1
0 BGP AS-PATH entries
0 BGP Community entries
0 BGP Prefix entries (Maximum-prefix:4294967295)

Neighbor        V     AS   MsgRcvd   MsgSent   TblVer   InQ   OutQ   Up/Down      State/PfxRcd
10.0.24.2       4    100      9         8        1       0     0    00:06:23          0
10.0.34.3       4    100      9         7        1       0     0    00:06:13          0

Total number of neighbors 2, established neighbors 2

R4#
```

3. 检查 BGP 路由通告的路由信息，此时，20.0.1.1/32 的路由选择 BGP 邻居 R2 设备通告。

```
R1#show    ip bgp
BGP table version is 2, local router ID is 1.1.1.1
Status codes: s suppressed, d damped, h history, * valid, > best, i - internal,
              S Stale, b - backup entry, m - multipath, f Filter, a additional-path
Origin codes: i - IGP, e - EGP, ? - incomplete

     Network         Next Hop         Metric      LocPrf      Weight Path
*bi 20.0.1.1/32      3.3.3.3           0           100         0 200 i
*>i                  2.2.2.2           0           100         0 200 i
```

4. 检查修改 IGP 的 Metric 值后，优先选择 Metric 值小的 BGP 邻居 R3 设备传递的路由。

```
R1#show    ip bgp
BGP table version is 2, local router ID is 1.1.1.1
Status codes: s suppressed, d damped, h history, * valid, > best, i - internal,
              S Stale, b - backup entry, m - multipath, f Filter, a additional-path
Origin codes: i - IGP, e - EGP, ? - incomplete

     Network         Next Hop         Metric      LocPrf      Weight Path
*>i 20.0.1.1/32      3.3.3.3           0           100         0 200 i
*bi                  2.2.2.2           0           100         0 200 i

R1#show    ip bgp    20.0.1.1
BGP routing table entry for 20.0.1.1/32(#0x7f6631beddd0)
Paths: (2 available, best #1, table Default-IP-Routing-Table)
  Not advertised to any peer

  200
      3.3.3.3 (metric 100) from 3.3.3.3 (3.3.3.3)
         Origin IGP, metric 0, localpref 100, valid, internal, best
         Last update: Wed Jun 28 09:24:47 2023
         RX ID: 0,TX ID: 0

  200
      2.2.2.2 (metric 200) from 2.2.2.2 (2.2.2.2)
         Origin IGP, metric 0, localpref 100, valid, internal, backup
         Last update: Wed Jun 28 09:24:42 2023
         RX ID: 0,TX ID: 0
R1#
```

➢ 问题与思考

根据 IGP 协议的 Metric 值选择最优路径，任务拓扑是根据 IGP 的 OSPF 路由协议 Metric 值优选最佳路径，以上述配置为例设置 BGP 路由根据 IGP 的 IS-IS 路由协议 Metric 值选择最佳路径。

1.2.8 选路原则 8——等价负载

➢ 原理

当到达同一目的地址存在多条等价路由时，可以通过 BGP 等价负载分担实现流量负载的目的。形成等价负载分担的条件是"BGP 选路原则"前 8 条原则需要比较的属性完全相同。

在公网中到达同一目的地的路由形成负载分担时，系统会首先判断最优路由的类型。若最优路由为 IBGP 路由则只是 IBGP 路由参与负载分担，若最优路由为 EBGP 路由则只是 EBGP 路由参与负载分担，若公网中到达同一目的地的 BGP 路由中同时包含 IBGP 路由和 EBGP 路由，则不能形成负载分担。

BGP 的等价负载默认不开启，需要在 BGP 进程下配置 maximum-paths [ibgp/ebgp] n，n 的取值为 2~32，BGP 路由才会执行等价负载均衡；

（1）设置关联 IBGP 关键字，会对 IBGP 路由执行等价负载均衡；
（2）设置关联 EBGP 关键字，会对 EBGP 路由执行等价负载均衡。

如果不配置该命令，则自动进入下一条选路原则的比较。

默认情况下设备只会对 AS-Path 列表完全相同的路由进行负载分担，否则需要通过配置命令 load-balancing as-path-ignore 忽略 AS-Path 路径不一致。

➢ 任务拓扑

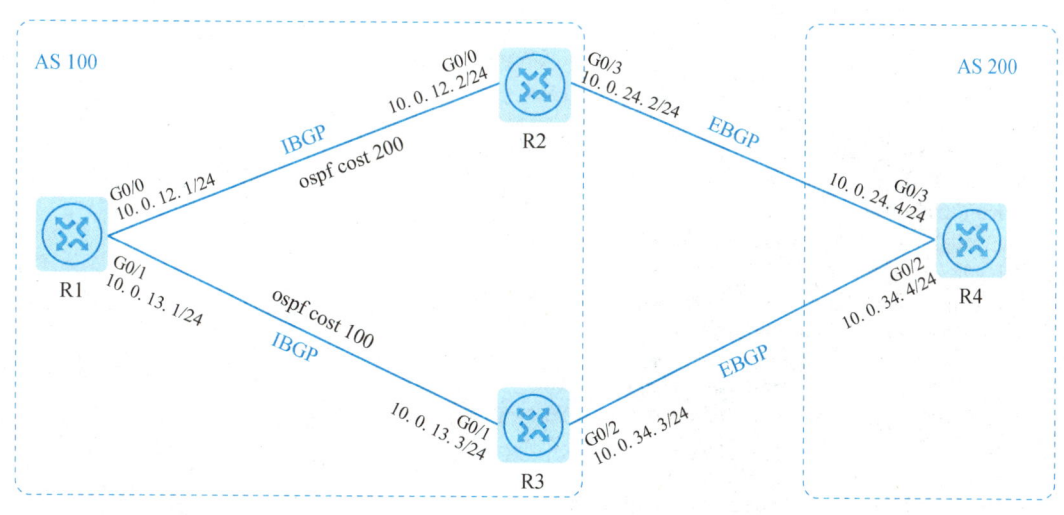

➢ 实施步骤

1. 根据任务拓扑配置各设备接口 IP 地址。

```
Ruijie>enable
Password:ruijie
```

```
Ruijie#configure terminal
Ruijie(config)#hostname   R1
R1(config)#interface   gigabitEthernet 0/0
R1(config-if-GigabitEthernet 0/0)#no   switchport
R1(config-if-GigabitEthernet 0/0)#ip address    10.0.12.1 24
R1(config-if-GigabitEthernet 0/0)#exit
R1(config)#interface   gigabitEthernet 0/1
R1(config-if-GigabitEthernet 0/1)#no   switchport
R1(config-if-GigabitEthernet 0/1)#ip address    10.0.13.1 24
R1(config-if-GigabitEthernet 0/1)#exit
R1(config)#interface   loopback 0
R1(config-if-Loopback 0)#ip address    1.1.1.1 32
R1(config-if-Loopback 0)#exit
R1(config)#interface loopback 1
R1(config-if-Loopback 1)#ip address    20.0.1.1 32
R1(config-if-Loopback 1)#exit
R1(config)#

Ruijie>enable
Password:ruijie
Ruijie#configure terminal
Ruijie(config)#hostname   R2
R2(config)#interface   gigabitEthernet 0/0
R2(config-if-GigabitEthernet 0/0)#no   switchport
R2(config-if-GigabitEthernet 0/0)#ip address    10.0.12.2 24
R2(config-if-GigabitEthernet 0/0)#exit
R2(config)#interface   gigabitEthernet 0/3
R2(config-if-GigabitEthernet 0/3)#no   switchport
R2(config-if-GigabitEthernet 0/3)#ip address    10.0.24.2 24
R2(config-if-GigabitEthernet 0/3)#exit
R2(config)#interface loopback 0
R2(config-if-Loopback 0)#ip address    2.2.2.2 32
R2(config-if-Loopback 0)#exit
R2(config)#

Ruijie>enable
Password:ruijie
Ruijie#configure terminal
Ruijie(config)#hostname   R1
R1(config)#hostname   R3
R3(config)#interface   gigabitEthernet 0/1
R3(config-if-GigabitEthernet 0/1)#no   switchport
R3(config-if-GigabitEthernet 0/1)#ip address    10.0.13.3 24
R3(config-if-GigabitEthernet 0/1)#exit
R3(config)#interface   gigabitEthernet 0/2
R3(config-if-GigabitEthernet 0/2)#no   switchport
R3(config-if-GigabitEthernet 0/2)#ip address    10.0.34.3 24
R3(config-if-GigabitEthernet 0/2)#exit
R3(config)#interface   loopback 0
R3(config-if-Loopback 0)#ip address    3.3.3.3 32
R3(config-if-Loopback 0)#exit
R3(config)#

Ruijie>enable
Password:ruijie
```

```
Ruijie#configure terminal
Ruijie(config)#hostname   R4
R4(config)#interface   gigabitEthernet 0/2
R4(config-if-GigabitEthernet 0/2)#no   switchport
R4(config-if-GigabitEthernet 0/2)#ip address   10.0.34.4 24
R4(config-if-GigabitEthernet 0/2)#exit
R4(config)#interface   gigabitEthernet 0/3
R4(config-if-GigabitEthernet 0/3)#no   switchport
R4(config-if-GigabitEthernet 0/3)#ip address   10.0.24.4 24
R4(config-if-GigabitEthernet 0/3)#exit
R4(config)#
```

2. 在 AS 100 内部建立 OSPF 邻居关系并通告 loopback 0 接口。

```
R1(config)#router ospf   10
R1(config-router)#router-id   1.1.1.1
Change router-id and update OSPF process! [yes/no]:y
R1(config-router)#network   1.1.1.1 0.0.0.0 area   0
R1(config-router)#network   10.0.12.0 0.0.0.255 area   0
R1(config-router)#network   10.0.13.0 0.0.0.255 area   0
R1(config-router)#exit
R1(config)#

R2(config)#router   ospf   10
R2(config-router)#router-id   2.2.2.2
Change router-id and update OSPF process! [yes/no]:y
R2(config-router)#network   2.2.2.2 0.0.0.0 area   0
R2(config-router)#network   10.0.12.0 0.0.0.255 area   0
R2(config-router)#exit
R2(config)#

R3(config)#router   ospf   10
R3(config-router)#router-id   3.3.3.3
Change router-id and update OSPF process! [yes/no]:y
R3(config-router)#network   3.3.3.3 0.0.0.0 area   0
R3(config-router)#network   10.0.13.0 0.0.0.255 area   0
R3(config-router)#exit
R3(config)#
```

3. 在各设备之间建立 BGP 邻居关系，在 AS 100 内建立 loopback 0 接口建立 IBGP 邻居关系。

```
R1(config)#router   bgp   100
R1(config-router)#bgp   router-id   1.1.1.1
R1(config-router)#neighbor   2.2.2.2 remote-as   100
R1(config-router)#neighbor   2.2.2.2 update-source loopback 0
R1(config-router)#neighbor   3.3.3.3 remote-as   100
R1(config-router)#neighbor   3.3.3.3 update-source loopback 0
R1(config-router)#exit
R1(config)#

R2(config)#router   bgp   100
R2(config-router)#bgp   router-id   2.2.2.2
R2(config-router)#neighbor   1.1.1.1 remote-as   100
R2(config-router)#neighbor   1.1.1.1 update-source   loopback 0
R2(config-router)#neighbor   10.0.24.4 remote-as   200
```

```
R2(config-router)#neighbor    1.1.1.1 next-hop-self
R2(config-router)#exit
R2(config)#

R3(config)#router   bgp   100
R3(config-router)#bgp    router-id   3.3.3.3
R3(config-router)#neighbor    1.1.1.1 remote-as    100
R3(config-router)#neighbor    1.1.1.1 update-source loopback 0
R3(config-router)#neighbor    1.1.1.1 next-hop-self
R3(config-router)#neighbor    10.0.34.4 remote-as    200
R3(config-router)#exit
R3(config)#

R4(config)#router bgp   200
R4(config-router)#bgp    router-id 4.4.4.4
R4(config-router)#neighbor    10.0.24.2 remote-as    100
R4(config-router)#neighbor    10.0.34.3 remote-as    100
R4(config-router)#exit
R4(config)#
```

4. 在 R1 设备将 loopback 1 接口通告到 BGP 路由表中。

```
R1(config)#router   bgp   100
R1(config-router)#network    20.0.1.1 mask    255.255.255.255
R1(config-router)#exit
R1(config)#
```

5. 在 R4 设备针对 20.0.1.1/32 网段设置等价负载均衡，根据需求设置 EBGP 或者 IBGP 邻居设置多路径负载分担。

```
R4(config)#router   bgp   200
R4(config-router)#maximum-paths ebgp  2
R4(config-router)#exit
R4(config)#exit
R4#
```

> 任务验证

1. 检查 OSPF 路由通告的路由信息。

```
R1#show ip route
Codes:   C - Connected, L - Local, S - Static
         R - RIP, O - OSPF, B - BGP, I - IS-IS, V - Overflow route
         N1 - OSPF NSSA external type 1, N2 - OSPF NSSA external type 2
         E1 - OSPF external type 1, E2 - OSPF external type 2
         SU - IS-IS summary, L1 - IS-IS level-1, L2 - IS-IS level-2
         IA - Inter area, EV - BGP EVPN, A - Arp to host
         LA - Local aggregate route
         * - candidate default

Gateway of last resort is no set
C     1.1.1.1/32 is local host.
O     2.2.2.2/32 [110/1] via 10.0.12.2, 00:11:58, GigabitEthernet 0/0
O     3.3.3.3/32 [110/1] via 10.0.13.3, 00:12:15, GigabitEthernet 0/1
C     10.0.12.0/24 is directly connected, GigabitEthernet 0/0
C     10.0.12.1/32 is local host.
C     10.0.13.0/24 is directly connected, GigabitEthernet 0/1
```

```
C       10.0.13.1/32 is local host.
C       20.0.1.1/32 is local host.
R1#
```

2. 检查 BGP 邻居关系建立情况。

```
R1#show ip bgp    summary
For address family: IPv4 Unicast
BGP router identifier 1.1.1.1, local AS number 100
BGP table version is 1
0 BGP AS-PATH entries
0 BGP Community entries
0 BGP Prefix entries (Maximum-prefix:4294967295)

Neighbor         V     AS MsgRcvd MsgSent    TblVer    InQ OutQ Up/Down    State/PfxRcd
2.2.2.2          4     100    26       25         1      0    0 00:22:57        0
3.3.3.3          4     100    25       23         1      0    0 00:21:29        0

Total number of neighbors 2, established neighbors 2

R1#


R4#show    ip bgp    summary
For address family: IPv4 Unicast
BGP router identifier 4.4.4.4, local AS number 200
BGP table version is 1
0 BGP AS-PATH entries
0 BGP Community entries
0 BGP Prefix entries (Maximum-prefix:4294967295)

Neighbor         V     AS MsgRcvd MsgSent    TblVer    InQ OutQ Up/Down    State/PfxRcd
10.0.24.2        4     100    22       21         1      0    0 00:19:12        0
10.0.34.3        4     100    22       20         1      0    0 00:19:07        0

Total number of neighbors 2, established neighbors 2

R4#
```

3. 检查 BGP 路由通告的路由信息，此时路径选择根据默认选路原则选择下一跳地址是 10.0.24.2，即 R2 设备传递路由。

```
R4#show    ip bgp
BGP table version is 2, local router ID is 4.4.4.4
Status codes: s suppressed, d damped, h history, * valid, > best, i - internal,
              S Stale, b - backup entry, m - multipath, f Filter, a additional-path
Origin codes: i - IGP, e - EGP, ? - incomplete

     Network           Next Hop          Metric     LocPrf     Weight Path
*b   20.0.1.1/32       10.0.34.3            0                    0 100 i
*>                     10.0.24.2            0                    0 100 i

Total number of prefixes 1
R4#

=====优先选择 R2 的路径=====
```

```
R4#show    ip route   bgp
B     20.0.1.1/32 [20/0] via 10.0.24.2, 00:05:44
R4#
```

4．在 BGP 进程下设置等价负载分担路径之后，此时 20.0.1.1/32 的路由为等价负载。

```
R4#show    ip bgp
BGP table version is 2, local router ID is 4.4.4.4
Status codes: s suppressed, d damped, h history, * valid, > best, i - internal,
              S Stale, b - backup entry, m - multipath, f Filter, a additional-path
Origin codes: i - IGP, e - EGP, ? - incomplete

      Network          Next Hop          Metric       LocPrf      Weight Path
*m   20.0.1.1/32       10.0.34.3         0                         0 100 i
*>                     10.0.24.2         0                         0 100 i

Total number of prefixes 1
R4#

R4#show    ip bgp    20.0.1.1
BGP routing table entry for 20.0.1.1/32(#0x7f107fbf3cf0)
Paths: (2 available, best #2, table Default-IP-Routing-Table)
  Advertised to update-groups:
   1

  100
     10.0.34.3 from 10.0.34.3 (3.3.3.3)
        Origin IGP, metric 0, localpref 100, valid, external, multipath
        Last update: Fri Jun 30 02:44:06 2023
        RX ID: 0,TX ID: 0

  100
     10.0.24.2 from 10.0.24.2 (2.2.2.2)
        Origin IGP, metric 0, localpref 100, valid, external, multipath, best
        Last update: Fri Jun 30 02:44:06 2023
        RX ID: 0,TX ID: 0
R4#

R4#show    ip route bgp
B     20.0.1.1/32 [20/0] via 10.0.24.2, 00:12:03
                  [20/0] via 10.0.34.3, 00:12:03
R4#
```

➢ 问题与思考

根据任务拓扑设置 IBGP 邻居之间的 BGP 路由负载分担。

1.2.9　选路原则 9——优选 Router ID 值小

➢ 原理

BGP 选路原则第 8 条为等价负载，默认 BGP 的等价负载不执行，则会执行 Router ID 值的比较；比较 BGP 协议的 Router ID 值，Router ID 值越小，路径越优先。

➢ 任务拓扑

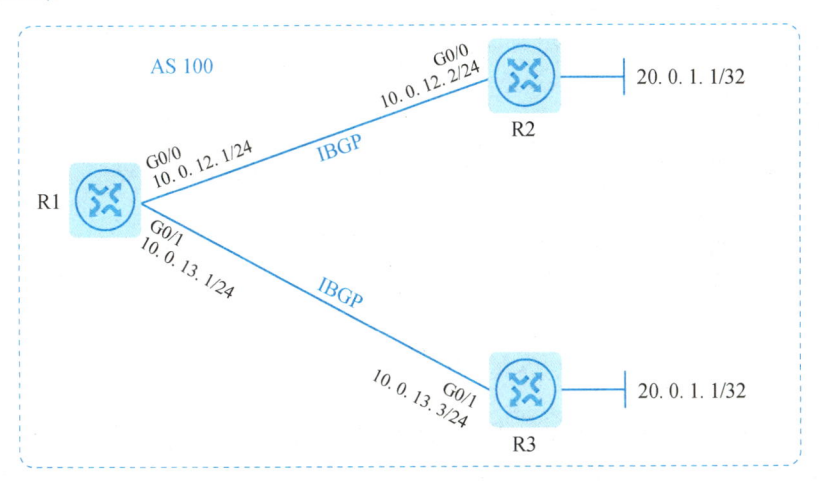

```
                    AS 100
                                G0/0
                              10.0.12.2/24
                                              R2 ──── 20.0.1.1/32
              G0/0
             10.0.12.1/24
                          IBGP
        R1
              G0/1
             10.0.13.1/24
                          IBGP
                              10.0.13.3/24
                                  G0/1
                                              R3 ──── 20.0.1.1/32
```

➢ 实施步骤

1. 根据任务拓扑配置各设备接口 IP 地址。

```
Ruijie>enable
Password:ruijie
Ruijie#configure terminal
Ruijie(config)#hostname    R1
R1(config)#interface    loopback 0
R1(config-if-Loopback 0)#ip address    1.1.1.1 32
R1(config-if-Loopback 0)#exit
R1(config)#interface    gigabitEthernet 0/0
R1(config-if-GigabitEthernet 0/0)#no    switchport
R1(config-if-GigabitEthernet 0/0)#ip address    10.0.12.1 24
R1(config-if-GigabitEthernet 0/0)#exit
R1(config)#interface    gigabitEthernet 0/1
R1(config-if-GigabitEthernet 0/1)#no    switchport
R1(config-if-GigabitEthernet 0/1)#ip address    10.0.13.1 24
R1(config-if-GigabitEthernet 0/1)#exit
R1(config)#

Ruijie>enable
Password:ruijie
Ruijie#configure terminal
Ruijie(config)#hostname    R2
R2(config)#interface    loopback 0
R2(config-if-Loopback 0)#ip address    2.2.2.2 32
R2(config-if-Loopback 0)#exit
R2(config)#interface    gigabitEthernet 0/0
R2(config-if-GigabitEthernet 0/0)#no    switchport
R2(config-if-GigabitEthernet 0/0)#ip address    10.0.12.2 24
R2(config-if-GigabitEthernet 0/0)#exit
R2(config)#

R2(config)#interface    loopback 1
R2(config-if-Loopback 1)#ip address    20.0.1.1 32
R2(config-if-Loopback 1)#exit
R2(config)#
```

```
Ruijie>enable
Password:ruijie
Ruijie#configure terminal
Ruijie(config)#hostname   R3
R3(config)#interface   loopback 0
R3(config-if-Loopback 0)#ip address   3.3.3.3 32
R3(config-if-Loopback 0)#exit
R3(config)#interface   gigabitEthernet 0/1
R3(config-if-GigabitEthernet 0/1)#no   switchport
R3(config-if-GigabitEthernet 0/1)#ip address   10.0.13.3 24
R3(config-if-GigabitEthernet 0/1)#exit
R3(config)#interface   loopback 1
R3(config-if-Loopback 1)#ip address   20.0.1.1 32
R3(config-if-Loopback 1)#exit
R3(config)#
```

2．在 AS 100 内配置 OSPF 协议并通告 loopback 0 接口。

```
R1(config)#router   ospf   10
R1(config-router)#router-id 1.1.1.1
Change router-id and update OSPF process! [yes/no]:y
R1(config-router)#network   1.1.1.1 0.0.0.0 area   0
R1(config-router)#network   10.0.12.0 0.0.0.255 area   0
R1(config-router)#network   10.0.13.0 0.0.0.255 area   0
R1(config-router)#exit
R1(config)#

R2(config)#router   ospf   10
R2(config-router)#router-id   2.2.2.2
Change router-id and update OSPF process! [yes/no]:y
R2(config-router)#network   2.2.2.2 0.0.0.0 area   0
R2(config-router)#network   10.0.12.0 0.0.0.255 area   0
R2(config-router)#exit
R2(config)#

R3(config)#router   ospf   10
R3(config-router)#router-id   3.3.3.3
Change router-id and update OSPF process! [yes/no]:y
R3(config-router)#network   3.3.3.3 0.0.0.0 area   0
R3(config-router)#network   10.0.13.0 0.0.0.255 area   0
R3(config-router)#exit
R3(config)#
```

3．在 AS 100 内使用 loopback 0 接口建立 IBGP 邻居关系。

```
R1(config)#router   bgp   100
R1(config-router)#bgp   router-id 1.1.1.1
R1(config-router)#neighbor   2.2.2.2 remote-as   100
R1(config-router)#neighbor   2.2.2.2 update-source loopback 0
R1(config-router)#neighbor   3.3.3.3 remote-as   100
R1(config-router)#neighbor   3.3.3.3 update-source   loopback 0
R1(config-router)#exit
R1(config)#

R2(config)#router   bgp   100
R2(config-router)#bgp   router-id   2.2.2.2
```

```
R2(config-router)#neighbor   1.1.1.1 remote-as   100
R2(config-router)#neighbor   1.1.1.1 update-source loopback 0
R2(config-router)#exit
R2(config)#


R3(config)#router   bgp   100
R3(config-router)#bgp   router-id   3.3.3.3
R3(config-router)#neighbor   1.1.1.1 remote-as   100
R3(config-router)#neighbor   1.1.1.1 update-source   loopback 0
R3(config-router)#exit
R3(config)#
```

4. 在 R2 设备、R3 设备通告 loopback 1 接口。

```
R2(config)#router   bgp 100
R2(config-router)#network   20.0.1.1 mask   255.255.255.255
R2(config-router)#exit
R2(config)#


R3(config)#router   bgp   100
R3(config-router)#network   20.0.1.1 mask   255.255.255.255
R3(config-router)#exit
R3(config)#
```

5. 将 R2 设备的 BGP 协议 Router ID 设置为 10.0.1.1，根据选路原则选择 Router ID 值小的路径。

```
R2(config)#router   bgp   100
R2(config-router)#bgp router-id 10.0.1.1
R2(config-router)#exit
R2(config)#
```

> 任务验证

1. 检查 OSPF 协议邻居建立情况。

```
R1#show   ip ospf   neighbor

OSPF process 10, 2 Neighbors, 2 is Full:
Neighbor ID   Pri   State       BFD State   Dead Time   Address       Interface
2.2.2.2       1     Full/DR     -           00:00:32    10.0.12.2     GigabitEthernet 0/0
3.3.3.3       1     Full/DR     -           00:00:36    10.0.13.3     GigabitEthernet 0/1
R1#
```

2. 检查 BGP 邻居关系建立情况。

```
R1#show   ip bgp   summary
For address family: IPv4 Unicast
BGP router identifier 1.1.1.1, local AS number 100
BGP table version is 2
0 BGP AS-PATH entries
0 BGP Community entries
0 BGP Prefix entries (Maximum-prefix:4294967295)

Neighbor        V     AS   MsgRcvd MsgSent   TblVer   InQ OutQ Up/Down    State/PfxRcd
2.2.2.2         4     100     7        5        1      0    0  00:03:39        0
3.3.3.3         4     100     7        4        1      0    0  00:03:22        0
```

Total number of neighbors 2, established neighbors 2

R1#

3. 检查 BGP 路由通告的路由信息，此时 20.0.1.1/32 的路径优选 R2 设备。

R1#show   ip bgp
BGP table version is 2, local router ID is 1.1.1.1
Status codes: s suppressed, d damped, h history, * valid, > best, i - internal,
              S Stale, b - backup entry, m - multipath, f Filter, a additional-path
Origin codes: i - IGP, e - EGP, ? - incomplete

Network	Next Hop	Metric	LocPrf	Weight	Path
*bi 20.0.1.1/32	3.3.3.3	0	100	0	i
*>i	2.2.2.2	0	100	0	i

Total number of prefixes 1
R1#

4. 将 R2 设备的 BGP 协议的 Router ID 设置为 10.0.1.1，根据选路原则优选 Router ID 值小的 R3 设备传递的路由。

R1#show   ip bgp
BGP table version is 2, local router ID is 1.1.1.1
Status codes: s suppressed, d damped, h history, * valid, > best, i - internal,
              S Stale, b - backup entry, m - multipath, f Filter, a additional-path
Origin codes: i - IGP, e - EGP, ? - incomplete

Network	Next Hop	Metric	LocPrf	Weight	Path
*bi 20.0.1.1/32	2.2.2.2	0	100	0	i
*>i	3.3.3.3	0	100	0	i

R1#show   ip bgp   20.0.1.1
BGP routing table entry for 20.0.1.1/32(#0x7efe77bfbd60)
Paths: (2 available, best #2, table Default-IP-Routing-Table)
  Not advertised to any peer

  Local
    2.2.2.2 (metric 1) from 2.2.2.2 (10.0.1.1)
      Origin IGP, metric 0, localpref 100, valid, internal, backup
      Last update: Fri Jun 30 03:37:30 2023
      RX ID: 0,TX ID: 0

  Local
    3.3.3.3 (metric 1) from 3.3.3.3 (3.3.3.3)
      Origin IGP, metric 0, localpref 100, valid, internal, **best**
      Last update: Fri Jun 30 03:36:32 2023
      RX ID: 0,TX ID: 0
R1#

> 问题与思考

BGP 路由通过路由器的 Router ID 选择最优路径，如果 BGP 路由中携带了 Originator_ID 属性，那么 BGP 路由如何选择最优路径？

1.2.10 选路原则 10——优选 Cluster List 短

➢ 原理

在 BGP 选路原则中通过 Cluster List 执行路径的优选仅在 BGP 路由反射器环境中，且有多个路由反射器的情况下才会进行 Cluster List 长度的比较，反射器在执行路由反射动作时会将自己的 Cluster ID 添加在 Cluster List 中，Cluster List 越短，路径越优先。

➢ 任务拓扑

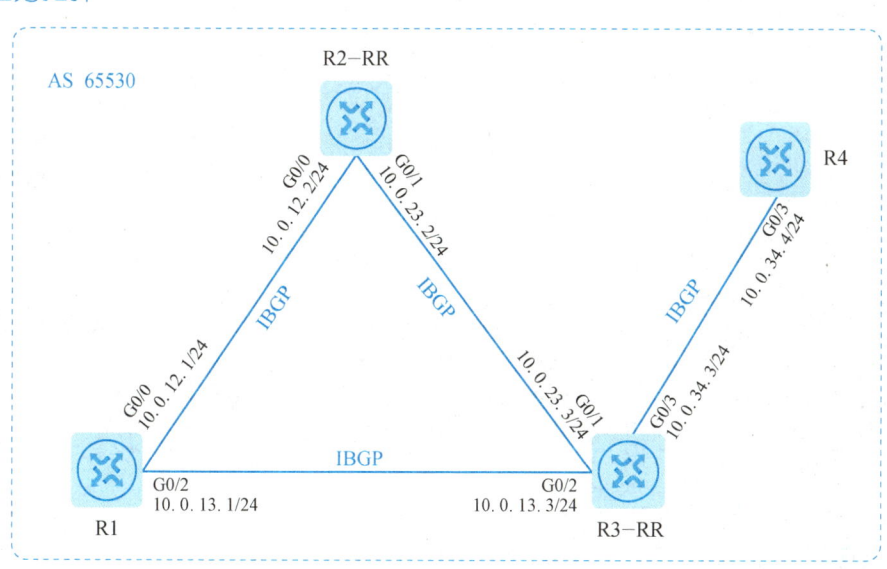

➢ 实施步骤

1. 根据任务拓扑配置各设备接口 IP 地址，分别在各设备创建 loopback 0 接口并用于建立 BGP 邻居关系。

```
Ruijie>enable
Password:ruijie
Ruijie#configure terminal
Ruijie(config)#hostname    R1
R1(config)#interface    gigabitEthernet 0/0
R1(config-if-GigabitEthernet 0/0)#no    switchport
R1(config-if-GigabitEthernet 0/0)#ip address    10.0.12.1 24
R1(config-if-GigabitEthernet 0/0)#exit
R1(config)#interface    gigabitEthernet 0/1
R1(config-if-GigabitEthernet 0/1)#no    switchport
R1(config-if-GigabitEthernet 0/1)#ip address    10.0.13.1 24
R1(config-if-GigabitEthernet 0/1)#exit
R1(config)#interface    gigabitEthernet 0/2
R1(config-if-GigabitEthernet 0/2)#no    switchport
R1(config-if-GigabitEthernet 0/2)#ip address    10.0.13.1 24
R1(config-if-GigabitEthernet 0/2)#exit
R1(config)#interface    loopback 0
R1(config-if-Loopback 0)#ip address    10.0.1.1 32
R1(config-if-Loopback 0)#exit
```

R1(config)#

Ruijie>enable
Password:ruijie
Ruijie#configure terminal
Ruijie(config)#hostname   R2
R2(config)#interface   gigabitEthernet 0/0
R2(config-if-GigabitEthernet 0/0)#no   switchport
R2(config-if-GigabitEthernet 0/0)#ip address   10.0.12.2 24
R2(config-if-GigabitEthernet 0/0)#exit
R2(config)#interface   gigabitEthernet 0/1
R2(config-if-GigabitEthernet 0/1)#no   switchport
R2(config-if-GigabitEthernet 0/1)#ip address   10.0.23.2 24
R2(config-if-GigabitEthernet 0/1)#exit
R2(config)#interface   loopback 0
R2(config-if-Loopback 0)#ip address   10.0.2.2 32
R2(config-if-Loopback 0)#exit
R2(config)#

Ruijie>enable
Password:ruijie
Ruijie#configure terminal
Ruijie(config)#hostname   R3
R3(config)#interface   gigabitEthernet 0/1
R3(config-if-GigabitEthernet 0/1)#no   switchport
R3(config-if-GigabitEthernet 0/1)#ip address   10.0.23.3 24
R3(config-if-GigabitEthernet 0/1)#exit
R3(config)#interface   gigabitEthernet 0/2
R3(config-if-GigabitEthernet 0/2)#no   switchport
R3(config-if-GigabitEthernet 0/2)#ip address   10.0.13.3 24
R3(config-if-GigabitEthernet 0/2)#exit
R3(config)#interface   gigabitEthernet 0/3
R3(config-if-GigabitEthernet 0/3)#no   switchport
R3(config-if-GigabitEthernet 0/3)#ip address   10.0.34.3 24
R3(config-if-GigabitEthernet 0/3)#exit
R3(config)#interface   loopback 0
R3(config-if-Loopback 0)#ip address   10.0.3.3 32
R3(config-if-Loopback 0)#exit
R3(config)#

Ruijie>enable
Password:ruijie
Ruijie#configure terminal
Ruijie(config)#hostname   R4
R4(config)#interface   gigabitEthernet 0/3
R4(config-if-GigabitEthernet 0/3)#no   switchport
R4(config-if-GigabitEthernet 0/3)#ip address   10.0.34.4 24
R4(config-if-GigabitEthernet 0/3)#exit
R4(config)#interface   loopback 0
R4(config-if-Loopback 0)#ip address   10.0.4.4 32
R4(config-if-Loopback 0)#exit
R4(config)#

2. 在 AS 65530 内配置 OSPF 协议，将各 loopback 0 接口通告到 OSPF 协议中。

R1(config)#router   ospf   10

```
R1(config-router)#router-id    1.1.1.1
Change router-id and update OSPF process! [yes/no]:y
R1(config-router)#network    10.0.1.1 0.0.0.0 area 0
R1(config-router)#network    10.0.12.0 0.0.0.255 area   0
R1(config-router)#network    10.0.13.0 0.0.0.255 area   0
R1(config-router)#exit
R1(config)#

R2(config)#router   ospf   10
R2(config-router)#router-id    2.2.2.2
Change router-id and update OSPF process! [yes/no]:y
R2(config-router)#network    10.0.2.2 0.0.0.0 area   0
R2(config-router)#network    10.0.12.0 0.0.0.255 area   0
R2(config-router)#network    10.0.23.0 0.0.0.255 area   0
R2(config-router)#exit
R2(config)#

R3(config)#router   ospf   10
R3(config-router)#router-id    3.3.3.3
Change router-id and update OSPF process! [yes/no]:y
R3(config-router)#network    10.0.23.0 0.0.0.255 area   0
R3(config-router)#network    10.0.13.0 0.0.0.255 area   0
R3(config-router)#network    10.0.34.0 0.0.0.255 are 0
R3(config-router)#network    10.0.3.3 0.0.0.0 area   0
R3(config-router)#exit
R3(config)#

R4(config)#router   ospf   10
R4(config-router)#router-id    4.4.4.4
Change router-id and update OSPF process! [yes/no]:y
R4(config-router)#network    10.0.4.4 0.0.0.0 area 0
R4(config-router)#network    10.0.34.0 0.0.0.255 area   0
R4(config-router)#exit
R4(config)#
```

3. 在 AS 65530 内各设备之间建立 IBGP 邻居关系。

```
R1(config)#router   bgp   65530
R1(config-router)#bgp   router-id   1.1.1.1
R1(config-router)#neighbor    10.0.2.2 remote-as    65530
R1(config-router)#neighbor    10.0.2.2 update-source    loopback 0
R1(config-router)#neighbor    10.0.3.3 remote-as 65530
R1(config-router)#neighbor    10.0.3.3 update-source loopback 0
R1(config-router)#exit
R1(config)#

R2(config)#router   bgp   65530
R2(config-router)#bgp   router-id   2.2.2.2
R2(config-router)#neighbor    10.0.1.1 remote-as    65530
R2(config-router)#neighbor    10.0.1.1 update-source loopback 0
R2(config-router)#neighbor    10.0.3.3 remote-as    65530
R2(config-router)#neighbor    10.0.3.3 update-source loopback 0
R2(config-router)#exit
R2(config)#

R3(config)#router bgp    65530
```

```
R3(config-router)#bgp    router-id    3.3.3.3
R3(config-router)#neighbor    10.0.1.1 remote-as    65530
R3(config-router)#neighbor    10.0.1.1 update-source loopback 0
R3(config-router)#neighbor    10.0.2.2 remote-as 65530
R3(config-router)#neighbor    10.0.2.2 update-source loopback 0
R3(config-router)#neighbor    10.0.4.4 remote-as    65530
R3(config-router)#neighbor    10.0.4.4 update-source    loopback 0
R3(config-router)#exit
R3(config)#

R4(config)#router    bgp 65530
R4(config-router)#bgp router-id    4.4.4.4
R4(config-router)#neighbor    10.0.3.3 remote-as    65530
R4(config-router)#neighbor    10.0.3.3 update-source loopback 0
R4(config-router)#exit
R4(config)#
```

4. 将 R2 设备和 R3 设备指定为 RR 设备，即将 R1 设备和 R2 设备、R4 设备设置为客户机，并将 R4 设备的路由反射给 R1 设备。

```
R2(config)#router    bgp    65530
R2(config-router)#neighbor    10.0.1.1 route-reflector-client
R2(config-router)#exit
R2(config)#

R3(config)#router    bgp    65530
R3(config-router)#neighbor 10.0.2.2 route-reflector-client
R3(config-router)#neighbor 10.0.4.4 route-reflector-client
R3(config-router)#exit
R3(config)#
```

5. 在 R4 设备创建 loopback 1 接口地址 20.0.1.1/32，并通告到 BGP 路由表中。

```
R4(config)#interface Loopback 1
R4(config-if- Loopback1)#ip address 20.0.1.1 32
R4(config-if- Loopback1)#exit
R4(config)#i

R4(config)#router    bgp    65530
R4(config-router)#network 20.0.1.1 mask 255.255.255.255
R4(config-router)#exit
R4(config)#
```

➢ 任务验证

1. 检查全局 IP 路由表的 OSPF 路由传递情况。

```
R1#show    ip route    ospf
O      10.0.2.2/32 [110/1] via 10.0.12.2, 00:56:42, GigabitEthernet 0/0
O      10.0.3.3/32 [110/1] via 10.0.13.3, 00:49:35, GigabitEthernet 0/2
O      10.0.4.4/32 [110/2] via 10.0.13.3, 00:55:41, GigabitEthernet 0/2
O      10.0.23.0/24 [110/2] via 10.0.13.3, 00:57:01, GigabitEthernet 0/2
                    [110/2] via 10.0.12.2, 00:57:01, GigabitEthernet 0/0
O      10.0.34.0/24 [110/2] via 10.0.13.3, 00:56:42, GigabitEthernet 0/2
R1#
```

2. 检查 BGP 邻居关系建立情况。

```
R1#show    ip bgp    summary
For address family: IPv4 Unicast
BGP router identifier 1.1.1.1, local AS number 65530
BGP table version is 1
0 BGP AS-PATH entries
0 BGP Community entries
0 BGP Prefix entries (Maximum-prefix:4294967295)

Neighbor      V      AS  MsgRcvd MsgSent   TblVer   InQ OutQ Up/Down     State/PfxRcd
10.0.2.2      4    65530     6       4        1     0    0  00:03:32        0
10.0.3.3      4    65530    52      51        1     0    0  00:49:54        0

Total number of neighbors 2, established neighbors 2

R1#

R3#show    ip bgp    summary
For address family: IPv4 Unicast
BGP router identifier 3.3.3.3, local AS number 65530
BGP table version is 1
0 BGP AS-PATH entries
0 BGP Community entries
0 BGP Prefix entries (Maximum-prefix:4294967295)

Neighbor      V      AS  MsgRcvd MsgSent   TblVer   InQ OutQ Up/Down     State/PfxRcd
10.0.1.1      4    65530    53      51        1     0    0  00:50:03        0
10.0.2.2      4    65530    15      15        1     0    0  00:12:25        0
10.0.4.4      4    65530    15      14        1     0    0  00:12:11        0

Total number of neighbors 3, established neighbors 3

R3#
```

3. 检查 BGP 路由通告的路由信息，优先选择 Cluster List 短的路径。

```
R1#show    ip bgp
BGP table version is 1, local router ID is 1.1.1.1
Status codes: s suppressed, d damped, h history, * valid, > best, i - internal,
              S Stale, b - backup entry, m - multipath, f Filter, a additional-path
Origin codes: i - IGP, e - EGP, ? - incomplete

    Network            Next Hop         Metric      LocPrf      Weight Path
* i 20.0.1.1/32        10.0.4.4           0          100           0    i


R1#show    ip bgp    20.0.1.1
BGP routing table entry for 20.0.1.1/32(#0x7f4165becdd0)
Paths: (2 available, best #2, table Default-IP-Routing-Table)
    Not advertised to any peer

  Local
      10.0.4.4 (metric 2) from 10.0.2.2 (4.4.4.4)
         Origin IGP, metric 0, localpref 100, valid, internal
         Originator: 4.4.4.4, Cluster list: 2.2.2.2 3.3.3.3
         Last update: Fri Jun 30 08:27:29 2023
         RX ID: 0,TX ID: 0
```

```
      Local
        10.0.4.4 (metric 2) from 10.0.3.3 (4.4.4.4)
          Origin IGP, metric 0, localpref 100, valid, internal, best
          Originator: 4.4.4.4, Cluster list: 3.3.3.3
          Last update: Fri Jun 30 08:27:24 2023
          RX ID: 0,TX ID: 0
R1#
```

> 问题与思考

BGP 路由反射器将反射的 BGP 路由经过多个反射簇反射之后，以最后一个设备接收到的反射路由为终点，终点设备接收到 BGP 路由反射器发送的路由下一跳地址是哪台设备？

1.2.11 选路原则 11——优选邻居 IP 地址小

> 原理

BGP 选路原则最后一条是通过比较邻居 IP 地址大小，该地址是配置 neighbor 命令指定的邻居地址，BGP 邻居 IP 地址越小，路径越优先。

> 任务拓扑

> 实施步骤

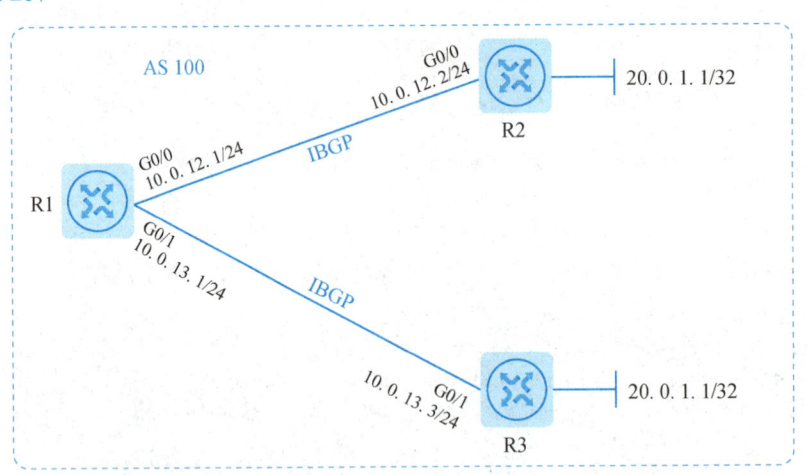

1. 根据任务拓扑配置各设备接口 IP 地址。

```
Ruijie>enable
Password:ruijie
Ruijie#configure terminal
Ruijie(config)#hostname   R1
R1(config)#interface    gigabitEthernet 0/0
R1(config-if-GigabitEthernet 0/0)#no    switchport
R1(config-if-GigabitEthernet 0/0)#ip address    10.0.12.1 24
R1(config-if-GigabitEthernet 0/0)#exit
R1(config)#interface    gigabitEthernet 0/1
R1(config-if-GigabitEthernet 0/1)#no    switchport
R1(config-if-GigabitEthernet 0/1)#ip address    10.0.13.1 24
```

```
R1(config-if-GigabitEthernet 0/1)#exit
R1(config)#interface    loopback 0
R1(config-if-Loopback 0)#ip address    10.0.1.1 32
R1(config-if-Loopback 0)#exit
R1(config)#


Password:ruijie
Ruijie#configure terminal
Ruijie(config)#hostname    R2
R2(config)#interface    gigabitEthernet 0/0
R2(config-if-GigabitEthernet 0/0)#no    switchport
R2(config-if-GigabitEthernet 0/0)#ip address    10.0.12.2 24
R2(config-if-GigabitEthernet 0/0)#exit
R2(config)#interface    loopback 0
R2(config-if-Loopback 0)#ip address    10.0.2.2 32
R2(config-if-Loopback 0)#exit
R2(config)#interface    loopback 1
R2(config-if-Loopback 1)#ip address    20.0.1.1 32
R2(config-if-Loopback 1)#exit
R2(config)#


Ruijie>enable
Password:ruijie
Ruijie#configure terminal
Ruijie(config)#hostname    R3
R3(config)#interface    gigabitEthernet 0/1
R3(config-if-GigabitEthernet 0/1)#no    switchport
R3(config-if-GigabitEthernet 0/1)#ip address    10.0.13.3 24
R3(config-if-GigabitEthernet 0/1)#exit
R3(config)#interface    loopback 0
R3(config-if-Loopback 0)#ip address    10.0.3.3 32
R3(config-if-Loopback 0)#exit
R3(config)#interface    loopback 1
R3(config-if-Loopback 1)#ip address    20.0.1.1 32
R3(config-if-Loopback 1)#exit
R3(config)#
```

2. 在 AS 100 内配置 OSPF 协议并通告 loopback 0 接口。

```
R1(config)#router    ospf    10
R1(config-router)#router-id    1.1.1.1
Change router-id and update OSPF process! [yes/no]:y
R1(config-router)#network    10.0.12.0 0.0.0.255 area    0
R1(config-router)#network    10.0.13.0 0.0.0.255 area    0
R1(config-router)#network    10.0.1.1 0.0.0.0 area 0
R1(config-router)#exit
R1(config)


R2(config)#router    ospf    10
R2(config-router)#router-id    2.2.2.2
Change router-id and update OSPF process! [yes/no]:y
R2(config-router)#network    10.0.2.2 0.0.0.0 area    0
R2(config-router)#network    10.0.12.0 0.0.0.255 area    0
R2(config-router)#exit
R2(config)#
```

```
R3(config)#router   ospf   10
R3(config-router)#router-id   3.3.3.3
Change router-id and update OSPF process! [yes/no]:y
R3(config-router)#network   10.0.3.3 0.0.0.0 area   0
R3(config-router)#network   10.0.13.0 0.0.0.255 area   0
R3(config-router)#exit
R3(config)#
```

3．在 AS 100 内通过 loopback 0 接口建立 IBGP 邻居关系。

```
R1(config)#router   bgp   100
R1(config-router)#bgp router-id   1.1.1.1
R1(config-router)#neighbor   10.0.2.2 remote-as   100
R1(config-router)#neighbor   10.0.2.2 update-source loopback 0
R1(config-router)#neighbor   10.0.3.3 remote-as   100
R1(config-router)#neighbor   10.0.3.3 update-source loopback 0
R1(config-router)#exit
R1(config)#

R2(config)#router   bgp   100
R2(config-router)#bgp   router-id   2.2.2.2
R2(config-router)#neighbor   10.0.1.1 remote-as   100
R2(config-router)#neighbor   10.0.1.1 update-source loopback 0
R2(config-router)#exit
R2(config)#

R3(config)#router   bgp   100
R3(config-router)#bgp   router-id   3.3.3.3
R3(config-router)#neighbor   10.0.1.1 remote-as   100
R3(config-router)#neighbor   10.0.1.1 update-source loopback 0
R3(config-router)#exit
R3(config)#
```

4．在 R2 设备与 R3 设备将 loopback 1 接口通告到 BGP 路由表中。

```
R2(config)#router   bgp   100
R2(config-router)#network   20.0.1.1   mask   255.255.255.255
R2(config-router)#exit
R2(config)#

R3(config)#router   bgp   100
R3(config-router)#network   20.0.1.1 mask   255.255.255.255
R3(config-router)#exit
R3(config)#
```

5．将 R2 设备的 BGP 协议 Router ID 设置与 R3 设备的 Router ID 一致，使其通过 IP 地址执行路径优选。

```
R2(config)#router   bgp   100
R2(config-router)#bgp   router-id   3.3.3.3
R2(config-router)#exit
R2(config)#
```

➢ 任务验证

1．检查 OSPF 协议邻居建立情况。

```
R1#show    ip ospf    neighbor

OSPF process 10, 2 Neighbors, 2 is Full:
Neighbor ID   Pri   State      BFD State   Dead Time    Address       Interface
2.2.2.2       1     Full/BDR   -           00:00:39     10.0.12.2     GigabitEthernet 0/0
3.3.3.3       1     Full/BDR   -           00:00:36     10.0.13.3     GigabitEthernet 0/1
R1#
```

2. 检查 BGP 邻居关系建立情况。

```
R1#show    ip bgp   summary
For address family: IPv4 Unicast
BGP router identifier 1.1.1.1, local AS number 100
BGP table version is 2
0 BGP AS-PATH entries
0 BGP Community entries
0 BGP Prefix entries (Maximum-prefix:4294967295)

Neighbor      V     AS MsgRcvd MsgSent    TblVer   InQ OutQ Up/Down    State/PfxRcd
10.0.2.2      4     100     37      36        1    0   0 00:33:06        0
10.0.3.3      4     100     36      34        1    0   0 00:32:47        0

Total number of neighbors 2, established neighbors 2

R1#
```

3. 检查 BGP 路由通告的路由信息。

```
R1#show    ip bgp
BGP table version is 2, local router ID is 1.1.1.1
Status codes: s suppressed, d damped, h history, * valid, > best, i - internal,
              S Stale, b - backup entry, m - multipath, f Filter, a additional-path
Origin codes: i - IGP, e - EGP, ? - incomplete

    Network          Next Hop           Metric      LocPrf      Weight Path
*bi 20.0.1.1/32      10.0.3.3           0           100         0      i
*>i                  10.0.2.2           0           100         0      i

Total number of prefixes 1
R1#
```

4. 将 R2 设备的 Router ID 设置与 R3 设备的 Router ID 相同后，选择 IP 地址小的邻居通告的路由。

```
R1#show    ip bgp   20.0.1.1
BGP routing table entry for 20.0.1.1/32(#0x7f9ce4fecd98)
Paths: (2 available, best #1, table Default-IP-Routing-Table)
  Not advertised to any peer

  Local
    10.0.2.2 (metric 1) from 10.0.2.2 (3.3.3.3)
      Origin IGP, metric 0, localpref 100, valid, internal, best
      Last update: Fri Jun 30 09:51:46 2023
      RX ID: 0,TX ID: 0

  Local
    10.0.3.3 (metric 1) from 10.0.3.3 (3.3.3.3)
```

```
        Origin IGP, metric 0, localpref 100, valid, internal, backup
        Last update: Fri Jun 30 09:50:27 2023
        RX ID: 0,TX ID: 0
R1#
```

> 问题与思考

将 BGP 协议引入其他路由协议的路由信息后，在 BGP 路由表中显示的下一跳地址为 0.0.0.0 的原因？

1.3 BGP 高级特性

1.3.1 BGP 反射器

> 原理

BGP 的 AS 内部支持跨设备建立 IBGP 对等体，因此会出现路径上的某些设备未运行 BGP 的情况；未运行 BGP 协议则不会收到 BGP 传递的路由信息，有数据经过该设备时，由于不存在路由信息，会导致数据包丢失，形成路由黑洞。

以下是解决 BGP 路由黑洞的几种方式：

（1）建立 IBGP 全互联，但建立物理的全互联结构，N 个 IBGP 对等体就要建立 N（N-1）/2 对 IBGP 邻居关系，会加大维护量，不推荐使用。

（2）配置路由反射器，使用路由反射器建立逻辑上的 IBGP 全互联。

（3）配置 BGP 联盟，通过 BGP 联盟建立逻辑上的 IBGP 全互联。

BGP 路由反射器是减少自治系统内的 IBGP 邻居连接数量的方法，将一台 BGP 路由器设置为 BGP 路由反射器，将自治系统内的 IBGP 邻居分为两类：客户端和非客户端。

BGP 路由反射器会将学习到的路由反射出去，从而使得 IBGP 路由在 AS 内传播时无须建立 IBGP 的全互联；指定 BGP 路由反射器，还需要指定客户端，而客户端也无须任何配置，对网络中是否存在 RR 设备并不知情。

在 AS 内实现路由反射器，规则如下：

配置路由反射器，并指定其客户端，路由反射器与客户端形成一个集群，路由反射器与客户端之间建立邻居关系。

在 AS 内，非客户端的 IBGP 邻居之间建立全互联，非客户端的 IBGP 邻居包括以下几种情况：一个集群内的多个路由反射器之间，集群内的路由反射器和集群外不参与路由反射器功能的 BGP 邻居（通常这些 BGP 路由器不支持路由反射器功能），集群内的路由反射器和其他集群的路由反射器之间。

路由反射器接收到一条路由的处理规则：

（1）从 EBGP 邻居接收到的路由更新，将发送给所有的客户端和非客户端；

（2）从客户端接收到的路由更新，将发送给其他客户端和所有非客户端；

（3）从 IBGP 非客户端接收到的路由更新，将发送给所有客户端。

路由反射器的处理规则打破了"AS 内部 IBGP 邻居水平分割原则",因此在有 RR 场景下可能会产生路由环路,为了防止 RR 产生路由环路,BGP 引入了两个可选非传递的路径属性:

(1) Originator_ID,RR 在反射出去的路由中增加 Originator_ID 属性,其值为本地 AS 中通告该路由的路由器的 BGP Router ID,若 AS 内存在多个 RR,则 Originator_ID 属性由第一个 RR 创建,且不被后续的 RR 更改,但 BGP 路由器收到一条协议携带 Originator_ID 属性的 IBGP 路由,并且 Originator_ID 属性值与自身的 Router ID 相同,则会忽略关于该路由的更新以达到防环的目的。

(2) Cluster_List,路由反射簇包括反射器 RR 及其客户端,一个 AS 内允许存在多个路由反射簇,每个反射簇都有一个唯一的 Cluster_ID,缺省为 RR 的 BGP Router ID,当一条路由被 RR 反射后,该反射簇内的 Cluster_ID 就会被添加至路由的 Cluster_ID 属性中;当 RR 收到一条携带 Cluster_ID 属性的 BGP 路由,且该属性值中包含该 RR 所在反射簇的 Cluster_ID 时,RR 认为该条路由存在环路,因此将忽略关于该条路由的更新。

➢ 任务拓扑

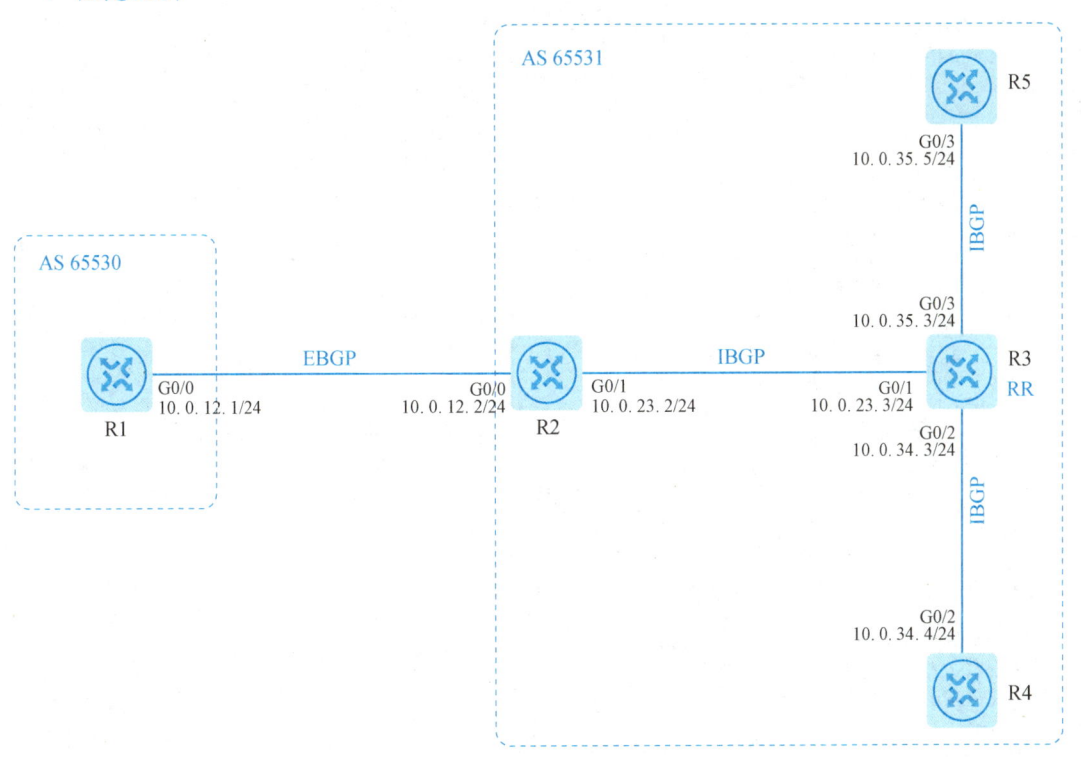

➢ 实施步骤

1. 根据任务拓扑配置各设备接口 IP 地址。

```
Ruijie>enable
Password:ruijie
Ruijie#configure terminal
Ruijie(config)#hostname    R1
R1(config)#interface    gigabitEthernet 0/0
R1(config-if-GigabitEthernet 0/0)#no    switchport
R1(config-if-GigabitEthernet 0/0)#ip address    10.0.12.1 24
```

```
R1(config-if-GigabitEthernet 0/0)#exit
R1(config)#interface    loopback 1
R1(config-if-Loopback 1)#ip address    20.0.1.1 32
R1(config-if-Loopback 1)#exit
R1(config)#

Ruijie>enable
Password:ruijie
Ruijie#configure terminal
Ruijie(config)#hostname    R2
R2(config)#interface    gigabitEthernet 0/0
R2(config-if-GigabitEthernet 0/0)#no    switchport
R2(config-if-GigabitEthernet 0/0)#ip address    10.0.12.2 24
R2(config-if-GigabitEthernet 0/0)#exit
R2(config)#interface    gigabitEthernet 0/1
R2(config-if-GigabitEthernet 0/1)#no    switchport
R2(config-if-GigabitEthernet 0/1)#ip address    10.0.23.2 24
R2(config-if-GigabitEthernet 0/1)#exit
R2(config)#interface    loopback 0
R2(config-if-Loopback 0)#ip address    10.0.2.2 32
R2(config-if-Loopback 0)#exit
R2(config)#

Ruijie>enable
Password:ruijie
Ruijie#configure terminal
Ruijie(config)#hostname    R3
R3(config)#interface    gigabitEthernet 0/1
R3(config-if-GigabitEthernet 0/1)#no    switchport
R3(config-if-GigabitEthernet 0/1)#ip address    10.0.23.3 24
R3(config-if-GigabitEthernet 0/1)#exit
R3(config)#interface    gigabitEthernet 0/02
R3(config-if-GigabitEthernet 0/2)#no    switchport
R3(config-if-GigabitEthernet 0/2)#ip address    10.0.34.3 24
R3(config-if-GigabitEthernet 0/2)#exit
R3(config)#interface    loopback 0
R3(config-if-Loopback 0)#ip address    10.0.3.3 32
R3(config-if-Loopback 0)#exit
R3(config)#interface    gigabitEthernet 0/3
R3(config-if-GigabitEthernet 0/3)#no    switchport
R3(config-if-GigabitEthernet 0/3)#ip address    10.0.35.3 24
R3(config-if-GigabitEthernet 0/3)#exit
R3(config)#

Ruijie>enable
Password:ruijie
Ruijie#configure terminal
Ruijie(config)#hostname    R4
R4(config)#interface    gigabitEthernet 0/2
R4(config-if-GigabitEthernet 0/2)#no    switchport
R4(config-if-GigabitEthernet 0/2)#ip address    10.0.34.4 24
R4(config-if-GigabitEthernet 0/2)#exit
R4(config)#interface    loopback 0
R4(config-if-Loopback 0)#ip address    10.0.4.4 32
R4(config-if-Loopback 0)#exit
```

```
R4(config)#

Ruijie>enable
Password:ruijie
Ruijie#configure terminal
Ruijie(config)#hostname   R5
R5(config)#interface gigabitEthernet 0/3
R5(config-if-GigabitEthernet 0/3)#no    switchport
R5(config-if-GigabitEthernet 0/3)#ip address    10.0.35.5 24
R5(config-if-GigabitEthernet 0/3)#exit
R5(config)#interface   loopback 0
R5(config-if-Loopback 0)#ip address    10.0.5.5 32
R5(config-if-Loopback 0)#exit
R5(config)#
```

2. 在 AS 65531 内部配置 OSPF 协议，并将 loopback 0 接口通告。

```
R2(config)#router   ospf   10
R2(config-router)#router-id   2.2.2.2
Change router-id and update OSPF process! [yes/no]:y
R2(config-router)#network    10.0.23.0 0.0.0.255 area   0
R2(config-router)#network    10.0.2.2 0.0.0.0 area    0
R2(config-router)#exit
R2(config)#

R3(config)#router   ospf   10
R3(config-router)#router-id   3.3.3.3
Change router-id and update OSPF process! [yes/no]:y
R3(config-router)#network    10.0.3.3 0.0.0.0 area    0
R3(config-router)#network    10.0.35.0 0.0.0.255 area    0
R3(config-router)#network    10.0.34.0 0.0.0.255 area    0
R3(config-router)#network    10.0.23.0 0.0.0.255 area    0
R3(config-router)#exit
R3(config)#

R4(config)#router   ospf   10
R4(config-router)#router-id 4.4.4.4
Change router-id and update OSPF process! [yes/no]:y
R4(config-router)#network    10.0.4.4 0.0.0.0 area    0
R4(config-router)#network    10.0.34.0 0.0.0.255 area    0
R4(config-router)#exit
R4(config)#

R5(config)#router   ospf   10
R5(config-router)#router-id   5.5.5.5
Change router-id and update OSPF process! [yes/no]:y
R5(config-router)#network    10.0.35.0 0.0.0.255 area    0
R5(config-router)#network    10.0.5.5 0.0.0.0 area    0
R5(config-router)#exit
R5(config)#
```

3. 在各设备之间建立 BGP 邻居关系，在 R1 设备与 R2 设备之间建立 EBGP 邻居关系，在 AS 65531 内部建立 IBGP 邻居关系，无须建立全互联。

```
R1(config)#router   bgp   65530
R1(config-router)#bgp   router-id    1.1.1.1
```

```
R1(config-router)#neighbor    10.0.12.2 remote-as    65531
R1(config-router)#exit
R1(config)#

R2(config)#router   bgp   65531
R2(config-router)#bgp   router-id    2.2.2.2
R2(config-router)#neighbor    10.0.12.1 remote-as    65530
R2(config-router)#neighbor    10.0.3.3 remote-as    65531
R2(config-router)#neighbor    10.0.3.3 update-source loopback 0
R2(config-router)#neighbor    10.0.3.3 next-hop-self
R2(config-router)#exit
R2(config)#

R3(config)#router   bgp   65531
R3(config-router)#bgp   router-id    3.3.3.3
R3(config-router)#neighbor    10.0.2.2 remote-as    65531
R3(config-router)#neighbor    10.0.2.2 update-source loopback 0
R3(config-router)#neighbor    10.0.4.4 remote-as    65531
R3(config-router)#neighbor    10.0.4.4 update-source loopback 0
R3(config-router)#neighbor    10.0.5.5 remote-as    65531
R3(config-router)#neighbor    10.0.5.5 update-source loopback 0
R3(config-router)#exit
R3(config)#

R4(config)#router   bgp   65531
R4(config-router)#bgp   router-id    4.4.4.4
R4(config-router)#neighbor    10.0.3.3 remote-as    65531
R4(config-router)#neighbor    10.0.3.3 update-source loopback 0
R4(config-router)#exit
R4(config)#

R4(config)#router   bgp   65531
R4(config-router)#bgp   router-id    4.4.4.4
R4(config-router)#neighbor    10.0.3.3 remote-as    65531
R4(config-router)#neighbor    10.0.3.3 update-source loopback 0
R4(config-router)#exit
R4(config)#
```

4. 在 R1 设备通告 BGP 路由。

```
R1(config)#router   bgp   65530
R1(config-router)#network    20.0.1.1 mask    255.255.255.255
R1(config-router)#exit
R1(config)#
```

5. 将 R3 设备设置为 RR 设备，即将 R4 设备和 R5 设备指定为 RR 客户端。

```
R3(config)#router   bgp   65531
R3(config-router)#neighbor    10.0.4.4 route-reflector-client
R3(config-router)#neighbor    10.0.5.5 route-reflector-client
R3(config-router)#exit
R3(config)#
```

➢ 任务验证

1. 检查 OSPF 协议邻居建立情况。

```
R3#show   ip ospf   neighbor
```

```
OSPF process 10, 3 Neighbors, 3 is Full:
Neighbor ID    Pri    State       BFD State    Dead Time    Address       Interface
2.2.2.2        1      Full/DR     -            00:00:36     10.0.23.2     GigabitEthernet 0/1
4.4.4.4        1      Full/DR     -            00:00:36     10.0.34.4     GigabitEthernet 0/2
5.5.5.5        1      Full/BDR    -            00:00:39     10.0.35.5     GigabitEthernet 0/3
R3#
```

2. 检查 OSPF 路由传递情况。

```
R2#show  ip route  ospf
O      10.0.3.3/32 [110/1] via 10.0.23.3, 03:20:21, GigabitEthernet 0/1
O      10.0.4.4/32 [110/2] via 10.0.23.3, 03:19:30, GigabitEthernet 0/1
O      10.0.5.5/32 [110/2] via 10.0.23.3, 03:19:53, GigabitEthernet 0/1
O      10.0.34.0/24 [110/2] via 10.0.23.3, 03:20:21, GigabitEthernet 0/1
O      10.0.35.0/24 [110/2] via 10.0.23.3, 03:20:21, GigabitEthernet 0/1
R2#
```

3. 检查 BGP 邻居关系建立情况。

```
R2#show  ip bgp  summary
For address family: IPv4 Unicast
BGP router identifier 2.2.2.2, local AS number 65531
BGP table version is 1
0 BGP AS-PATH entries
0 BGP Community entries
0 BGP Prefix entries (Maximum-prefix:4294967295)

Neighbor      V     AS MsgRcvd MsgSent    TblVer   InQ OutQ Up/Down    State/PfxRcd
10.0.3.3      4     65531    188     186      1    0    0 03:04:26         0
10.0.12.1     4     65530    187     186      1    0    0 03:05:02         0

Total number of neighbors 2, established neighbors 2

R2#


R3#show  ip bgp  summary
For address family: IPv4 Unicast
BGP router identifier 3.3.3.3, local AS number 65531
BGP table version is 1
0 BGP AS-PATH entries
0 BGP Community entries
0 BGP Prefix entries (Maximum-prefix:4294967295)

Neighbor      V     AS MsgRcvd MsgSent    TblVer   InQ OutQ Up/Down    State/PfxRcd
10.0.2.2      4     65531    187     186      1    0    0 03:04:35         0
10.0.4.4      4     65531    186     185      1    0    0 03:03:48         0
10.0.5.5      4     65531    186     184      1    0    0 03:03:29         0

Total number of neighbors 3, established neighbors 3

R3#
```

4. 检查 BGP 路由通告的路由信息，此时 BGP 路由只能传递到 R3 设备，由于 IBGP 水平分割的原因，从 IBGP 邻居接收的路由不传递给其他 IBGP 邻居。

```
R3#show    ip bgp
BGP table version is 2, local router ID is 3.3.3.3
Status codes: s suppressed, d damped, h history, * valid, > best, i - internal,
             S Stale, b - backup entry, m - multipath, f Filter, a additional-path
Origin codes: i - IGP, e - EGP, ? - incomplete

    Network            Next Hop         Metric      LocPrf       Weight Path
*>i 20.0.1.1/32        10.0.2.2         0           100          0 65530 i

Total number of prefixes 1
R3#
```

5. 设置 RR 之后，BGP 路由能正常传递给 R4 设备和 R5 设备。

```
R4#show    ip bgp
BGP table version is 2, local router ID is 4.4.4.4
Status codes: s suppressed, d damped, h history, * valid, > best, i - internal,
             S Stale, b - backup entry, m - multipath, f Filter, a additional-path
Origin codes: i - IGP, e - EGP, ? - incomplete

    Network            Next Hop         Metric      LocPrf       Weight Path
*>i 20.0.1.1/32        10.0.2.2         0           100          0 65530 i

Total number of prefixes 1
R4#


R5#show    ip bgp
BGP table version is 2, local router ID is 5.5.5.5
Status codes: s suppressed, d damped, h history, * valid, > best, i - internal,
             S Stale, b - backup entry, m - multipath, f Filter, a additional-path
Origin codes: i - IGP, e - EGP, ? - incomplete

    Network            Next Hop         Metric      LocPrf       Weight Path
*>i 20.0.1.1/32        10.0.2.2         0           100          0 65530 i

Total number of prefixes 1
R5#
```

> 问题与思考

1. BGP 协议在使用路由反射器的情况下如何实现环路防护？
2. 该 BGP 路由反射器实验 IBGP 邻居采用 loopback 接口来建立，若切换使用物理接口建立 IBGP 邻居，BGP 路由传递是否正常？请写出原因。

1.3.2 BGP 联盟

> 原理

BGP 联盟就是把一个大的 AS 划分成多个子 AS，这样被划分的 AS 就是一个联盟，子 AS 就是成员 AS。子 AS 之间建立联盟 EBGP 邻居关系，这样就打破了 IBGP 的水平分割，有效避免路由黑洞的同时，也减少了 AS 内部 IBGP 邻居激增的问题。

BGP 联盟内的成员 AS 间需要建立 EBGP 邻居关系，该 EBGP 邻居关系被称为联盟

EBGP，该 EBGP 邻居关系结合了正常 EBGP 和 IBGP 的特征。

联盟内 BGP 路由路径属性的传递规则：

Next-hop、Multi_Exit_DISC、Local_Pref 路径属性在传递过程中把整个联盟看作一个普通的 AS，因此可以保留原有的 IBGP 属性；

成员 AS 的 AS Number 只在联盟内部有效，只用来在联盟内部避免环路发生，不参与路由的选路决策，BGP 联盟相关属性只在联盟内部有效，传出联盟时会被自动删除。

➤ 任务拓扑

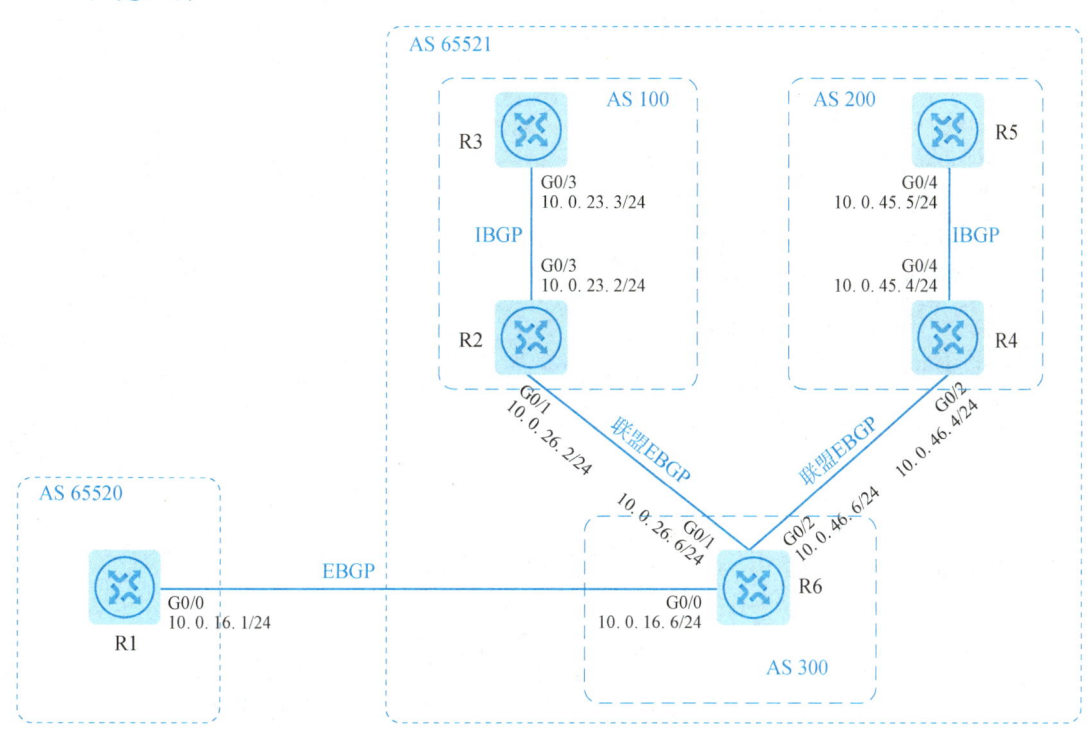

➤ 实施步骤

1. 根据任务拓扑配置各设备接口 IP 地址。

```
Ruijie>enable
Password:ruijie
Ruijie#configure terminal
Ruijie(config)#hostname    R1
R1(config)#interface    gigabitEthernet 0/0
R1(config-if-GigabitEthernet 0/0)#no    switchport
R1(config-if-GigabitEthernet 0/0)#ip address    10.0.16.1 24
R1(config-if-GigabitEthernet 0/0)#exit
R1(config)#interface loopback 0
R1(config-if-Loopback 0)#ip address    20.0.1.1 32
R1(config-if-Loopback 0)#exit
R1(config)#

Ruijie>enable
Password:ruijie
Ruijie#configure terminal
```

```
Ruijie(config)#hostname    R2
R2(config)#interface    gigabitEthernet 0/1
R2(config-if-GigabitEthernet 0/1)#no    switchport
R2(config-if-GigabitEthernet 0/1)#ip address    10.0.26.2 24
R2(config-if-GigabitEthernet 0/1)#exit
R2(config)#interface    gigabitEthernet 0/3
R2(config-if-GigabitEthernet 0/3)#no    switchport
R2(config-if-GigabitEthernet 0/3)#ip address    10.0.23.2 24
R2(config-if-GigabitEthernet 0/3)#exit
R2(config)#interface    loopback 0
R2(config-if-Loopback 0)#ip address    10.0.2.2 32
R2(config-if-Loopback 0)#exit
R2(config)#

Ruijie>enable
Password:ruijie
Ruijie#configure terminal
Ruijie(config)#hostname    R3
R3(config)#interface    gigabitEthernet 0/3
R3(config-if-GigabitEthernet 0/3)#no    switchport
R3(config-if-GigabitEthernet 0/3)#ip address    10.0.23.3 24
R3(config-if-GigabitEthernet 0/3)#exit
R3(config)#interface    loopback 0
R3(config-if-Loopback 0)#ip address    10.0.3.3 32
R3(config-if-Loopback 0)#exit
R3(config)#

Ruijie>enable
Password:ruijie
Ruijie#configure terminal
Ruijie(config)#hostname    R4
R4(config)#interface    gigabitEthernet 0/4
R4(config-if-GigabitEthernet 0/4)#no    switchport
R4(config-if-GigabitEthernet 0/4)#ip address    10.0.45.4 24
R4(config-if-GigabitEthernet 0/4)#exit
R4(config)#interface    gigabitEthernet 0/2
R4(config-if-GigabitEthernet 0/2)#no    switchport
R4(config-if-GigabitEthernet 0/2)#ip address    10.0.46.4 24
R4(config-if-GigabitEthernet 0/2)#exit
R4(config)#interface loopback 0
R4(config-if-Loopback 0)#ip address    10.0.4.4 32
R4(config-if-Loopback 0)#exit
R4(config)#

Ruijie>enable
Password:ruijie
Ruijie#configure terminal
Ruijie(config)#hostname    R5
R5(config)#interface    gigabitEthernet 0/4
R5(config-if-GigabitEthernet 0/4)#no    switchport
R5(config-if-GigabitEthernet 0/4)#ip address    10.0.45.5 24
R5(config-if-GigabitEthernet 0/4)#exit
R5(config)#interface    loopback 0
R5(config-if-Loopback 0)#ip address    10.0.5.5 32
R5(config-if-Loopback 0)#exit
```

```
R5(config)#interface    loopback 1
R5(config-if-Loopback 1)#ip address    20.0.2.1 32
R5(config-if-Loopback 1)#exit
R5(config)#


Ruijie>enable
Password:ruijie
Ruijie#configure terminal
Ruijie(config)#hostname    R6
R6(config)#interface    gigabitEthernet 0/0
R6(config-if-GigabitEthernet 0/0)#no    switchport
R6(config-if-GigabitEthernet 0/0)#ip address    10.0.16.6 24
R6(config-if-GigabitEthernet 0/0)#exit
R6(config)#interface    gigabitEthernet 0/1
R6(config-if-GigabitEthernet 0/1)#no    switchport
R6(config-if-GigabitEthernet 0/1)#ip address    10.0.26.6 24
R6(config-if-GigabitEthernet 0/1)#exit
R6(config)#interface    gigabitEthernet 0/2
R6(config-if-GigabitEthernet 0/2)#no    switchport
R6(config-if-GigabitEthernet 0/2)#ip address    10.0.46.6 24
R6(config-if-GigabitEthernet 0/2)#exit
R6(config)#interface    loopback 0
R6(config-if-Loopback 0)#ip address    10.0.6.6 32
R6(config-if-Loopback 0)#exit
R6(config)#
```

2. 在 AS 65521 内部建立 OSPF 协议，通告 loopback 0 接口。

```
R2(config)#router ospf    10
R2(config-router)#router-id 2.2.2.2
Change router-id and update OSPF process! [yes/no]:y
R2(config-router)#network    10.0.23.0 0.0.0.255 area    0
R2(config-router)#network    10.0.2.2 0.0.0.0 area    0
R2(config-router)#network    10.0.26.0 0.0.0.255 area    0
R2(config-router)#exit
R2(config)#


R3(config)#router    ospf    10
R3(config-router)#router-id 3.3.3.3
Change router-id and update OSPF process! [yes/no]:y
R3(config-router)#network    10.0.23.0 0.0.0.255 area    0
R3(config-router)#network    10.0.3.3 0.0.0.0 area    0
R3(config-router)#exit
R3(config)#


R4(config)#router    ospf    10
R4(config-router)#router-id    4.4.4.4
Change router-id and update OSPF process! [yes/no]:y
R4(config-router)#network    10.0.45.0 0.0.0.255 area    0
R4(config-router)#network    10.0.4.4 0.0.0.0 area    0
R4(config-router)#network    10.0.46.0 0.0.0.255 area    0
R4(config-router)#exit
R4(config)#


R5(config)#router ospf    10
R5(config-router)#router-id 5.5.5.5
```

```
Change router-id and update OSPF process! [yes/no]:y
R5(config-router)#network    10.0.45.0 0.0.0.255 are 0
R5(config-router)#network    10.0.5.5 0.0.0.0 area    0
R5(config-router)#exit
R5(config)#


R6(config)#router   ospf   10
R6(config-router)#router-id    6.6.6.6
Change router-id and update OSPF process! [yes/no]:y
R6(config-router)#network    10.0.46.0 0.0.0.255 area   0
R6(config-router)#network    10.0.6.6 0.0.0.0 area    0
R6(config-router)#network    10.0.26.0 0.0.0.255 area   0
R6(config-router)#exit
R6(config)#
```

3. BGP 邻居关系建立，AS 65521 内部配置 BGP 联盟减少 IBGP 全互联的配置，联盟 EBGP 之间需要指定联盟 EBGP 邻居的 ID（联盟 AS 号），BGP 联盟内直接配置 IBGP 邻居关系。

```
R1(config)#router   bgp   65520
R1(config-router)#bgp   router-id   1.1.1.1
R1(config-router)#neighbor   10.0.16.6 remote-as   65521
R1(config-router)#exit
R1(config)#


R2(config)#router   bgp   100
R2(config-router)#bgp   confederation   identifier   65521
R2(config-router)#bgp   confederation   peers   300
R2(config-router)#neighbor   10.0.6.6 remote-as   300
R2(config-router)#neighbor   10.0.6.6 update-source loopback 0
R2(config-router)#neighbor   10.0.6.6 ebgp-multihop
R2(config-router)#exit
R2(config)#


R3(config)#router   bgp   100
R3(config-router)#bgp   confederation   identifier   65521
R3(config-router)#bgp   router-id   3.3.3.3
R3(config-router)#neighbor   10.0.2.2 remote-as   100
R3(config-router)#neighbor   10.0.2.2 update-source loopback 0
R3(config-router)#exit
R3(config)#


R4(config)#router   bgp   200
R4(config-router)#bgp   router-id   4.4.4.4
R4(config-router)#bgp   confederation   identifier   65521
R4(config-router)#neighbor   10.0.5.5 remote-as   200
R4(config-router)#neighbor   10.0.5.5 update-source loopback 0
R4(config-router)#bgp   confederation   peers   300
R4(config-router)#neighbor   10.0.6.6 remote-as   300
R4(config-router)#neighbor   10.0.6.6 update-source loopback 0
R4(config-router)#neighbor   10.0.6.6 ebgp-multihop
R4(config-router)#exit
R4(config)#


R5(config)#router   bgp   200
R5(config-router)#bgp   confederation   identifier   65521
R5(config-router)#neighbor   10.0.4.4 remote-as   200
```

```
R5(config-router)#neighbor    10.0.4.4 update-source loopback 0
R5(config-router)#exit
R5(config)#

R6(config)#router   bgp   300
R6(config-router)#bgp    router-id 6.6.6.6
R6(config-router)#bgp    confederation   identifier   65521
R6(config-router)#bgp    confederation   peers   100
R6(config-router)#bgp    confederation   peers   200
R6(config-router)#neighbor    10.0.2.2 remote-as    100
R6(config-router)#neighbor    10.0.2.2 update-source   loopback 0
R6(config-router)#neighbor    10.0.2.2 ebgp-multihop
R6(config-router)#neighbor    10.0.2.2 next-hop-self
R6(config-router)#neighbor    10.0.4.4 remote-as    200
R6(config-router)#neighbor    10.0.4.4 update-source loopback 0
R6(config-router)#neighbor    10.0.4.4 ebgp-multihop
R6(config-router)#neighbor    10.0.4.4 next-hop-self
R6(config-router)#neighbor    10.0.16.1 remote-as    65520
R6(config-router)#exit
R6(config)#
```

4. 在 R1 设备通告 BGP 路由，观察在 AS 65521 内传递的路由情况。

```
R1(config)#router bgp    65520
R1(config-router)#network    20.0.1.1 mask    255.255.255.255
R1(config-router)#exit
R1(config)#
```

5. 在 R5 设备通告 BGP 路由，观察路由在 AS 65521 内和在 AS 65520 内传递的情况。

```
R5(config)#router   bgp   200
R5(config-router)#network    20.0.2.1 mask    255.255.255.255
R5(config-router)#exit
R5(config)#
```

> 任务验证

1. 检查 OSPF 协议邻居建立情况及路由通告的路由信息。

```
R6#show    ip ospf   neighbor

OSPF process 10, 2 Neighbors, 2 is Full:
Neighbor ID    Pri    State      BFD State    Dead Time    Address         Interface
2.2.2.2        1      Full/DR    -            00:00:33     10.0.26.2       GigabitEthernet 0/1
4.4.4.4        1      Full/DR    -            00:00:38     10.0.46.4       GigabitEthernet 0/2
R6#

R6#show    ip route   ospf
O      10.0.2.2/32 [110/1] via 10.0.26.2, 00:30:53, GigabitEthernet 0/1
O      10.0.3.3/32 [110/2] via 10.0.26.2, 00:30:53, GigabitEthernet 0/1
O      10.0.4.4/32 [110/1] via 10.0.46.4, 00:31:09, GigabitEthernet 0/2
O      10.0.5.5/32 [110/2] via 10.0.46.4, 00:31:09, GigabitEthernet 0/2
O      10.0.23.0/24 [110/2] via 10.0.26.2, 00:30:53, GigabitEthernet 0/1
O      10.0.45.0/24 [110/2] via 10.0.46.4, 00:31:09, GigabitEthernet 0/2
R6#
```

2. 检查 BGP 邻居关系建立情况。

```
R6#show    ip bgp    summary
For address family: IPv4 Unicast
BGP router identifier 6.6.6.6, local AS number 300
BGP table version is 1
0 BGP AS-PATH entries
0 BGP Community entries
0 BGP Prefix entries (Maximum-prefix:4294967295)

Neighbor        V       AS   MsgRcvd MsgSent    TblVer   InQ OutQ Up/Down    State/PfxRcd
10.0.2.2        4      100      19      18         1      0    0  00:16:58         0
10.0.4.4        4      200      19      17         1      0    0  00:16:50         0
10.0.16.1       4    65520      19      18         1      0    0  00:16:26         0

Total number of neighbors 3, established neighbors 3

R6#
```

3. 检查 R1 设备通告的 BGP 路由，路由传递到 AS 65521 内根据 BGP 联盟规则记录子 AS 号。

```
R6#show    ip bgp
BGP table version is 2, local router ID is 6.6.6.6
Status codes: s suppressed, d damped, h history, * valid, > best, i - internal,
              S Stale, b - backup entry, m - multipath, f Filter, a additional-path
Origin codes: i - IGP, e - EGP, ? - incomplete

    Network          Next Hop          Metric     LocPrf     Weight Path
*>  20.0.1.1/32      10.0.16.1           0                      0 65520 i

Total number of prefixes 1
R6#


R3#show    ip bgp
BGP table version is 2, local router ID is 3.3.3.3
Status codes: s suppressed, d damped, h history, * valid, > best, i - internal,
              S Stale, b - backup entry, m - multipath, f Filter, a additional-path
Origin codes: i - IGP, e - EGP, ? - incomplete

    Network          Next Hop          Metric     LocPrf     Weight Path
*>i 20.0.1.1/32      10.0.6.6            0         100          0 (300) 65520 i

Total number of prefixes 1
R3#
```

4. 检查 R5 设备通告的 BGP 路由，在 AS 65521 内传递会记录 BGP 联盟内的子 AS 号，传递出联盟 AS 后，不再记录联盟内的子 AS 号。

```
R3#show    ip bgp
BGP table version is 2, local router ID is 3.3.3.3
Status codes: s suppressed, d damped, h history, * valid, > best, i - internal,
              S Stale, b - backup entry, m - multipath, f Filter, a additional-path
Origin codes: i - IGP, e - EGP, ? - incomplete

    Network          Next Hop          Metric     LocPrf     Weight Path
*>i 20.0.1.1/32      10.0.6.6            0         100          0 (300) 65520 i
```

```
 *>i 20.0.2.1/32        10.0.6.6                 0             100           0 (300 200) i

Total number of prefixes 2
R3#

R6#show    ip bgp
BGP table version is 5, local router ID is 6.6.6.6
Status codes: s suppressed, d damped, h history, * valid, > best, i - internal,
              S Stale, b - backup entry, m - multipath, f Filter, a additional-path
Origin codes: i - IGP, e - EGP, ? - incomplete

    Network            Next Hop         Metric        LocPrf        Weight Path
 *> 20.0.1.1/32        10.0.16.1         0                          0 65520 i
 *> 20.0.2.1/32        10.0.5.5          0             100          0 (200) i

Total number of prefixes 2
R6#


R1#show    ip bgp
BGP table version is 5, local router ID is 1.1.1.1
Status codes: s suppressed, d damped, h history, * valid, > best, i - internal,
              S Stale, b - backup entry, m - multipath, f Filter, a additional-path
Origin codes: i - IGP, e - EGP, ? - incomplete

    Network            Next Hop         Metric        LocPrf        Weight Path
 *> 20.0.1.1/32        0.0.0.0           0                          32768      i
 *> 20.0.2.1/32        10.0.16.6         0                          0 65521 i

Total number of prefixes 2
R1#
```

➢ 问题与思考

BGP 路由反射器与 BGP 联盟的区别？

1.3.3　BGP 汇总

➢ 原理

为了减少路由震荡，保证网络稳定性，通常使用路由汇总的方式来实现，BGP 支持对路由进行手工汇总，并控制汇总路由的属性，并且决定是否向 BGP 对等体发布明细路由。

network：手工配置指向 null 0 的静态汇总路由，然后通过 network 方式通告到 BGP 路由表中。

aggregate-address：在 BGP 进程下进行手工路由汇总。

为了避免路由汇总可能引起的路由环路，BGP 设计了 as-set 属性，标明了汇总路由所经过的 AS 号。当汇总路由重新进入了 as-set 属性中所列出的任何一个 AS 时，BGP 将会检测到这样的情况，丢弃该汇总路由，从而避免路由环路。

aggregate 方式扩展参数：

summary-only 仅将汇总路由通告给 BGP 邻居，明细路由将不再通告。

suppress-map 可以将某些路由选择性地抑制掉而不通告给邻居。

attribute-map 可以设置汇总路由的相关属性。

➢ 任务拓扑

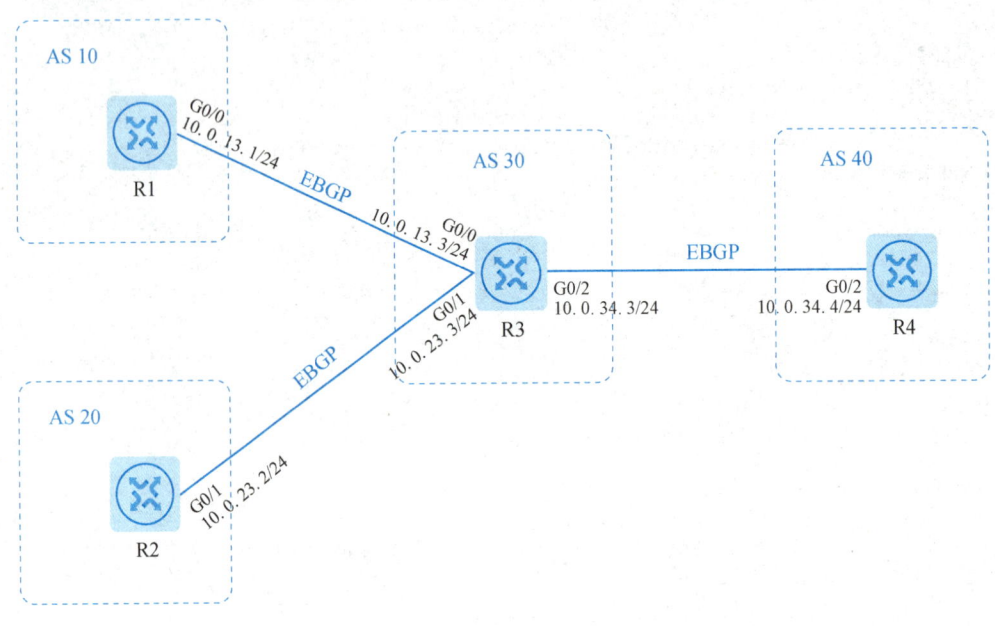

```
                    AS 10
                         G0/0
                         10.0.13.1/24
                    R1              EBGP
                                         10.0.13.3/24  G0/0
                                                      AS 30         AS 40
                                                                    EBGP
                                                    R3  G0/2              G0/2
                                                    10.0.34.3/24          10.0.34.4/24
                                                G0/1                                  R4
                                                10.0.23.3/24
                              EBGP
                    AS 20
                         G0/1
                         10.0.23.2/24
                    R2
```

➢ 实施步骤

1. 根据任务拓扑配置各设备接口 IP 地址。

```
Ruijie>enable
Password:ruijie
Ruijie#configure terminal
Ruijie(config)#hostname    R1
R1(config)#interface    gigabitEthernet 0/0
R1(config-if-GigabitEthernet 0/0)#no    switchport
R1(config-if-GigabitEthernet 0/0)#ip address    10.0.13.1 24
R1(config-if-GigabitEthernet 0/0)#exit
R1(config)#

Ruijie>enable
Password:ruijie
Ruijie#configure terminal
Ruijie(config)#hostname    R2
R2(config)#interface    gigabitEthernet 0/1
R2(config-if-GigabitEthernet 0/1)#no    switchport
R2(config-if-GigabitEthernet 0/1)#ip address    10.0.23.2 24
R2(config-if-GigabitEthernet 0/1)#exit
R2(config)#

Ruijie>enable
Password:ruijie
Ruijie#configure terminal
Ruijie(config)#hostname    R3
R3(config)#interface    gigabitEthernet 0/0
R3(config-if-GigabitEthernet 0/0)#no    switchport
R3(config-if-GigabitEthernet 0/0)#ip address    10.0.13.3 24
```

```
R3(config-if-GigabitEthernet 0/0)#exit
R3(config)#interface   gigabitEthernet 0/1
R3(config-if-GigabitEthernet 0/1)#no   switchport
R3(config-if-GigabitEthernet 0/1)#ip address    10.0.23.3 24
R3(config-if-GigabitEthernet 0/1)#exit
R3(config)#interface   gigabitEthernet 0/2
R3(config-if-GigabitEthernet 0/2)#no   switchport
R3(config-if-GigabitEthernet 0/2)#ip address    10.0.34.3 24
R3(config-if-GigabitEthernet 0/2)#exit
R3(config)#

Ruijie>enable
Password:ruijie
Ruijie#configure terminal
Ruijie(config)#hostname   R4
R4(config)#interface   gigabitEthernet 0/2
R4(config-if-GigabitEthernet 0/2)#no   switchport
R4(config-if-GigabitEthernet 0/2)#ip address    10.0.34.4 24
R4(config-if-GigabitEthernet 0/2)#exit
R4(config)#
```

2. 在各 AS 之间建立 EBGP 邻居关系。

```
R1(config)#router   bgp   10
R1(config-router)#bgp   router-id   1.1.1.1
R1(config-router)#neighbor   10.0.13.3 remote-as   30
R1(config-router)#exit
R1(config)#

R2(config)#router   bgp   20
R2(config-router)#bgp   router-id   2.2.2.2
R2(config-router)#neighbor   10.0.23.3 remote-as   30
R2(config-router)#exit
R2(config)#

R3(config)#router   bgp   30
R3(config-router)#bgp router-id   3.3.3.3
R3(config-router)#neighbor   10.0.13.1 remote-as   10
R3(config-router)#neighbor   10.0.23.2 remote-as   20
R3(config-router)#neighbor   10.0.34.4 remote-as   40
R3(config-router)#exit
R3(config)#

R4(config)#router   bgp   40
R4(config-router)#bgp   router-id   4.4.4.4
R4(config-router)#neighbor   10.0.34.3 remote-as   30
R4(config-router)#exit
R4(config)#
```

3. 在 R1 设备与 R2 设备分别创建两个 loopback 接口，并将这些接口通告到 BGP 路由表中。

```
R1(config)#interface   loopback 0
R1(config-if-Loopback 0)#ip address    20.0.1.1 32
R1(config-if-Loopback 0)#exit
R1(config)#interface   loopback 1
```

```
R1(config-if-Loopback 1)#ip address    20.0.1.2 32
R1(config-if-Loopback 1)#exit
R1(config)#
R1(config)#router   bgp   10
R1(config-router)#network   20.0.1.1 mask    255.255.255.255
R1(config-router)#network   20.0.1.2 mask    255.255.255.255
R1(config-router)#exit
R1(config)#

R2(config)#interface   loopback 0
R2(config-if-Loopback 0)#ip address    20.0.1.3 32
R2(config-if-Loopback 0)#exit
R2(config)#interface   loopback 1
R2(config-if-Loopback 1)#ip address    20.0.1.4 32
R2(config-if-Loopback 1)#exit
R2(config)#router   bgp   20
R2(config-router)#network   20.0.1.3 mask 255.255.255.255
R2(config-router)#network   20.0.1.4 mask    255.255.255.255
R2(config-router)#exit
R2(config)#
```

4. 在 R3 设备通过手工配置指向 null 0 的静态汇总路由，然后通过 network 命令将汇总的静态路由通告到 BGP 路由表中。

```
R3(config)#ip route 20.0.1.0 255.255.255.0 null 0
R3(config)#router   bgp   30
R3(config-router)#network   20.0.1.0 mask    255.255.255.0
R3(config-router)#exit
R3(config)#
```

5. 在 R3 设备通过 aggregate-address 命令在 BGP 进程下进行手工路由汇总。

```
R3(config)#router   bgp   30
R3(config-router)#aggregate-address    20.0.1.0 255.255.255.0
R3(config-router)#exit
R3(config)#
```

6. 由于前两种路由汇总方式 BGP 协议会将明细路由与汇总路由同时传递给邻居，因此在汇总路由添加扩展 summary-only 参数，此参数表示仅将汇总路由通告给 BGP 邻居，明细路由不通告。

```
R3(config)#router   bgp   30
R3(config-router)#aggregate-address    20.0.1.0 255.255.255.0 summary-only
R3(config-router)#exit
R3(config)#
```

7. as-set 参数，BGP 路由汇总添加 as-set 路径属性，使汇总的路由继承明细路由的 AS 号，避免将汇总路由通告给明细路由的通告者。

```
R3(config)#router   bgp   30
R3(config-router)#aggregate-address    20.0.1.0 255.255.255.0 summary-only   as-set
R3(config-router)#exit
R3(config)#
```

8. 设置 BGP 路由汇总的 suppress-map 参数，通过该命令可以将某些路由选择性地抑制掉而不通告给邻居。

```
R3(config)#ip prefix-list ruijie permit    20.0.1.1/32
R3(config)#ip prefix-list   ruijie permit 20.0.1.2/32
R3(config)#route-map ruijie permit    10
R3(config-route-map)#match    ip address    prefix-list    ruijie
R3(config-route-map)#exit
R3(config)#route-map ruijie deny    20
R3(config-route-map)#exit
R3(config)#router bgp    30
R3(config-router)#aggregate-address    20.0.1.0 255.255.255.0 summary-only as-set    suppress-map ruijie
R3(config-router)#exit
R3(config)#
```

9. 设置 BGP 路由汇总的 attribute-map 参数，此参数可以将汇总路由的相关属性修改，比如将起源属性从【i】改为【?】。

```
R3(config)#ip access-list standard    1
R3(config-std-nacl)#permit    20.0.1.3 0.0.0.0
R3(config-std-nacl)#permit    20.0.1.4 0.0.0.0
R3(config-std-nacl)#exit
R3(config)#route-map Att permit    10
R3(config-route-map)#match    ip address    1
R3(config-route-map)#set    origin    incomplete
R3(config-route-map)#exit
R3(config)#route-map    Att permit    20
R3(config-route-map)#exit
R3(config)#router    bgp    30
R3(config-router)#aggregate-address    20.0.1.0 255.255.255.0 as-set    summary-only attribute-map Att
R3(config-router)#exit
R3(config)
```

> 任务验证

1. 检查 BGP 邻居关系建立情况。

```
R3#show   ip bgp   summary
For address family: IPv4 Unicast
BGP router identifier 3.3.3.3, local AS number 30
BGP table version is 7
0 BGP AS-PATH entries
0 BGP Community entries
0 BGP Prefix entries (Maximum-prefix:4294967295)

Neighbor        V     AS MsgRcvd MsgSent    TblVer  InQ OutQ Up/Down    State/PfxRcd
10.0.13.1       4     10    41      41         7    0    0 00:31:07         0
10.0.23.2       4     20    41      40         7    0    0 00:31:02         0
10.0.34.4       4     40    40      39         7    0    0 00:30:47         0

Total number of neighbors 3, established neighbors 3

R3#
```

2. 检查 BGP 路由通告的路由信息。

```
R4#show    ip bgp
BGP table version is 9, local router ID is 4.4.4.4
Status codes: s suppressed, d damped, h history, * valid, > best, i - internal,
              S Stale, b - backup entry, m - multipath, f Filter, a additional-path
```

Origin codes: i - IGP, e - EGP, ? - incomplete

Network	Next Hop	Metric	LocPrf	Weight Path
*> 20.0.1.1/32	10.0.34.3	0		0 30 10 i
*> 20.0.1.2/32	10.0.34.3	0		0 30 10 i
*> 20.0.1.3/32	10.0.34.3	0		0 30 20 i
*> 20.0.1.4/32	10.0.34.3	0		0 30 20 i

Total number of prefixes 4
R4#

3．检查手工静态汇总路由，在 R4 设备能查看到 20.0.1.0/24 的汇总路由，其中也接收到了其余的明细路由。

R4#show   ip bgp
BGP table version is 10, local router ID is 4.4.4.4
Status codes: s suppressed, d damped, h history, * valid, > best, i - internal,
              S Stale, b - backup entry, m - multipath, f Filter, a additional-path
Origin codes: i - IGP, e - EGP, ? - incomplete

Network	Next Hop	Metric	LocPrf	Weight Path
*> 20.0.1.0/24	10.0.34.3	0		0 30 i
*> 20.0.1.1/32	10.0.34.3	0		0 30 10 i
*> 20.0.1.2/32	10.0.34.3	0		0 30 10 i
*> 20.0.1.3/32	10.0.34.3	0		0 30 20 i
*> 20.0.1.4/32	10.0.34.3	0		0 30 20 i

Total number of prefixes 5
R4#

4．BGP 的路由汇总与静态手工汇总方式的路由一致，明细路由与汇总路由同时传递。

R4#show   ip bgp
BGP table version is 12, local router ID is 4.4.4.4
Status codes: s suppressed, d damped, h history, * valid, > best, i - internal,
              S Stale, b - backup entry, m - multipath, f Filter, a additional-path
Origin codes: i - IGP, e - EGP, ? - incomplete

Network	Next Hop	Metric	LocPrf	Weight Path
*> 20.0.1.0/24	10.0.34.3	0		0 30 i
*> 20.0.1.1/32	10.0.34.3	0		0 30 10 i
*> 20.0.1.2/32	10.0.34.3	0		0 30 10 i
*> 20.0.1.3/32	10.0.34.3	0		0 30 20 i
*> 20.0.1.4/32	10.0.34.3	0		0 30 20 i

Total number of prefixes 5
R4#

5．检查添加了 summary-only 参数的 BGP 汇总路由，R3 设备将明细路由做抑制操作，仅将汇总后的路由通告给邻居，由于汇总路由在 R3 设备执行，会将汇总的路由发送给明细路由通告者。

R3(config)#show   ip bgp
BGP table version is 13, local router ID is 3.3.3.3
Status codes: s suppressed, d damped, h history, * valid, > best, i - internal,
              S Stale, b - backup entry, m - multipath, f Filter, a additional-path

```
Origin codes: i - IGP, e - EGP, ? - incomplete

     Network          Next Hop         Metric     LocPrf     Weight Path
*>   20.0.1.0/24      0.0.0.0                                32768      i
s>   20.0.1.1/32      10.0.13.1        0                     0 10 i
s>   20.0.1.2/32      10.0.13.1        0                     0 10 i
s>   20.0.1.3/32      10.0.23.2        0                     0 20 i
s>   20.0.1.4/32      10.0.23.2        0                     0 20 i

Total number of prefixes 5
R3(config)#

R4#show   ip bgp
BGP table version is 13, local router ID is 4.4.4.4
Status codes: s suppressed, d damped, h history, * valid, > best, i - internal,
              S Stale, b - backup entry, m - multipath, f Filter, a additional-path
Origin codes: i - IGP, e - EGP, ? - incomplete

     Network          Next Hop         Metric     LocPrf     Weight Path
*>   20.0.1.0/24      10.0.34.3        0                     0 30 i

Total number of prefixes 1
R4#

R1#show   ip bgp
BGP table version is 16, local router ID is 1.1.1.1
Status codes: s suppressed, d damped, h history, * valid, > best, i - internal,
              S Stale, b - backup entry, m - multipath, f Filter, a additional-path
Origin codes: i - IGP, e - EGP, ? - incomplete

     Network          Next Hop         Metric     LocPrf     Weight Path
*>   20.0.1.0/24      10.0.13.3        0                     0 30 i
*>   20.0.1.1/32      0.0.0.0          0                     32768      i
*>   20.0.1.2/32      0.0.0.0          0                     32768      i

Total number of prefixes 3
R1#
```

6. 检查设置 as-set 参数的汇总路由继承明细路由的 AS 号，避免出现路由环路。

```
R4#show   ip bgp
BGP table version is 16, local router ID is 4.4.4.4
Status codes: s suppressed, d damped, h history, * valid, > best, i - internal,
              S Stale, b - backup entry, m - multipath, f Filter, a additional-path
Origin codes: i - IGP, e - EGP, ? - incomplete

     Network          Next Hop         Metric     LocPrf     Weight Path
*>   20.0.1.0/24      10.0.34.3        0                     0 30 {10,20} i

Total number of prefixes 1
R4#
```

7. BGP 路由汇总使用 suppress-map 参数，此时将 20.0.1.1/32 和 20.0.1.2/32 明细路由抑制，通告其余的明细路由。

```
R4#show   ip bgp
BGP table version is 17, local router ID is 4.4.4.4
```

```
Status codes: s suppressed, d damped, h history, * valid, > best, i - internal,
              S Stale, b - backup entry, m - multipath, f Filter, a additional-path
Origin codes: i - IGP, e - EGP, ? - incomplete

     Network          Next Hop         Metric     LocPrf      Weight Path
*>   20.0.1.0/24      10.0.34.3        0                      0 30 {10,20} i
*>   20.0.1.3/32      10.0.34.3        0                      0 30 20 i
*>   20.0.1.4/32      10.0.34.3        0                      0 30 20 i

Total number of prefixes 3
R4#


R3#show    ip bgp
BGP table version is 17, local router ID is 3.3.3.3
Status codes: s suppressed, d damped, h history, * valid, > best, i - internal,
              S Stale, b - backup entry, m - multipath, f Filter, a additionai-path
Origin codes: i - IGP, e - EGP, ? - incomplete

     Network          Next Hop         Metric     LocPrf      Weight Path
*>   20.0.1.0/24      0.0.0.0                                 32768 {10,20} i
s>   20.0.1.1/32      10.0.13.1        0                      0 10 i
s>   20.0.1.2/32      10.0.13.1        0                      0 10 i
*>   20.0.1.3/32      10.0.23.2        0                      0 20 i
*>   20.0.1.4/32      10.0.23.2        0                      0 20 i

Total number of prefixes 5
R3#
```

8．根据 attribute-map 参数，修改 BGP 汇总路由的属性。

```
---------未修改前属性为【i】---------
R4#show    ip bgp
BGP table version is 18, local router ID is 4.4.4.4
Status codes: s suppressed, d damped, h history, * valid, > best, i - internal,
              S Stale, b - backup entry, m - multipath, f Filter, a additional-path
Origin codes: i - IGP, e - EGP, ? - incomplete

     Network          Next Hop         Metric     LocPrf      Weight Path
*>   20.0.1.0/24      10.0.34.3        0                      0 30 {10,20} i
Total number of prefixes 1

---------修改后属性为【?】---------
R4#show ip bgp
BGP table version is 33, local router ID is 4.4.4.4
Status codes: s suppressed, d damped, h history, * valid, > best, i - internal,
              S Stale, b - backup entry, m - multipath, f Filter, a additional-path
Origin codes: i - IGP, e - EGP, ? - incomplete

     Network          Next Hop         Metric     LocPrf      Weight Path
*>   20.0.1.0/24      10.0.34.3        0                      0 30 {10,20} ?

Total number of prefixes 1
```

> 问题与思考

1. BGP 路由汇总配置添加 as-set 参数有什么作用？
2. BGP 路由汇总传递给邻居时，如何将明细路由屏蔽？

第 2 章　IS-IS 路由协议

2.1　IS-IS 协议基础

2.1.1　IS-IS 协议邻居建立

> 原理

IS-IS（Intermediate System to Intermediate System，中间系统到中间系统）是一种路由选择协议，适用于 IP 和 ISO CLNS 的双环境网络，是一种可扩展的、健壮的、易使用的 IGP 协议。

IS-IS 作为链路状态协议，具有链路状态协议的共通性。通过发送 Hello 报文来发现和维护邻居关系，通过向邻居发送协议数据报文 LSP（Link State PDU），来通告自身的链路状态。

IS-IS 路由器类型：

（1）Level-1 路由器

Level-1 路由器负责区域内的路由，它只与属于同一区域的 Level-1 和 Level-1-2 路由器形成邻居关系，和属于不同区域的 Level-1 路由器不能形成邻居关系。Level-1 路由器只负责维护本区域 Level-1 的链路状态数据库 LSDB（Link State Database），该 LSDB 包含本区域的路由信息，将到达本区域外的报文转发给最近的 Level-1-2 路由器。

（2）Level-2 路由器

Level-2 路由器负责区域间的路由，它可以与同一区域或者不同区域的 Level-2 路由器或者其他区域的 Level-1-2 路由器形成邻居关系。Level-2 路由器只负责维护一个 Level-2 的 LSDB，该 LSDB 包含区域间的路由信息。

所有 Level-2 级别（即形成 Level-2 邻居关系）的路由器组成路由域的骨干网，负责在不同区域间通信。路由域中 Level-2 级别的路由器必须是物理连续的，以保证骨干网的连续性。只有 Level-2 级别的路由器才能直接与区域外的路由器交换数据报文或路由信息。

（3）Level-1-2 路由器

同时属于 Level-1 和 Level-2 的路由器被称为 Level-1-2 路由器，它可以与同一区域的 Level-1 和 Level-1-2 路由器形成 Level-1 邻居关系，也可以与其他区域的 Level-2 和 Level-1-2 路由器形成 Level-2 邻居关系。Level-1 路由器必须通过 Level-1-2 路由器才能连接至其他区域。

➢ 任务拓扑

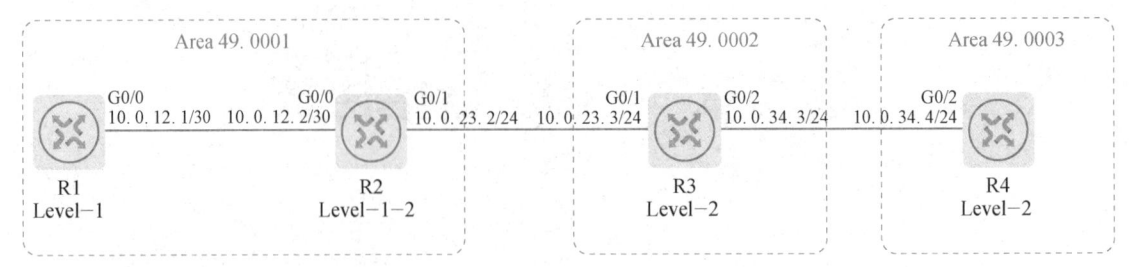

```
        Area 49.0001              |            Area 49.0002           |           Area 49.0003
  G0/0            G0/0   G0/1              G0/1         G0/2                  G0/2
10.0.12.1/30  10.0.12.2/30  10.0.23.2/24  10.0.23.3/24  10.0.34.3/24       10.0.34.4/24
     R1              R2                       R3                              R4
   Level-1       Level-1-2                  Level-2                        Level-2
```

➢ 实施步骤

1. 根据任务拓扑配置各设备接口 IP 地址，在 R1 设备和 R4 设备分别创建 loopback 0 接口，并用来做连通测试。

```
Ruijie>enable
Password:ruijie
Ruijie#configure terminal
Ruijie(config)#hostname   R1
R1(config)#interface    gigabitEthernet 0/0
R1(config-if-GigabitEthernet 0/0)# ip address    10.0.12.1 30
R1(config-if-GigabitEthernet 0/0)#exit
R1(config)#interface   loopback 0
R1(config-if-Loopback 0)#ip address   20.0.1.1 32
R1(config-if-Loopback 0)#exit
R1(config)#

Ruijie>enable
Password:ruijie
Ruijie#configure terminal
Ruijie(config)#hostname   R2
R2(config)# interface    gigabitEthernet 0/0
R2(config-if-GigabitEthernet 0/0)#ip address   10.0.12.2 30
R2(config-if-GigabitEthernet 0/0)#exit
R2(config)#interface    gigabitEthernet 0/1
R2(config-if-GigabitEthernet 0/1)#no   switchport
R2(config-if-GigabitEthernet 0/1)#ip address   10.0.23.2 24
R2(config-if-GigabitEthernet 0/1)#exit
R2(config)#

Ruijie>enable
Password:ruijie
Ruijie#configure terminal
Ruijie(config)#hostname   R3
R3(config)#interface    gigabitEthernet 0/1
R3(config-if-GigabitEthernet 0/1)#no   switchport
R3(config-if-GigabitEthernet 0/1)#ip address   10.0.23.3 24
R3(config-if-GigabitEthernet 0/1)#exit
R3(config)#interface    gigabitEthernet 0/2
R3(config-if-GigabitEthernet 0/2)#no   switchport
R3(config-if-GigabitEthernet 0/2)#ip address   10.0.34.3 24
R3(config-if-GigabitEthernet 0/2)#exit
R3(config)#
```

```
Ruijie>enable
Password:ruijie
Ruijie#configure terminal
Ruijie(config)#hostname   R4
R4(config)#interface    gigabitEthernet 0/2
R4(config-if-GigabitEthernet 0/2)#no switchport
R4(config-if-GigabitEthernet 0/2)#ip address    10.0.34.4 24
R4(config-if-GigabitEthernet 0/2)#exit
R4(config)#interface    loopback 0
R4(config-if-Loopback 0)#ip address    20.0.2.1 32
R4(config-if-Loopback 0)#exit
R4(config)#
```

2．配置 IS-IS 协议，各设备创建 IS-IS 协议进程，并在设备接口上开启 IS-IS 协议，将接口通告。

```
R1(config)#router    isis    1
R1(config-router)#net    49.0001.0000.0000.0001.00
R1(config-router)#is-type level-1
R1(config-router)#exit
R1(config)#interface    gigabitEthernet 0/0
R1(config-if-GigabitEthernet 0/0)#ip router    isis    1
R1(config-if-GigabitEthernet 0/0)#exit
R1(config)#interface    loopback 0
R1(config-if-Loopback 0)#ip router    isis    1
R1(config-if-Loopback 0)#exit
R1(config)#

R2(config)#router    isis    1
R2(config-router)#net    49.0001.0000.0000.0002.00
R2(config-router)#exit
R2(config)#interface    gigabitEthernet 0/0
R2(config-if-GigabitEthernet 0/0)#ip router    isis    1
R2(config-if-GigabitEthernet 0/0)#exit
R2(config)#interface    gigabitEthernet 0/1
R2(config-if-GigabitEthernet 0/1)#ip router    isis    1
R2(config-if-GigabitEthernet 0/1)#exit
R2(config)#

R3(config)#router    isis    1
R3(config-router)#net    49.0002.0000.0000.0003.00
R3(config-router)#is-type    level-2
R3(config-router)#exit
R3(config)#interface    gigabitEthernet 0/1
R3(config-if-GigabitEthernet 0/1)#ip router    isis    1
R3(config-if-GigabitEthernet 0/1)#exit
R3(config)#interface    gigabitEthernet 0/2
R3(config-if-GigabitEthernet 0/2)#ip router    isis    1
R3(config-if-GigabitEthernet 0/2)#exit
R3(config)#

R4(config)#router    isis    1
```

```
R4(config-router)#net    49.0003.0000.0000.0004.00
R4(config-router)#is-type    level-2
R4(config-router)#exit
R4(config)#interface    gigabitEthernet 0/2
R4(config-if-GigabitEthernet 0/2)#ip router    isis    1
R4(config-if-GigabitEthernet 0/2)#exit
R4(config)#interface    loopback 0
R4(config-if-Loopback 0)#ip router    isis    1
R4(config-if-Loopback 0)#exit
R4(config)#
```

> 任务验证

1．通过 show isis neighbors 命令检查 IS-IS 协议邻居关系建立情况。

```
R2#show   isis   neighbors

Area 1:
System Id     Type   IP Address       State    Holdtime    Circuit       Interface
R1            L1     10.0.12.1        Up       23          R2.01         GigabitEthernet 0/0
R3            L2     10.0.23.3        Up       8           R3.01         GigabitEthernet 0/1

R2#


R3#show   isis   neighbors

Area 1:
System Id     Type   IP Address       State    Holdtime    Circuit       Interface
R2            L2     10.0.23.2        Up       28          R3.01         GigabitEthernet 0/1
R4            L2     10.0.34.4        Up       8           R4.01         GigabitEthernet 0/2

R3#
```

2．检查 IS-IS 路由传递情况。

```
R1#show    ip route

Codes:   C - Connected, L - Local, S - Static
         R - RIP, O - OSPF, B - BGP, I - IS-IS, V - Overflow route
         N1 - OSPF NSSA external type 1, N2 - OSPF NSSA external type 2
         E1 - OSPF external type 1, E2 - OSPF external type 2
         SU - IS-IS summary, L1 - IS-IS level-1, L2 - IS-IS level-2
         IA - Inter area, EV - BGP EVPN, A - Arp to host
         LA - Local aggregate route
         * - candidate default

Gateway of last resort is 10.0.12.2 to network 0.0.0.0
I*L1   0.0.0.0/0 [115/1] via 10.0.12.2, 04:50:18, GigabitEthernet 0/0
C      10.0.12.0/30 is directly connected, GigabitEthernet 0/0
C      10.0.12.1/32 is local host.
I L1   10.0.23.0/24 [115/2] via 10.0.12.2, 04:50:29, GigabitEthernet 0/0
C      20.0.1.1/32 is local host.
R1#
```

```
R4#show    ip route

Codes:    C - Connected, L - Local, S - Static
          R - RIP, O - OSPF, B - BGP, I - IS-IS, V - Overflow route
          N1 - OSPF NSSA external type 1, N2 - OSPF NSSA external type 2
          E1 - OSPF external type 1, E2 - OSPF external type 2
          SU - IS-IS summary, L1 - IS-IS level-1, L2 - IS-IS level-2
          IA - Inter area, EV - BGP EVPN, A - Arp to host
          LA - Local aggregate route
          * - candidate default

Gateway of last resort is no set
I L2    10.0.12.0/30 [115/3] via 10.0.34.3, 04:50:36, GigabitEthernet 0/2
I L2    10.0.23.0/24 [115/2] via 10.0.34.3, 04:50:44, GigabitEthernet 0/2
C       10.0.34.0/24 is directly connected, GigabitEthernet 0/2
C       10.0.34.4/32 is local host.
I L2    20.0.1.1/32 [115/3] via 10.0.34.3, 04:27:18, GigabitEthernet 0/2
C       20.0.2.1/32 is local host.
R4#
```

3．测试 R1 设备和 R4 设备 loopback 0 接口连通情况。

```
R1#ping    20.0.2.1
Sending 5, 100-byte ICMP Echoes to 20.0.2.1, timeout is 2 seconds:
   < press Ctrl+C to break >
!!!!!
Success rate is 100 percent (5/5), round-trip min/avg/max = 3/10/36 ms.
R1#
```

> 问题与思考

1．IS-IS 协议的 NET 地址的长度默认值是多少？
2．IS-IS 协议不同级别路由器的区别？
3．IS-IS 协议的点到点网络类型和广播网络类型的建立过程有什么不同？

2.1.2　IS-IS 协议路由渗透

> 原理

通常情况下，Level-1 区域内的路由通过 Leve-1 路由器进行管理。所有的 Level-2 和 Level-1-2 路由器构成一个连续的骨干区域。Level-1 区域必须且只能与骨干区域相连，不同的 Level-1 区域之间并不相连。

在 RFC1195 中规定的集成 IS-IS 只能将 Level-1 的区域做类似 OSPF 的特殊区域处理，Level-2 中的路由不能发布到 Level-1 中，Level-1 路由器只能选择最近的一个 Level-1-2 路由器作为本区域的所有流量的出口（根据设置的 ATT bit 产生缺省路由），显然这样很容易造成次优路径。

在 RFC2966 中定义了路由泄露：可以将 Level-2 的 IP 路由引入到 Level-1 中，这样就可以允许 Level-1 路由器对某些或全部的 Level-2 路由选择出区域的最佳路径。

> **任务拓扑**

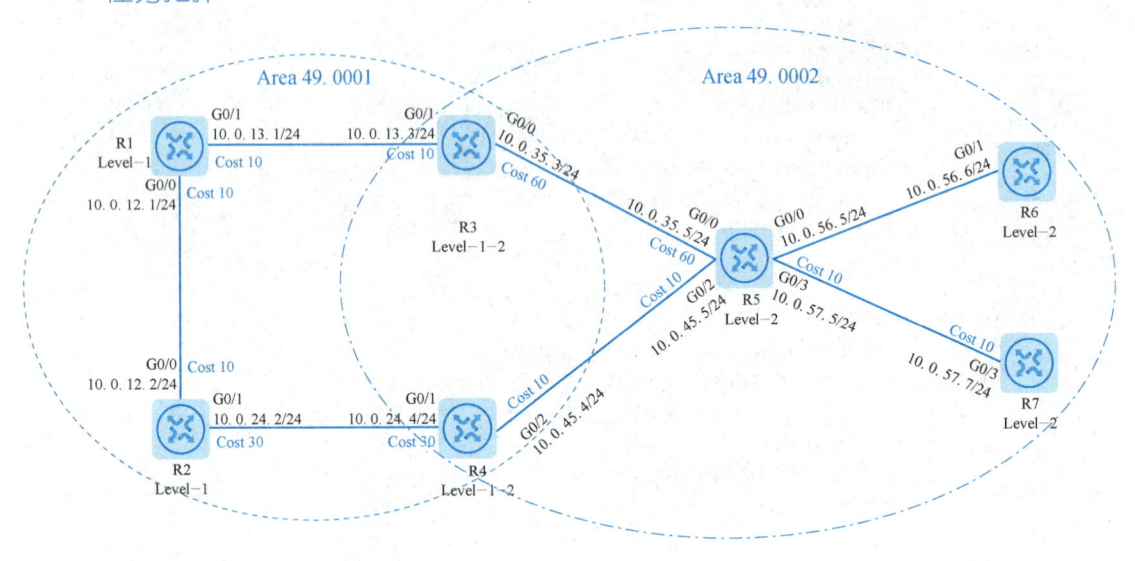

[拓扑图：Area 49.0001 包含 R1 (Level-1)、R2 (Level-1)、R3 (Level-1-2)、R4 (Level-1-2)；Area 49.0002 包含 R5 (Level-2)、R6 (Level-2)、R7 (Level-2)。

接口及 Cost 信息：
- R1 G0/1 10.0.13.1/24 Cost 10 —— R3 G0/1 10.0.13.3/24 Cost 10
- R1 G0/0 10.0.12.1/24 Cost 10
- R2 G0/0 10.0.12.2/24 Cost 10
- R2 G0/1 10.0.24.2/24 Cost 30 —— R4 G0/1 10.0.24.4/24 Cost 30
- R3 G0/0 10.0.35.3/24 Cost 60 —— R5 G0/0 10.0.35.5/24 Cost 60
- R4 G0/0 10.0.45.4/24 Cost 10 —— R5 G0/2 10.0.45.5/24 Cost 10
- R5 G0/0 10.0.56.5/24 Cost 10 —— R6 G0/1 10.0.56.6/24
- R5 G0/3 10.0.57.5/24 Cost 10 —— R7 G0/3 10.0.57.7/24]

> **实施步骤**

1. 根据任务拓扑配置各设备接口 IP 地址。

```
Ruijie>enable
Password:ruijie
Ruijie#configure terminal
Ruijie(config)#hostname    R1
R1(config)#interface    gigabitEthernet 0/0
R1(config-if-GigabitEthernet 0/0)#no    switchport
R1(config-if-GigabitEthernet 0/0)#ip address    10.0.12.1 24
R1(config-if-GigabitEthernet 0/0)#exit
R1(config)#interface    gigabitEthernet 0/1
R1(config-if-GigabitEthernet 0/1)#no    switchport
R1(config-if-GigabitEthernet 0/1)#ip address    10.0.13.1 24
R1(config-if-GigabitEthernet 0/1)#exit
R1(config)#

Ruijie>enable
Password:ruijie
Ruijie#configure terminal
Ruijie(config)#hostname    R2
R2(config)#interface    gigabitEthernet 0/0
R2(config-if-GigabitEthernet 0/0)#no    switchport
R2(config-if-GigabitEthernet 0/0)#ip address    10.0.12.2 24
R2(config-if-GigabitEthernet 0/0)#exit
R2(config)#interface    gigabitEthernet 0/1
R2(config-if-GigabitEthernet 0/1)#no    switchport
R2(config-if-GigabitEthernet 0/1)#ip address    10.0.24.2 24
R2(config-if-GigabitEthernet 0/1)#exit
R2(config)#

Ruijie>enable
Password:ruijie
Ruijie#configure terminal
```

```
Ruijie(config)#hostname    R3
R3(config)#interface    gigabitEthernet.0/1
R3(config-if-GigabitEthernet 0/1)#no    switchport
R3(config-if-GigabitEthernet 0/1)#ip address    10.0.13.3 24
R3(config-if-GigabitEthernet 0/1)#exit
R3(config)#interface    gigabitEthernet 0/0
R3(config-if-GigabitEthernet 0/0)#no    switchport
R3(config-if-GigabitEthernet 0/0)#ip address    10.0.35.3 24
R3(config-if-GigabitEthernet 0/0)#exit
R3(config)#

Ruijie>enable
Password:ruijie
Ruijie#configure terminal
Ruijie(config)#hostname    R4
R4(config)#interface    gigabitEthernet 0/1
R4(config-if-GigabitEthernet 0/1)#no    switchport
R4(config-if-GigabitEthernet 0/1)#ip address    10.0.24.4 24
R4(config-if-GigabitEthernet 0/1)#exit
R4(config)#interface    gigabitEthernet 0/2
R4(config-if-GigabitEthernet 0/2)#no    switchport
R4(config-if-GigabitEthernet 0/2)#ip address    10.0.45.4 24
R4(config-if-GigabitEthernet 0/2)#exit
R4(config)#

Ruijie>enable
Password:ruijie
Ruijie#configure terminal
Ruijie(config)#hostname    R5
R5(config)#interface    gigabitEthernet 0/0
R5(config-if-GigabitEthernet 0/0)#no    switchport
R5(config-if-GigabitEthernet 0/0)#ip address    10.0.35.5 24
R5(config-if-GigabitEthernet 0/0)#exit
R5(config)#interface    gigabitEthernet 0/1
R5(config-if-GigabitEthernet 0/1)#no    switchport
R5(config-if-GigabitEthernet 0/1)#ip address    10.0.56.5 24
R5(config-if-GigabitEthernet 0/1)#exit
R5(config)#interface    gigabitEthernet 0/2
R5(config-if-GigabitEthernet 0/2)#no    switchport
R5(config-if-GigabitEthernet 0/2)#ip address    10.0.45.5 24
R5(config-if-GigabitEthernet 0/2)#exit
R5(config)#interface    gigabitEthernet 0/3
R5(config-if-GigabitEthernet 0/3)#no    switchport
R5(config-if-GigabitEthernet 0/3)#ip address    10.0.57.5 24
R5(config-if-GigabitEthernet 0/3)#exit
R5(config)#

Ruijie>enable
Password:ruijie
Ruijie#configure terminal
Ruijie(config)#hostname    R6
R6(config)#interface    gigabitEthernet 0/1
R6(config-if-GigabitEthernet 0/1)#no    switchport
R6(config-if-GigabitEthernet 0/1)#ip address    10.0.56.6 24
R6(config-if-GigabitEthernet 0/1)#exit
```

R6(config)#

Ruijie>enable
Password:ruijie
Ruijie#configure terminal
Ruijie(config)#hostname   R7
R7(config)#interface   gigabitEthernet 0/3
R7(config-if-GigabitEthernet 0/3)#no   switchport
R7(config-if-GigabitEthernet 0/3)#ip address   10.0.57.7 24
R7(config-if-GigabitEthernet 0/3)#exit
R7(config)#

2. 配置 IS-IS 路由协议，R1 设备、R2 设备、R3 设备、R4 设备属于 Area 49.0001，R5 设备、R6 设备、R7 设备属于 Area 49.0002。

R1(config)#router   isis   1
R1(config-router)#net   49.0001.0000.0000.0001.00
R1(config-router)#is-type   level-1
R1(config-router)#exit
R1(config)#interface   gigabitEthernet 0/0
R1(config-if-GigabitEthernet 0/0)#ip router   isis   1
R1(config-if-GigabitEthernet 0/0)#exit
R1(config)#interface   gigabitEthernet 0/1
R1(config-if-GigabitEthernet 0/1)#ip router isis   1
R1(config-if-GigabitEthernet 0/1)#exit
R1(config)#

R2(config)#router   isis   1
R2(config-router)#net   49.0001.0000.0000.0002.00
R2(config-router)#is-type   level-1
R2(config-router)#exit
R2(config)#interface   gigabitEthernet 0/0
R2(config-if-GigabitEthernet 0/0)#ip router   isis   1
R2(config-if-GigabitEthernet 0/0)#exit
R2(config)#interface   gigabitEthernet 0/1
R2(config-if-GigabitEthernet 0/1)#ip router   isis   1
R2(config-if-GigabitEthernet 0/1)#exit
R2(config)#

R3(config)#router   isis   1
R3(config-router)#net   49.0001.0000.0000.0003.00
R3(config-router)#exit
R3(config)#interface   gigabitEthernet 0/1
R3(config-if-GigabitEthernet 0/1)#ip router   isis   1
R3(config-if-GigabitEthernet 0/1)#exit
R3(config)#interface   gigabitEthernet 0/0
R3(config-if-GigabitEthernet 0/0)#ip router   isis   1
R3(config-if-GigabitEthernet 0/0)#exit
R3(config)#

R4(config)#router   isis   1
R4(config-router)#net   49.0001.0000.0000.0004.00
R4(config-router)#exit
R4(config)#interface   gigabitEthernet 0/1
R4(config-if-GigabitEthernet 0/1)#ip router   isis   1
R4(config-if-GigabitEthernet 0/1)#exit

```
R4(config)#interface   gigabitEthernet 0/2
R4(config-if-GigabitEthernet 0/2)#ip router   isis   1
R4(config-if-GigabitEthernet 0/2)#exit
R4(config)#

R5(config)#router   isis   1
R5(config-router)#net   49.0002.0000.0000.0005.00
R5(config-router)#is-type   level-2
R5(config-router)#exit
R5(config)#interface   gigabitEthernet 0/0
R5(config-if-GigabitEthernet 0/0)#ip router   isis   1
R5(config-if-GigabitEthernet 0/0)#exit
R5(config)#interface   gigabitEthernet 0/1
R5(config-if-GigabitEthernet 0/1)#ip router   isis   1
R5(config-if-GigabitEthernet 0/1)#exit
R5(config)#interface   gigabitEthernet 0/2
R5(config-if-GigabitEthernet 0/2)#ip router   isis   1
R5(config-if-GigabitEthernet 0/2)#exit
R5(config)#interface   gigabitEthernet 0/3
R5(config-if-GigabitEthernet 0/3)#ip router   isis   1
R5(config-if-GigabitEthernet 0/3)#exit
R5(config)#

R6(config)#router   isis   1
R6(config-router)#net   49.0002.0000.0000.0006.00
R6(config-router)#is-type level-2
R6(config-router)#exit
R6(config)#interface   gigabitEthernet 0/1
R6(config-if-GigabitEthernet 0/1)#ip router   isis   1
R6(config-if-GigabitEthernet 0/1)#exit
R6(config)#

R7(config)#router   isis   1
R7(config-router)#net   49.0002.0000.0000.0007.00
R7(config-router)#is-type   level-2
R7(config-router)#exit
R7(config)#interface   gigabitEthernet 0/3
R7(config-if-GigabitEthernet 0/3)#ip router   isis   1
R7(config-if-GigabitEthernet 0/3)#exit
R7(config)#
```

3．根据任务拓扑，修改 IS-IS 的接口 Metric 值。

```
R1(config)#interface   gigabitEthernet 0/0
R1(config-if-GigabitEthernet 0/0)#isis   metric 10
R1(config-if-GigabitEthernet 0/0)#exit
R1(config)#
R1(config)#interface   gigabitEthernet 0/1
R1(config-if-GigabitEthernet 0/1)#isis metric   10
R1(config-if-GigabitEthernet 0/1)#exit

R2(config)#interface   gigabitEthernet 0/0
R2(config-if-GigabitEthernet 0/0)#isis   metric   10
R2(config-if-GigabitEthernet 0/0)#exit
R2(config)#interface   gigabitEthernet 0/1
R2(config-if-GigabitEthernet 0/1)#isis   metric   30
```

```
R2(config-if-GigabitEthernet 0/1)#exit
R2(config)#

R3(config)#interface   gigabitEthernet 0/1
R3(config-if-GigabitEthernet 0/1)#isis   metric   10
R3(config-if-GigabitEthernet 0/1)#exit
R3(config)#interface   gigabitEthernet 0/0
R3(config-if-GigabitEthernet 0/0)#isis   metric   60
R3(config-if-GigabitEthernet 0/0)#exit
R3(config)#

R4(config)#interface   gigabitEthernet 0/1
R4(config-if-GigabitEthernet 0/1)#isis   metric   30
R4(config-if-GigabitEthernet 0/1)#exit
R4(config)#interface   gigabitEthernet 0/2
R4(config-if-GigabitEthernet 0/2)#isis   metric   10
R4(config-if-GigabitEthernet 0/2)#exit
R4(config)#

R5(config)#interface   gigabitEthernet 0/0
R5(config-if-GigabitEthernet 0/0)#isis   metric   60
R5(config-if-GigabitEthernet 0/0)#exit
R5(config)#interface   gigabitEthernet 0/2
R5(config-if-GigabitEthernet 0/2)#isis   metric   10
R5(config-if-GigabitEthernet 0/2)#exit
R5(config)#interface   gigabitEthernet 0/1
R5(config-if-GigabitEthernet 0/1)#exit
R5(config)#interface   gigabitEthernet 0/3
R5(config-if-GigabitEthernet 0/3)#isis   metric 10
R5(config-if-GigabitEthernet 0/3)#exit
R5(config)#

R7(config)#interface   gigabitEthernet 0/3
R7(config-if-GigabitEthernet 0/3)#isis   metric   10
R7(config-if-GigabitEthernet 0/3)#exit
R7(config)#
```

4. 在 R7 设备创建 loopback 0 接口，同时在接口上开启 IS-IS 协议功能通告路由。

```
R7(config)#interface   loopback 0
R7(config-if-Loopback 0)#ip address   20.0.2.1 32
R7(config-if-Loopback 0)#exit
R7(config)#interface   loopback 0
R7(config-if-Loopback 0)#ip router   isis   1
R7(config-if-Loopback 0)#exit
R7(config)#
```

5. R3 设备和 R4 设备同为 Level-1-2 路由器，且都有 Level-1 和 Level-2 的数据库。Level-1 区域的路由器通过 ATT 置位 LSP 产生默认路由，通过默认路由访问其他区域目标网段时可能会出现次优路径的情况。为防止次优路径的产生，在 Level-1-2 路由器设置路由渗透，将 Level-2 区域的明细路由渗透到 Level-1 区域中。通过调用 route-map，针对部分目标网段设置路由渗透。

```
R3(config)#ip prefix-list ruijie permit   20.0.2.1/32
R3(config)#route-map ruijie permit   10
```

```
R3(config-route-map)#match   ip address   prefix-list   ruijie
R3(config-route-map)#exit
R3(config)#route-map ruijie permit   20
R3(config-route-map)#exit
R3(config)#
R3(config)#router   isis   1
R3(config-router)#redistribute   isis   1 level-2 into   level-1   route-map ruijie
R3(config-router)#exit
R3(config)#

R4(config)#ip prefix-list   ruijie permit   20.0.2.1/32
R4(config)#route-map   ruijie permit   10
R4(config-route-map)#match   ip address   prefix-list   ruijie
R4(config-route-map)#exit
R4(config)#route-map ruijie permit   20
R4(config-route-map)#exit
R4(config)#router   isis   1
R4(config-router)#redistribute   isis   1 level-2 into   level-1   route-map ruijie
R4(config-router)#exit
R4(config)#
```

> 任务验证

1. 通过 show   isis   neighbors 命令或者 show   clns neighbors 命令检查 IS-IS 协议邻居关系建立情况。

```
R2#show   isis   neighbors

Area 1:
System Id        Type   IP Address         State   Holdtime   Circuit         Interface
R1               L1     10.0.12.1          Up      28         R2.01           GigabitEthernet 0/0
R4               L1     10.0.24.4          Up      7          R4.01           GigabitEthernet 0/1

R2#

R4#show   clns neighbors

Area 1:
System Id         SNPA                    State   Holdtime   Type Protocol Interface
R2                5000.0002.0002          Up      27         L1   IS-IS    GigabitEthernet 0/1
R5                5000.0005.0003          Up      8          L2   IS-IS    GigabitEthernet 0/2

R4#

R5#show   isis   neighbors

Area 1:
System Id        Type   IP Address         State   Holdtime   Circuit         Interface
R3               L2     10.0.35.3          Up      21         R5.01           GigabitEthernet 0/0
R6               L2     10.0.56.6          Up      7          R6.01           GigabitEthernet 0/1
R4               L2     10.0.45.4          Up      22         R5.03           GigabitEthernet 0/2
R7               L2     10.0.57.7          Up      8          R7.01           GigabitEthernet 0/3

R5#
```

2. 通过 show   isis   database 命令查看数据库信息，Level-1 区域的路由器根据 ATT 置位条件产生一条去往最近的 Level-1-2 路由器的默认路由访问其他区域网段。

```
R3#show  isis  database
s
Area 1:
IS-IS Level-1 Link State Database:
LSPID                    LSP Seq Num      LSP Checksum    LSP Holdtime    ATT/P/OL
R1.00-00                 0x00000020       0x2D4C          476             0/0/0
R2.00-00                 0x00000021       0xBD83          726             0/0/0
R2.01-00                 0x00000019       0xDBE7          992             0/0/0
R3.00-00            *    0x0000001F       0xEDCB          502             1/0/0
R3.01-00            *    0x00000019       0xE3DB          1026            0/0/0
R4.00-00                 0x0000001E       0x2D73          513             1/0/0
R4.01-00                 0x00000019       0x03B9          1093            0/0/0

IS-IS Level-2 Link State Database:
LSPID                    LSP Seq Num      LSP Checksum    LSP Holdtime    ATT/P/OL
R3.00-00            *    0x0000008D       0xB5CE          1096            0/0/0
R4.00-00                 0x00000086       0x5C58          1024            0/0/0
R5.00-00                 0x00000025       0xFFCC          542             0/0/0
R5.01-00                 0x00000019       0x2297          1127            0/0/0
R5.03-00                 0x00000019       0x2D89          1136            0/0/0
R6.00-00                 0x0000001D       0x4920          1165            0/0/0
R6.01-00                 0x00000019       0x5A5B          1165            0/0/0
R7.00-00                 0x0000001D       0x3775          989             0/0/0
R7.01-00                 0x00000019       0x6053          1184            0/0/0

R3#
```

3. 检查 IS-IS 路由通告的路由信息，默认情况下 Level-1 区域无法接收其他区域传递的明细路由，根据 ATT 置位条件产生一条默认路由，此时能在 R1 设备或者 R2 设备查看的路由只有 Level-1 区域内的明细路由及一条默认路由。

```
R1#show   ip route
Codes:   C - Connected, L - Local, S - Static
         R - RIP, O - OSPF, B - BGP, I - IS-IS, V - Overflow route
         N1 - OSPF NSSA external type 1, N2 - OSPF NSSA external type 2
         E1 - OSPF external type 1, E2 - OSPF external type 2
         SU - IS-IS summary, L1 - IS-IS level-1, L2 - IS-IS level-2
         IA - Inter area, EV - BGP EVPN, A - Arp to host
         LA - Local aggregate route
         * - candidate default

Gateway of last resort is 10.0.13.3 to network 0.0.0.0
I*L1    0.0.0.0/0 [115/10] via 10.0.13.3, 00:39:24, GigabitEthernet 0/1
C       10.0.12.0/24 is directly connected, GigabitEthernet 0/0
C       10.0.12.1/32 is local host.
C       10.0.13.0/24 is directly connected, GigabitEthernet 0/1
C       10.0.13.1/32 is local host.
I L1    10.0.24.0/24 [115/20] via 10.0.12.2, 00:39:13, GigabitEthernet 0/0
I L1    10.0.35.0/24 [115/70] via 10.0.13.3, 00:38:59, GigabitEthernet 0/1
I L1    10.0.45.0/24 [115/30] via 10.0.12.2, 00:38:45, GigabitEthernet 0/0
R1#
```

```
R2#show    ip route
Codes:    C - Connected, L - Local, S - Static
          R - RIP, O - OSPF, B - BGP, I - IS-IS, V - Overflow route
          N1 - OSPF NSSA external type 1, N2 - OSPF NSSA external type 2
          E1 - OSPF external type 1, E2 - OSPF external type 2
          SU - IS-IS summary, L1 - IS-IS level-1, L2 - IS-IS level-2
          IA - Inter area, EV - BGP EVPN, A - Arp to host
          LA - Local aggregate route
          * - candidate default

Gateway of last resort is 10.0.12.1 to network 0.0.0.0
I*L1   0.0.0.0/0 [115/20] via 10.0.12.1, 00:10:01, GigabitEthernet 0/0
C      10.0.12.0/24 is directly connected, GigabitEthernet 0/0
C      10.0.12.2/32 is local host.
I L1   10.0.13.0/24 [115/20] via 10.0.12.1, 04:59:05, GigabitEthernet 0/0
C      10.0.24.0/24 is directly connected, GigabitEthernet 0/1
C      10.0.24.2/32 is local host.
I L1   10.0.35.0/24 [115/80] via 10.0.12.1, 04:58:47, GigabitEthernet 0/0
I L1   10.0.45.0/24 [115/40] via 10.0.24.4, 00:10:01, GigabitEthernet 0/1
R2#
```

4. 在 Area 49.0002 区域的设备查看路由，Level-2 区域属于骨干区域，Level-2 区域路由器能够计算出其他区域的明细路由，可以直接通过明细路由访问目标网段。

```
R6#show    ip route
Codes:    C - Connected, L - Local, S - Static
          R - RIP, O - OSPF, B - BGP, I - IS-IS, V - Overflow route
          N1 - OSPF NSSA external type 1, N2 - OSPF NSSA external type 2
          E1 - OSPF external type 1, E2 - OSPF external type 2
          SU - IS-IS summary, L1 - IS-IS level-1, L2 - IS-IS level-2
          IA - Inter area, EV - BGP EVPN, A - Arp to host
          LA - Local aggregate route
          * - candidate default

Gateway of last resort is no set
I L2   10.0.12.0/24 [115/51] via 10.0.56.5, 04:04:17, GigabitEthernet 0/1
I L2   10.0.13.0/24 [115/61] via 10.0.56.5, 04:04:17, GigabitEthernet 0/1
I L2   10.0.24.0/24 [115/41] via 10.0.56.5, 04:04:17, GigabitEthernet 0/1
I L2   10.0.35.0/24 [115/61] via 10.0.56.5, 04:04:27, GigabitEthernet 0/1
I L2   10.0.45.0/24 [115/11] via 10.0.56.5, 04:04:17, GigabitEthernet 0/1
C      10.0.56.0/24 is directly connected, GigabitEthernet 0/1
C      10.0.56.6/32 is local host.
I L2   10.0.57.0/24 [115/11] via 10.0.56.5, 04:04:07, GigabitEthernet 0/1
I L2   20.0.2.1/32 [115/11] via 10.0.56.5, 03:56:44, GigabitEthernet 0/1
R6#
```

5. 在 R2 设备通过 tracertroute 命令检查路径，R2 设备访问目标的路径选择 R1-R2-R3-R5-R7 路径，该路径为次优路径。因 Level-1 路由器根据产生的默认路由访问目标，使其选择去往目标最近的 Level-1-2 路由器，导致 R2 设备访问 20.0.2.1/32 时选择了次优路径。

```
R2#traceroute    20.0.2.1
  < press Ctrl+C to break >
Tracing the route to 20.0.2.1
```

```
1       10.0.12.1       <1 msec     <1 msec     <1 msec
2       10.0.12.1       <1 msec     <1 msec     <1 msec
3       10.0.13.3       3 msec
4       10.0.35.5       1 msec      <1 msec
5       20.0.2.1        2 msec
R2#
```

6. 在 R3 设备和 R4 设备设置完路由渗透后，选择 R2-R4-R5-R7 路径，此时选择明细路由访问目标网段为最优路径。

```
R2#traceroute   20.0.2.1
 < press Ctrl+C to break >
Tracing the route to 20.0.2.1

1       10.0.24.4       1 msec      <1 msec     <1 msec
2       10.0.45.5       5 msec
3       20.0.2.1        1 msec      1 msec
R2#

R2#show    ip route

Codes:    C - Connected, L - Local, S - Static
          R - RIP, O - OSPF, B - BGP, I - IS-IS, V - Overflow route
          N1 - OSPF NSSA external type 1, N2 - OSPF NSSA external type 2
          E1 - OSPF external type 1, E2 - OSPF external type 2
          SU - IS-IS summary, L1 - IS-IS level-1, L2 - IS-IS level-2
          IA - Inter area, EV - BGP EVPN, A - Arp to host
          LA - Local aggregate route
          * - candidate default

Gateway of last resort is 10.0.12.1 to network 0.0.0.0
I*L1    0.0.0.0/0 [115/20] via 10.0.12.1, 01:25:54, GigabitEthernet 0/0
C       10.0.12.0/24 is directly connected, GigabitEthernet 0/0
C       10.0.12.2/32 is local host.
I L1    10.0.13.0/24 [115/20] via 10.0.12.1, 06:14:58, GigabitEthernet 0/0
C       10.0.24.0/24 is directly connected, GigabitEthernet 0/1
C       10.0.24.2/32 is local host.
I L1    10.0.35.0/24 [115/80] via 10.0.12.1, 06:14:40, GigabitEthernet 0/0
I L1    10.0.45.0/24 [115/40] via 10.0.24.4, 01:25:54, GigabitEthernet 0/1
I IA    20.0.2.1/32 [115/50] via 10.0.24.4, 00:00:07, GigabitEthernet 0/1
R2#
```

> 问题与思考

1．IS-IS 协议与 OSPF 协议一般部署在网络拓扑中的哪个位置？
2．IS-IS 路由渗透的主要作用是什么？执行路由渗透后是否会导致环路产生？原因是什么？

2.2 IS-IS 特性

2.2.1 IS-IS 协议认证

➢ 原理

IS-IS 协议认证可以提高 IS-IS 网络的安全性。

接口认证在建立维护邻居关系中起作用。如果两台 IS-IS 设备配置了不同的接口认证密码，那么就无法建立邻居关系，这样可以避免未经授权或认证的 IS-IS 设备加入到需要认证的 IS-IS 网络中。接口认证的密码被封装在 Hello 报文中进行发送。

IS-IS 区域认证和路由域认证用于验证 LSP、CSNP、PSNP 报文，以避免未经授权或认证的路由信息被注入到 IS-IS 的链路状态数据库当中。在发送时，其认证密码被封装在相应的 LSP、CSNP、PSNP 报文中进行发送。

IS-IS 区域认证的密码被封装在 Level-1 的 LSP、CSNP、PSNP 报文中。接收这些报文时，也会验证密码是否一致。

IS-IS 路由域认证的密码被封装在 Level-2 的 LSP、CSNP、PSNP 报文中。接收这些报文时，也会验证密码是否一致。

目前提供两种认证方式：明文认证和 MD5 加密认证。

IS-IS 明文认证方式只提供有限的安全性，因为报文中传送的密码是可直观显示的。

IS-IS MD5 加密认证方式则可以提供较高的安全性，因为报文中传送的密码是经过 MD5 算法加密的。

➢ 任务拓扑

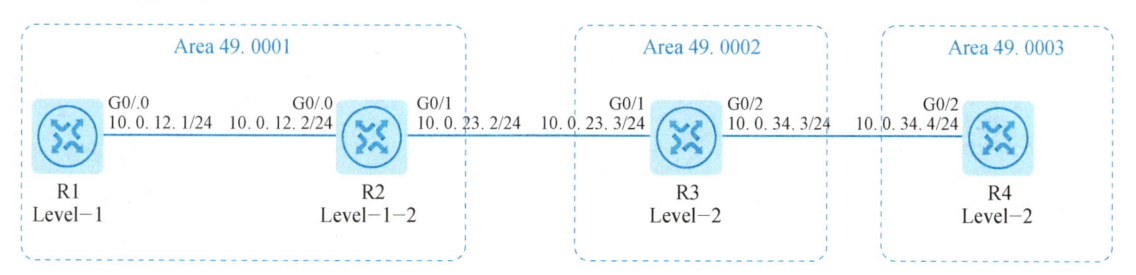

```
Area 49.0001                    Area 49.0002                    Area 49.0003
     G0/.0      G0/.0      G0/1      G0/1      G0/2      G0/2
  10.0.12.1/24  10.0.12.2/24  10.0.23.2/24  10.0.23.3/24  10.0.34.3/24  10.0.34.4/24
     R1              R2              R3              R4
  Level-1        Level-1-2        Level-2         Level-2
```

➢ 实施步骤

1. 根据任务拓扑配置各设备接口 IP 地址。

```
Ruijie>enable
Password:ruijie
Ruijie#configure terminal
Ruijie(config)#hostname   R1
R1(config)#interface   gigabitEthernet 0/0
R1(config-if-GigabitEthernet 0/0)#no   switchport
R1(config-if-GigabitEthernet 0/0)#ip address   10.0.12.1 24
R1(config-if-GigabitEthernet 0/0)#exit
R1(config)#
```

```
Ruijie>enable
Password:ruijie
Ruijie#configure terminal
Ruijie(config)#hostname    R2
R2(config)#interface    gigabitEthernet 0/0
R2(config-if-GigabitEthernet 0/0)#no    switchport
R2(config-if-GigabitEthernet 0/0)#ip address    10.0.12.2 24
R2(config-if-GigabitEthernet 0/0)#exit
R2(config)#interface    gigabitEthernet 0/1
R2(config-if-GigabitEthernet 0/1)#no    switchport
R2(config-if-GigabitEthernet 0/1)#ip address    10.0.23.2 24
R2(config-if-GigabitEthernet 0/1)#exit
R2(config)#

Ruijie>enable
Password:ruijie
Ruijie#configure terminal
Ruijie(config)#hostname    R3
R3(config)#interface    gigabitEthernet 0/1
R3(config-if-GigabitEthernet 0/1)#no    switchport
R3(config-if-GigabitEthernet 0/1)#ip address    10.0.23.3 24
R3(config-if-GigabitEthernet 0/1)#exit
R3(config)#interface    gigabitEthernet 0/2
R3(config-if-GigabitEthernet 0/2)#no    switchport
R3(config-if-GigabitEthernet 0/2)#ip address    10.0.34.3 24
R3(config-if-GigabitEthernet 0/2)#exit
R3(config)#

Ruijie>enable
Password:ruijie
Ruijie#configure terminal
Ruijie(config)#hostname    R4
R4(config)#interface    gigabitEthernet 0/2
R4(config-if-GigabitEthernet 0/2)#no    switchport
R4(config-if-GigabitEthernet 0/2)#ip address    10.0.34.4 24
R4(config-if-GigabitEthernet 0/2)#exit
R4(config)#
```

2. 配置 IS-IS 协议，R1 设备和 R2 设备属于 Area 49.0001，R3 设备属于 Area 49.0002，R4 设备属于 Area 49.0003。

```
R1(config)#router    isis    1
R1(config-router)#net    49.0001.0000.0000.0001.00
R1(config-router)#is-type    level-1
R1(config-router)#exit
R1(config)#int gigabitEthernet 0/0
R1(config-if-GigabitEthernet 0/0)#ip router isis    1
R1(config-if-GigabitEthernet 0/0)#exit
R1(config)#

R2(config)#router    isis    1
R2(config-router)#net 49.0001.0000.0000.0002.00
R2(config-router)#exit
R2(config)#interface    gigabitEthernet 0/0
R2(config-if-GigabitEthernet 0/0)#ip router    isis    1
```

```
R2(config-if-GigabitEthernet 0/0)#exit
R2(config)#interface   gigabitEthernet 0/1
R2(config-if-GigabitEthernet 0/1)#ip router   isis   1
R2(config-if-GigabitEthernet 0/1)#exit
R2(config)#

R3(config)#router isis   1
R3(config-router)#net   49.0002.0000.0000.0003.00
R3(config-router)#is-type   level-2
R3(config-router)#exit
R3(config)#interface   gigabitEthernet 0/1
R3(config-if-GigabitEthernet 0/1)#ip router   isis   1
R3(config-if-GigabitEthernet 0/1)#exit
R3(config)#interface   gigabitEthernet 0/2
R3(config-if-GigabitEthernet 0/2)#ip router isis   1
R3(config-if-GigabitEthernet 0/2)#exit
R3(config)#

R4(config)#router   isis   1
R4(config-router)#net   49.0003.0000.0000.0004.00
R4(config-router)#is-type   level-2
R4(config-router)#exit
R4(config)#interface   gigabitEthernet 0/2
R4(config-if-GigabitEthernet 0/2)#ip router   isis   1
R4(config-if-GigabitEthernet 0/2)#exit
R4(config)#
```

3．将全网互联接口配置接口认证，针对各个不同路由器级别发送 Hello 报文做验证，R3 设备和 R4 设备开启 MD5 加密认证，创建密码链。

```
R1(config)#interface   gigabitEthernet 0/0
R1(config-if-GigabitEthernet 0/0)#isis   password admin
R1(config-if-GigabitEthernet 0/0)#exit
R1(config)#

R2(config)#interface   gigabitEthernet 0/0
R2(config-if-GigabitEthernet 0/0)#isis   password   admin
R2(config-if-GigabitEthernet 0/0)#exit
R2(config)#interface   gigabitEthernet 0/1
R2(config-if-GigabitEthernet 0/1)#isis   password   admin
R2(config-if-GigabitEthernet 0/1)#exit
R2(config)#

R3(config-keychain)#key   1
R3(config-keychain-key)#key-string admin
R3(config-keychain-key)#exit
R3(config)#interface   gigabitEthernet 0/2
R3(config-if-GigabitEthernet 0/2)#isis   authentication   mode   md5
R3(config-if-GigabitEthernet 0/2)#isis   authentication   key-chain   isis
R3(config-if-GigabitEthernet 0/2)#exit
R3(config)#

R4(config)#key chain isis
R4(config-keychain)#key 1
R4(config-keychain-key)#key-string admin
R4(config-keychain-key)#exit
```

```
R4(config-keychain)#exit
R4(config)#interface  gigabitEthernet 0/2
R4(config-if-GigabitEthernet 0/2)#isis  authentication mode   md5
R4(config-if-GigabitEthernet 0/2)#isis  authentication  key-chain   isis
R4(config-if-GigabitEthernet 0/2)#exit
R4(config)#
```

➢ 任务验证

1. 检查 IS-IS 协议邻居关系建立情况。

```
R2#show isis   neighbors

Area 1:
System Id     Type    IP Address        State    Holdtime   Circuit        Interface
R1            L1      10.0.12.1    Up       23         R2.01          GigabitEthernet 0/0
R3            L2      10.0.23.3    Up        9         R3.01          GigabitEthernet 0/1

R2#


R3#show   isis   neighbors

Area 1:
System Id     Type    IP Address        State    Holdtime   Circuit        Interface
R2            L2      10.0.23.2    Up       22         R3.01          GigabitEthernet 0/1
R4            L2      10.0.34.4    Up        9         R4.01          GigabitEthernet 0/2

R3#
```

2. 配置接口认证后，IS-IS 协议邻居关系正常建立且正常传递路由信息。

```
R4#show   isis   neighbors

Area 1:
System Id     Type    IP Address        State    Holdtime   Circuit        Interface
R3            L2      10.0.34.3    Up       26         R4.01          GigabitEthernet 0/2

R4#

R4#show   ip route   isis
I L2    10.0.12.0/24 [115/3] via 10.0.34.3, 00:44:02, GigabitEthernet 0/2
I L2    10.0.23.0/24 [115/2] via 10.0.34.3, 00:44:02, GigabitEthernet 0/2
R4#
```

➢ 问题与思考

根据本节任务拓扑完成路由域认证配置，并说明路由域认证与接口、区域认证的不同点。

2.2.2 IS-IS 协议汇总

➢ 原理

IS-IS 路由汇总是指将具有相同前缀的多条路由信息汇总为一条路由，并将汇总后的路由（替代大量明细路由）发布给邻居。路由汇总有助于减轻协议交互的负担和减小路由表的规模，同时也能减小路由震荡对路由表的影响。

路由汇总后，只要汇总路由内存在任一可达地址或可达网段路由信息，就向外发布汇总路由，而不发布明细路由。

➢ 任务拓扑

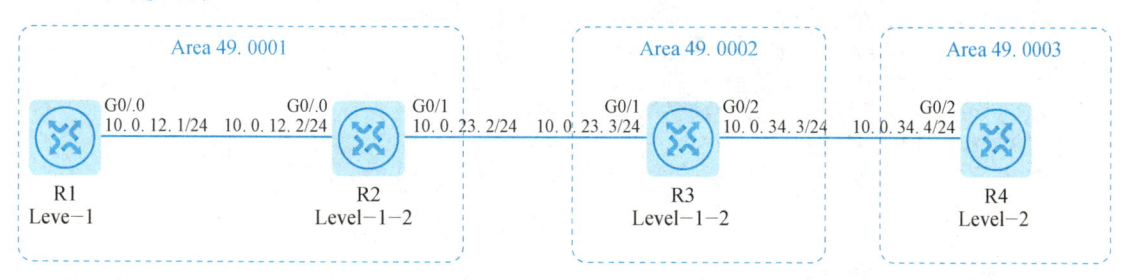

```
         Area 49.0001                          Area 49.0002              Area 49.0003
     G0/.0        G0/.0      G0/1        G0/1        G0/2       G0/2
  10.0.12.1/24  10.0.12.2/24 10.0.23.2/24 10.0.23.3/24 10.0.34.3/24 10.0.34.4/24
       R1              R2                R3                R4
     Leve-1        Level-1-2          Level-1-2         Level-2
```

➢ 实施步骤

1. 根据任务拓扑配置各设备接口 IP 地址。

```
Ruijie>enable
Password:ruijie
Ruijie#configure terminal
Ruijie(config)#hostname    R1
R1(config)#interface    gigabitEthernet 0/0
R1(config-if-GigabitEthernet 0/0)#no    switchport
R1(config-if-GigabitEthernet 0/0)#ip address    10.0.12.1 24
R1(config-if-GigabitEthernet 0/0)#exit
R1(config)#

Ruijie>enable
Password:ruijie
Ruijie#configure terminal
Ruijie(config)#hostname    R2
R2(config)#interface    gigabitEthernet 0/0
R2(config-if-GigabitEthernet 0/0)#no    switchport
R2(config-if-GigabitEthernet 0/0)#ip address    10.0.12.2 24
R2(config-if-GigabitEthernet 0/0)#exit
R2(config)#interface    gigabitEthernet 0/1
R2(config-if-GigabitEthernet 0/1)#no    switchport
R2(config-if-GigabitEthernet 0/1)#ip address    10.0.23.2 24
R2(config-if-GigabitEthernet 0/1)#exit
R2(config)#

Ruijie>enable
Password:ruijie
Ruijie#configure terminal
Ruijie(config)#hostname    R3
R3(config)#interface    gigabitEthernet 0/1
R3(config-if-GigabitEthernet 0/1)#no    switchport
R3(config-if-GigabitEthernet 0/1)#ip address    10.0.23.3 24
R3(config-if-GigabitEthernet 0/1)#exit
R3(config)#interface    gigabitEthernet 0/2
R3(config-if-GigabitEthernet 0/2)#no    switchport
R3(config-if-GigabitEthernet 0/2)#ip address    10.0.34.3 24
```

```
R3(config-if-GigabitEthernet 0/2)#exit
R3(config)#

Ruijie>enable
Password:ruijie
Ruijie#configure terminal
Ruijie(config)#hostname    R4
R4(config)#interface    gigabitEthernet 0/2
R4(config-if-GigabitEthernet 0/2)#no    switchport
R4(config-if-GigabitEthernet 0/2)#ip address    10.0.34.4 24
R4(config-if-GigabitEthernet 0/2)#exit
R4(config)#
```

2．配置 IS-IS 协议，各设备建立 IS-IS 邻居关系。

```
R1(config)#router   isis   1
R1(config-router)#net    49.0001.0000.0000.0001.00
R1(config-router)#is-type    level-1
R1(config-router)#exit
R1(config)#interface    gigabitEthernet 0/0
R1(config-if-GigabitEthernet 0/0)#ip router    isis   1
R1(config-if-GigabitEthernet 0/0)#exit
R1(config)#

R2(config)#router   isis   1
R2(config-router)#net    49.0001.0000.0000.0002.00
R2(config-router)#exit
R2(config)#interface    gigabitEthernet 0/0
R2(config-if-GigabitEthernet 0/0)#ip router    isis   1
R2(config-if-GigabitEthernet 0/0)#exit
R2(config)#interface    g R2(config)#interface    gigabitEthernet 0/1
R2(config-if-GigabitEthernet 0/1)#ip router    isis   1
R2(config-if-GigabitEthernet 0/1)#exit
R2(config)#

R3(config)#router   isis   1
R3(config-router)#net    49.0002.0000.0000.0003.00
R3(config-router)#exit
R3(config)#interface    gigabitEthernet 0/2
R3(config-if-GigabitEthernet 0/2)#ip router    isis   1
R3(config-if-GigabitEthernet 0/2)#exit
R3(config)#interface    gigabitEthernet 0/1
R3(config-if-GigabitEthernet 0/1)#ip router    isis   1
R3(config-if-GigabitEthernet 0/1)#exit
R3(config)#

R4(config)#router   isis   1
R4(config-router)#net    49.0003.0000.0000.0004.00
R4(config-router)#is-type    level-2
R4(config-router)#exit
R4(config)#interface    gigabitEthernet 0/2
R4(config-if-GigabitEthernet 0/2)#ip router    isis   1
R4(config-if-GigabitEthernet 0/2)#exit
R4(config)#
```

3．在 R1 设备和 R4 设备分别创建多个 loopback 接口，同时在 loopback 接口上开启 IS-IS

协议功能通告路由。

```
R1(config)#interface   loopback 0
R1(config-if-Loopback 0)#ip address   20.0.1.1 32
R1(config-if-Loopback 0)#ip router   isis   1
R1(config-if-Loopback 0)#exit
R1(config)#interface   loopback 1
R1(config-if-Loopback 1)#ip address   20.0.1.2 32
R1(config-if-Loopback 1)#ip router   isis   1
R1(config-if-Loopback 1)#exit
R1(config)#interface   loopback 2
R1(config-if-Loopback 2)#ip address   20.0.1.3 32
R1(config-if-Loopback 2)#ip router   isis   1
R1(config-if-Loopback 2)#exit
R1(config)#

R4(config)#interface   loopback 0
R4(config-if-Loopback 0)#ip address   20.0.2.1 32
R4(config-if-Loopback 0)#ip router   isis   1
R4(config-if-Loopback 0)#exit
R4(config)#interface   loopback 1
R4(config-if-Loopback 1)#ip address   20.0.2.2 32
R4(config-if-Loopback 1)#ip router   isis   1
R4(config-if-Loopback 1)#exit
R4(config)#interface   loopback 2
R4(config-if-Loopback 2)#ip address   20.0.2.3 32
R4(config-if-Loopback 2)#ip router   isis   1
R4(config-if-Loopback 2)#exit
R4(config)#
```

4. 在 R1 设备配置 Level-1 区域的路由汇总，汇总路由后需要添加参数 Level-1/Level-2。

```
R1(config)#router   isis   1
R1(config-router)#summary-address   20.0.1.0 255.255.255.0 level-1
R1(config-router)#exit
R1(config)#
```

5. 在 R4 设备配置 Level-2 区域的路由汇总，汇总路由后需要添加参数 Level-1/Level-2。

```
R4(config)#router   isis   1
R4(config-router)#summary-address   20.0.2.0 255.255.255.0 level-2
R4(config-router)#exit
R4(config)#
```

➢ 任务验证

1. 检查 IS-IS 协议邻居关系建立情况。

```
R2#show   isis   neighbors

Area 1:
System Id      Type    IP Address      State    Holdtime    Circuit      Interface
R1             L1      10.0.12.1       Up       29          R2.01        GigabitEthernet 0/0
R3             L2      10.0.23.3       Up       7           R3.02        GigabitEthernet 0/1

R2#
```

```
R3#show   isis   neighbors

Area 1:
System Id       Type    IP Address      State   Holdtime    Circuit     Interface
R2              L2      10.0.23.2       Up      23          R3.02       GigabitEthernet 0/1
R4              L2      10.0.34.4       Up      9           R4.01       GigabitEthernet 0/2

R3#
```

2. 检查 IS-IS 路由通告的路由信息，此时路由表只有明细路由。

```
R2#show   ip route

Codes:    C - Connected, L - Local, S - Static
          R - RIP, O - OSPF, B - BGP, I - IS-IS, V - Overflow route
          N1 - OSPF NSSA external type 1, N2 - OSPF NSSA external type 2
          E1 - OSPF external type 1, E2 - OSPF external type 2
          SU - IS-IS summary, L1 - IS-IS level-1, L2 - IS-IS level-2
          IA - Inter area, EV - BGP EVPN, A - Arp to host
          LA - Local aggregate route
          * - candidate default

Gateway of last resort is no set
C         10.0.12.0/24 is directly connected, GigabitEthernet 0/0
C         10.0.12.2/32 is local host.
C         10.0.23.0/24 is directly connected, GigabitEthernet 0/1
C         10.0.23.2/32 is local host.
I L2      10.0.34.0/24 [115/2] via 10.0.23.3, 00:17:02, GigabitEthernet 0/1
I L1      20.0.1.1/32 [115/1] via 10.0.12.1, 00:04:36, GigabitEthernet 0/0
I L1      20.0.1.2/32 [115/1] via 10.0.12.1, 00:04:32, GigabitEthernet 0/0
I L1      20.0.1.3/32 [115/1] via 10.0.12.1, 00:04:29, GigabitEthernet 0/0
I L2      20.0.2.1/32 [115/2] via 10.0.23.3, 00:04:14, GigabitEthernet 0/1
I L2      20.0.2.2/32 [115/2] via 10.0.23.3, 00:04:10, GigabitEthernet 0/1
I L2      20.0.2.3/32 [115/2] via 10.0.23.3, 00:04:06, GigabitEthernet 0/1
R2#

R4#show   ip route

Codes:    C - Connected, L - Local, S - Static
          R - RIP, O - OSPF, B - BGP, I - IS-IS, V - Overflow route
          N1 - OSPF NSSA external type 1, N2 - OSPF NSSA external type 2
          E1 - OSPF external type 1, E2 - OSPF external type 2
          SU - IS-IS summary, L1 - IS-IS level-1, L2 - IS-IS level-2
          IA - Inter area, EV - BGP EVPN, A - Arp to host
          LA - Local aggregate route
          * - candidate default

Gateway of last resort is no set
I L2      10.0.12.0/24 [115/3] via 10.0.34.3, 00:16:52, GigabitEthernet 0/2
I L2      10.0.23.0/24 [115/2] via 10.0.34.3, 00:16:59, GigabitEthernet 0/2
C         10.0.34.0/24 is directly connected, GigabitEthernet 0/2
C         10.0.34.4/32 is local host.
I L2      20.0.1.1/32 [115/3] via 10.0.34.3, 00:04:50, GigabitEthernet 0/2
I L2      20.0.1.2/32 [115/3] via 10.0.34.3, 00:04:46, GigabitEthernet 0/2
I L2      20.0.1.3/32 [115/3] via 10.0.34.3, 00:04:40, GigabitEthernet 0/2
C         20.0.2.1/32 is local host.
```

```
C       20.0.2.2/32 is local host.
C       20.0.2.3/32 is local host.
R4#
```

3. 在 R2 设备查看全局 IP 路由表，R1 设备将 loopback 接口汇总后，R2 设备只接收一条汇总路由 20.0.1.0/24。另外，R2 设备是 Level-1-2 路由器，会向 Level-1 区域发送 ATT 置位 LSP，Level-1 区域的路由器根据 ATT 置位条件产生一条默认路由，访问其他区域目标网段。

```
R2#show    ip route

Codes:    C - Connected, L - Local, S - Static
          R - RIP, O - OSPF, B - BGP, I - IS-IS, V - Overflow route
          N1 - OSPF NSSA external type 1, N2 - OSPF NSSA external type 2
          E1 - OSPF external type 1, E2 - OSPF external type 2
          SU - IS-IS summary, L1 - IS-IS level-1, L2 - IS-IS level-2
          IA - Inter area, EV - BGP EVPN, A - Arp to host
          LA - Local aggregate route
          * - candidate default

Gateway of last resort is no set
C       10.0.12.0/24 is directly connected, GigabitEthernet 0/0
C       10.0.12.2/32 is local host.
C       10.0.23.0/24 is directly connected, GigabitEthernet 0/1
C       10.0.23.2/32 is local host.
I L2    10.0.34.0/24 [115/2] via 10.0.23.3, 00:19:32, GigabitEthernet 0/1
I L1    20.0.1.0/24 [115/1] via 10.0.12.1, 00:00:56, GigabitEthernet 0/0
I L2    20.0.2.1/32 [115/2] via 10.0.23.3, 00:06:44, GigabitEthernet 0/1
I L2    20.0.2.2/32 [115/2] via 10.0.23.3, 00:06:40, GigabitEthernet 0/1
I L2    20.0.2.3/32 [115/2] via 10.0.23.3, 00:06:36, GigabitEthernet 0/1
R2#

R1#show    ip route

Codes:    C - Connected, L - Local, S - Static
          R - RIP, O - OSPF, B - BGP, I - IS-IS, V - Overflow route
          N1 - OSPF NSSA external type 1, N2 - OSPF NSSA external type 2
          E1 - OSPF external type 1, E2 - OSPF external type 2
          SU - IS-IS summary, L1 - IS-IS level-1, L2 - IS-IS level-2
          IA - Inter area, EV - BGP EVPN, A - Arp to host
          LA - Local aggregate route
          * - candidate default

Gateway of last resort is 10.0.12.2 to network 0.0.0.0
I*L1    0.0.0.0/0 [115/1] via 10.0.12.2, 00:39:13, GigabitEthernet 0/0
C       10.0.12.0/24 is directly connected, GigabitEthernet 0/0
C       10.0.12.1/32 is local host.
I L1    10.0.23.0/24 [115/2] via 10.0.12.2, 00:39:33, GigabitEthernet 0/0
I SU    20.0.1.0/24 [115/0] via 0.0.0.0, 00:20:38, Null 0
C       20.0.1.1/32 is local host.
C       20.0.1.2/32 is local host.
C       20.0.1.3/32 is local host.
R1#
```

5. 在 R4 设备执行路由汇总，R3 设备只能接收到汇总路由。

```
R4#show    ip route
```

```
Codes:   C - Connected, L - Local, S - Static
         R - RIP, O - OSPF, B - BGP, I - IS-IS, V - Overflow route
         N1 - OSPF NSSA external type 1, N2 - OSPF NSSA external type 2
         E1 - OSPF external type 1, E2 - OSPF external type 2
         SU - IS-IS summary, L1 - IS-IS level-1, L2 - IS-IS level-2
         IA - Inter area, EV - BGP EVPN, A - Arp to host
         LA - Local aggregate route
         * - candidate default

Gateway of last resort is no set
I L2    10.0.12.0/24 [115/3] via 10.0.34.3, 00:27:12, GigabitEthernet 0/2
I L2    10.0.23.0/24 [115/2] via 10.0.34.3, 00:27:19, GigabitEthernet 0/2
C       10.0.34.0/24 is directly connected, GigabitEthernet 0/2
C       10.0.34.4/32 is local host.
I L2    20.0.1.0/24 [115/3] via 10.0.34.3, 00:09:01, GigabitEthernet 0/2
I SU    20.0.2.0/24 [115/0] via 0.0.0.0, 00:03:20, Null 0
C       20.0.2.1/32 is local host.
C       20.0.2.2/32 is local host.
C       20.0.2.3/32 is local host.
R4#

R3#show   ip route

Codes:   C - Connected, L - Local, S - Static
         R - RIP, O - OSPF, B - BGP, I - IS-IS, V - Overflow route
         N1 - OSPF NSSA external type 1, N2 - OSPF NSSA external type 2
         E1 - OSPF external type 1, E2 - OSPF external type 2
         SU - IS-IS summary, L1 - IS-IS level-1, L2 - IS-IS level-2
         IA - Inter area, EV - BGP EVPN, A - Arp to host
         LA - Local aggregate route
         * - candidate default

Gateway of last resort is no set
I L2    10.0.12.0/24 [115/2] via 10.0.23.2, 00:35:22, GigabitEthernet 0/1
C       10.0.23.0/24 is directly connected, GigabitEthernet 0/1
C       10.0.23.3/32 is local host.
C       10.0.34.0/24 is directly connected, GigabitEthernet 0/2
C       10.0.34.3/32 is local host.
I L2    20.0.1.0/24 [115/2] via 10.0.23.2, 00:16:45, GigabitEthernet 0/1
I L2    **20.0.2.0/24 [115/1] via 10.0.34.4, 00:00:23, GigabitEthernet 0/2**
R3#
```

> 问题与思考

IS-IS 路由在 Level-1 区域上默认会由 Level-1-2 路由器产生 ATT 置位的 LSP 生成默认路由，Level-2 区域是否能自动生成默认路由？又如何生成默认路由？

第 3 章 MPLS 与 VPN 应用

3.1 MPLS 协议

3.1.1 静态 MPLS 协议

➤ 原理

MPLS（Multi-Protocol Label Switching）是多协议标签交换的简称，所谓多协议是指 MPLS 支持多种网络层协议，例如 IP、IPv6、IPX 等；而且兼容包括 ATM、帧中继、以太网、PPP 等在内的多种链路层技术；所谓标签交换就是给报文附上标签，根据标签进行转发。MPLS 使用的是面向无连接的控制面和面向连接的数据面，从而使面向无连接的 IP 网络中添加了面向连接的属性。最初 MPLS 技术是为了提高路由设备的转发速度，随着硬件技术和网络处理器的发展，这一优势已经不明显了。但由于 MPLS 将二层交换和三层路由技术结合起来的固有优势，它在解决 VPN（虚拟专用）和 TE（流量工程）问题时具有其他技术无可比拟的地方。在解决企业互联问题，提供各种新业务方面，MPLS 越来越被运营商看好，成为运营商在 IP 网络中提供增值业务的重要手段。因此 MPLS 技术获得了越来越多的关注，MPLS 的应用也逐步转向 MPLS VPN 和流量工程等应用。

MPLS 节点

一个运行 MPLS 功能的节点能够识别 MPLS 的信令协议（控制协议），能够运行一个或者多个的三层路由协议（包括静态路由），并且能够根据 MPLS 标签转发报文。通常一个 MPLS 节点也有能力转发原始的三层报文（如 IP 报文）。

转发等价类 FEC（Forwarding Equivalence Class）

转发等价类是指在转发过程中以等价的方式处理的一类数据报文，例如，目的地址前缀相同的数据报文。针对不同的应用，FEC 的归类方法也可以不尽相同。在 IP 单播路由应用中 FEC 可以按照目的地址前缀进行分类，也就是一条路由对应一个 FEC。属于相同转发等价类的报文在 MPLS 网络中将获得完全相同的处理。

LSR（Label Switching Router）标签交换路由器

LSR 是 MPLS 的网络的核心设备，它具有标签交换和标签分发功能。在 MPLS 体系文档 RFC3031 中讲到 LSR 同时是一个有能力转发原始的三层报文（如 IP 报文或者 IPv6 报文等）的 MPLS 节点，对于 MPLS 在 IP 网络中的应用，意味着 LSR 有能力同时执行正常的 IP 报文转发。

LER（Label Switching Edge Router）标签边缘路由器

在 MPLS 的网络边缘，进入到 MPLS 网络的流量由 LSR 区分为不同的 FEC，并为这些 FEC 请求相应的标签；离开 MPLS 网络的流量由 LER 弹出标签并还原为原始的报文。因此 LER 提供了流量分类、标签的映射和标签的移除功能。

LSP（Label Switched Path）标签交换路径

标签交换路径中一个 FEC 的数据流在不同的节点被赋予确定的标签，数据转发根据已经分配好的标签在各个节点上执行标签交换动作。数据流所走的路径就是 LSP，是一系列 LSR 的集合。可以将 LSP 看作类似穿越 MPLS 核心网络的一个隧道。

标签是一个长度固定、具有本地意义的短标识符。说它具有本地意义指的是标签只在相邻的两个 LSR 之间分发传递，因此也只在这两个 LSR 之间有效。一个标签用于标识一个 FEC，当报文到达 MPLS 网络入口时，它将按一定规则被划分给不同的 FEC，根据报文所属的 FEC 将相应的标签封装在报文中，在 MPLS 网络中按标签进行转发。

➢ 任务拓扑

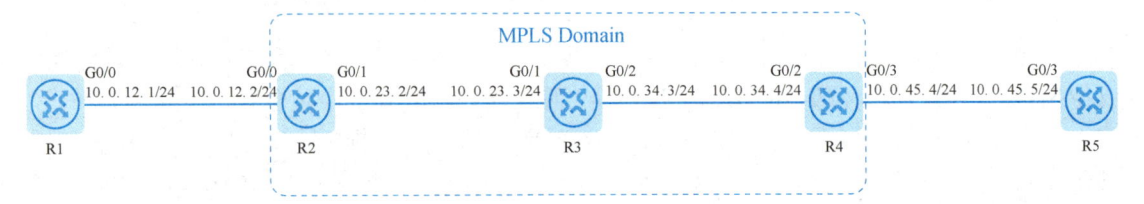

➢ 实施步骤

1. 根据任务拓扑配置各设备接口 IP 地址。

```
Ruijie>enable
Password:ruijie
Ruijie#configure terminal
Ruijie(config)#hostname    R1
R1(config)#interface    gigabitEthernet 0/0
R1(config-if-GigabitEthernet 0/0)#no    switchport
R1(config-if-GigabitEthernet 0/0)#ip address    10.0.12.1 24
R1(config-if-GigabitEthernet 0/0)#exit
R1(config)#interface    loopback 0
R1(config-if-Loopback 0)#ip address    10.0.1.1 32
R1(config-if-Loopback 0)#exit
R1(config)#

Ruijie>enable
Password:ruijie
Ruijie#configure terminal
Ruijie(config)#hostname    R2
R2(config)#interface    gigabitEthernet 0/0
R2(config-if-GigabitEthernet 0/0)#no    switchport
R2(config-if-GigabitEthernet 0/0)#ip address    10.0.12.2 24
R2(config-if-GigabitEthernet 0/0)#exit
R2(config)#interface    gigabitEthernet 0/1
R2(config-if-GigabitEthernet 0/1)#no    switchport
R2(config-if-GigabitEthernet 0/1)#ip address    10.0.23.2 24
R2(config-if-GigabitEthernet 0/1)#exit
```

```
R2(config)#interface    loopback 0
R2(config-if-Loopback 0)#ip address    10.0.2.2 32
R2(config-if-Loopback 0)#exit
R2(config)#

Ruijie>enable
Password:ruijie
Ruijie#configure terminal
Ruijie(config)#hostname   R3
R3(config)#interface    loopback 0
R3(config-if-Loopback 0)#ip address    10.0.3.3 32
R3(config-if-Loopback 0)#exit
R3(config)#interface    gigabitEthernet 0/1
R3(config-if-GigabitEthernet 0/1)#no    switchport
R3(config-if-GigabitEthernet 0/1)#ip address    10.0.23.3 24
R3(config-if-GigabitEthernet 0/1)#exit
R3(config)#interface    gigabitEthernet 0/2
R3(config-if-GigabitEthernet 0/2)#no    switchport
R3(config-if-GigabitEthernet 0/2)#ip address    10.0.34.3 24
R3(config-if-GigabitEthernet 0/2)#exit
R3(config)#

Ruijie>enable
Password:ruijie
Ruijie#configure terminal
Ruijie(config)#hostname   R4
R4(config)#interface    loopback 0
R4(config-if-Loopback 0)#ip address    10.0.4.4 32
R4(config-if-Loopback 0)#exit
R4(config)#interface    gigabitEthernet 0/2
R4(config-if-GigabitEthernet 0/2)#no    switchport
R4(config-if-GigabitEthernet 0/2)#ip address    10.0.34.4 24
R4(config-if-GigabitEthernet 0/2)#exit
R4(config)#interface    gigabitEthernet 0/3
R4(config-if-GigabitEthernet 0/3)#no    switchport
R4(config-if-GigabitEthernet 0/3)#ip address    10.0.45.4 24
R4(config-if-GigabitEthernet 0/3)#exit
R4(config)#

Ruijie>enable
Password:ruijie
Ruijie#configure terminal
Ruijie(config)#hostname   R5
R5(config)#interface    gigabitEthernet 0/3
R5(config-if-GigabitEthernet 0/3)#no    switchport
R5(config-if-GigabitEthernet 0/3)#ip address    10.0.45.5 24
R5(config-if-GigabitEthernet 0/3)#exit
R5(config)#interface    loopback 0
R5(config-if-Loopback 0)#ip address    10.0.5.5 32
R5(config-if-Loopback 0)#exit
R5(config)#
```

2．在 MPLS 域内配置 OSPF 协议实现内部路由互通。

```
R1(config)#router    ospf 1
R1(config-router)#router-id    1.1.1.1
```

```
Change router-id and update OSPF process! [yes/no]:y
R1(config-router)#network    10.0.12.0 0.0.0.255 area    0
R1(config-router)#network    10.0.1.1 0.0.0.0 area    0
R1(config-router)#exit
R1(config)#


R2(config)#router   ospf   1
R2(config-router)#router-id    2.2.2.2
Change router-id and update OSPF process! [yes/no]:y
R2(config-router)#network    10.0.2.2 0.0.0.0 area    0
R2(config-router)#network    10.0.12.0 0.0.0.255 area 0
R2(config-router)#network    10.0.23.0 0.0.0.255 area 0
R2(config-router)#exit
R2(config)#


R3(config)#router   ospf   1
R3(config-router)#router-id    3.3.3.3
Change router-id and update OSPF process! [yes/no]:y
R3(config-router)#network    10.0.3.3 0.0.0.0 area    0
R3(config-router)#network    10.0.23.0 0.0.0.255 area    0
R3(config-router)#network    10.0.34.0 0.0.0.255 area    0
R3(config-router)#exit
R3(config)#


R4(config)#router   ospf   1
R4(config-router)#router-id    4.4.4.4
Change router-id and update OSPF process! [yes/no]:y
R4(config-router)#network    10.0.34.0 0.0.0.255 area    0
R4(config-router)#network    10.0.45.0 0.0.0.255 area    0
R4(config-router)#network    10.0.4.4 0.0.0.0 area    0
R4(config-router)#exit
R4(config)#


R5(config)#router   ospf   1
R5(config-router)#router-id    5.5.5.5
Change router-id and update OSPF process! [yes/no]:y
R5(config-router)#network    10.0.45.0 0.0.0.255 area    0
R5(config-router)#network    10.0.5.5 0.0.0.0 area    0
R5(config-router)#exit
R5(config)#
```

3. 在各设备全局模式开启 mpls   ip 协议，同时在接口开启标签交换功能。

```
R1(config)#mpls  ip
R1(config)#interface    gigabitEthernet 0/0
R1(config-if-GigabitEthernet 0/0)#label-switching
R1(config-if-GigabitEthernet 0/0)#exit
R1(config)#

R2(config)#mpls ip
R2(config)#interface    gigabitEthernet 0/0
R2(config-if-GigabitEthernet 0/0)#label-switching
R2(config-if-GigabitEthernet 0/0)#exit
R2(config)#interface    gigabitEthernet 0/1
R2(config-if-GigabitEthernet 0/1)#label-switching
```

```
R2(config-if-GigabitEthernet 0/1)#exit
R2(config)#


R3(config)#mpls   ip
R3(config)#interface    gigabitEthernet 0/1
R3(config-if-GigabitEthernet 0/1)#label-switching
R3(config-if-GigabitEthernet 0/1)#exit
R3(config)#interface    gigabitEthernet 0/2
R3(config-if-GigabitEthernet 0/2)#label-switching
R3(config-if-GigabitEthernet 0/2)#exit
R3(config)#

R4(config)#mpls   ip
R4(config)#interface    gigabitEthernet 0/2
R4(config-if-GigabitEthernet 0/2)#label-switching
R4(config-if-GigabitEthernet 0/2)#exit
R4(config)#interface    gigabitEthernet 0/3
R4(config-if-GigabitEthernet 0/3)#label-switching
R4(config-if-GigabitEthernet 0/3)#exit
R4(config)#

R5(config)#interface    gigabitEthernet 0/3
R5(config-if-GigabitEthernet 0/3)#label-switching
R5(config-if-GigabitEthernet 0/3)#exit
R5(config)#
```

4. 在网络设备全局配置模式下创建 R1 设备到 R5 设备关于目的 10.0.5.5/32 的静态 LSP。

```
R1(config)#mpls    static ftn 10.0.5.5/32 out-label    102 nexthop gigabitEthernet 0/0 10.0.12.2

R2(config)#mpls    static ilm    in-label    102 forward-action    swap-label 203 nexthop gigabitEthernet 0/1 10.0.23.3 fec 10.0.5.5/32

R3(config)#mpls static ilm    in-label    203 forward-action    swap-label    304 nexthop gigabitEthernet 0/2 10.0.34.4 fec 10.0.5.5/32

R4(config)#mpls    static ilm    in-label    304 forward-action swap-label 3 nexthop gigabitEthernet 0/3 10.0.45.5 fec    10.0.5.5/32
```

5. 在网络设备全局模式下创建 R5 设备到 R1 设备关于目的 10.0.1.1/32 的静态 LSP。

```
R5(config)#mpls    static ftn    10.0.1.1/32 out-label    504 nexthop gigabitEthernet 0/3 10.0.45.4

R4(config)#mpls    static ilm    in-label    504 forward-action swap-label    403 nexthop gigabitEthernet 0/2 10.0.34.3 fec 10.0.1.1/32

R3(config)#mpls static ilm in-label    403 forward-action    swap-label    302 nexthop    gigabitEthernet 0/1 10.0.23.2 fec    10.0.1.1/32

R2(config)#mpls static ilm    in-label    302 forward-action swap-label 3 nexthop gigabitEthernet 0/0 10.0.12.1 fec    10.0.1.1/32
```

➢ 任务验证

1. 检查 OSPF 协议配置情况。

```
R2#show    ip ospf    neighbor

OSPF process 1, 2 Neighbors, 2 is Full:
Neighbor ID    Pri    State      BFD State    Dead Time    Address      Interface
1.1.1.1        1      Full/DR    -            00:00:35     10.0.12.1    GigabitEthernet 0/0
3.3.3.3        1      Full/DR    -            00:00:33     10.0.23.3    GigabitEthernet 0/1
R2#

R3#show    ip ospf    neighbor

OSPF process 1, 2 Neighbors, 2 is Full:
Neighbor ID    Pri    State       BFD State    Dead Time    Address      Interface
2.2.2.2        1      Full/BDR    -            00:00:39     10.0.23.2    GigabitEthernet 0/1
4.4.4.4        1      Full/DR     -            00:00:36     10.0.34.4    GigabitEthernet 0/2
R3#
```

2．检查各设备之间的互通情况。

```
R1#show    ip route    ospf
O    10.0.2.2/32 [110/1] via 10.0.12.2, 01:39:57, GigabitEthernet 0/0
O    10.0.3.3/32 [110/2] via 10.0.12.2, 01:14:55, GigabitEthernet 0/0
O    10.0.4.4/32 [110/3] via 10.0.12.2, 01:14:55, GigabitEthernet 0/0
O    10.0.5.5/32 [110/4] via 10.0.12.2, 00:35:28, GigabitEthernet 0/0
O    10.0.23.0/24 [110/2] via 10.0.12.2, 01:15:02, GigabitEthernet 0/0
O    10.0.34.0/24 [110/3] via 10.0.12.2, 01:14:55, GigabitEthernet 0/0
O    10.0.45.0/24 [110/4] via 10.0.12.2, 01:14:55, GigabitEthernet 0/0
R1#

R1#ping 10.0.3.3
Sending 5, 100-byte ICMP Echoes to 10.0.3.3, timeout is 2 seconds:
  < press Ctrl+C to break >
!!!!!
Success rate is 100 percent (5/5), round-trip min/avg/max = 1/2/4 ms.
R1#
R1#ping    10.0.4.4
Sending 5, 100-byte ICMP Echoes to 10.0.4.4, timeout is 2 seconds:
  < press Ctrl+C to break >
!!!!!
Success rate is 100 percent (5/5), round-trip min/avg/max = 2/2/4 ms.
R1#ping 10.0.5.5
Sending 5, 100-byte ICMP Echoes to 10.0.5.5, timeout is 2 seconds:
  < press Ctrl+C to break >
!!!!!
Success rate is 100 percent (5/5), round-trip min/avg/max = 3/4/5 ms.
R1#
```

3．检查 MPLS 协议配置情况。

```
R2#show    mpls    forwarding-table

Label Operation Code:
PH--PUSH label
PP--POP label
SW--SWAP label
SP--SWAP topmost label and push new label
DP--DROP packet
PC--POP label and continue lookup by IP or Label
```

PI--POP label and do ip lookup forward
PN--POP label and forward to nexthop
PM--POP label and do MAC lookup forward
PV--POP label and output to VC attach interface
IP--IP lookup forward
  s--stale

Local label	Outgoing label	OP	FEC	Outgoing interface	Nexthop
102	203	SW	10.0.5.5/32	Gi0/1	10.0.23.3
302	imp-null	PP	10.0.1.1/32	Gi0/0	10.0.12.1

R3#show mpls forwarding-table

Label Operation Code:
PH--PUSH label
PP--POP label
SW--SWAP label
SP--SWAP topmost label and push new label
DP--DROP packet
PC--POP label and continue lookup by IP or Label
PI--POP label and do ip lookup forward
PN--POP label and forward to nexthop
PM--POP label and do MAC lookup forward
PV--POP label and output to VC attach interface
IP--IP lookup forward
  s--stale

Local label	Outgoing label	OP	FEC	Outgoing interface	Nexthop
203	304	SW	10.0.5.5/32	Gi0/2	10.0.34.4
403	302	SW	10.0.1.1/32	Gi0/1	10.0.23.2

R3#

R2#show mpls forwarding-table detail
Label Operation Code:
PH--PUSH label
PP--POP label
SW--SWAP label
SP--SWAP topmost label and push new label
DP--DROP packet
PC--POP label and continue lookup by IP or Label
PI--POP label and do ip lookup forward
PN--POP label and forward to nexthop
PM--POP label and do MAC lookup forward
PV--POP label and output to VC attach interface
IP--IP lookup forward
  s--stale

Local label	Outgoing label	OP	FEC	Outgoing interface	Nexthop
102	203	SW	10.0.5.5/32	Gi0/1	10.0.23.3
	Added by Route(vrf Global), Tag Stack: { 203 }				
302	imp-null	PP	10.0.1.1/32	Gi0/0	10.0.12.1
	Added by Route(vrf Global), Tag Stack: { }				

R2#

4. 测试 R1 设备和 R5 设备之间的 loopback 0 接口互通情况。

```
R5#ping    10.0.1.1 source    10.0.5.5
Sending 5, 100-byte ICMP Echoes to 10.0.1.1, timeout is 2 seconds:
  < press Ctrl+C to break >
!!!!!
Success rate is 100 percent (5/5), round-trip min/avg/max = 3/4/6 ms.
R5#

R1#ping    10.0.5.5 source    10.0.1.1
Sending 5, 100-byte ICMP Echoes to 10.0.5.5, timeout is 2 seconds:
  < press Ctrl+C to break >
!!!!!
Success rate is 100 percent (5/5), round-trip min/avg/max = 3/3/4 ms.
R1#
```

5. 使用 Wireshark 抓包工具分析数据流量，对应的 FEC 在 MPLS 域内压入标签转发，中间节点负责将标签替换。

```
→ 1 0.0000... 10.0.1.1      10.0.5.5      ICMP    146 Echo (ping) request  id=0x0700, seq=15359/65339, ttl=64 (reply in 2)
  2 0.0036... 10.0.5.5      10.0.1.1      ICMP    146 Echo (ping) reply    id=0x0700, seq=15359/65339, ttl=64 (request in 1)
  3 0.1053... 10.0.1.1      10.0.5.5      ICMP    146 Echo (ping) request  id=0x0700, seq=18175/65350, ttl=64 (no response found!)
  4 0.1086... 10.0.5.5      10.0.1.1      ICMP    146 Echo (ping) reply    id=0x0700, seq=18175/65350, ttl=64 (request in 3)
  5 0.2103... 10.0.1.1      10.0.5.5      ICMP    146 Echo (ping) request  id=0x0700, seq=20735/65360, ttl=64 (reply in 6)
  6 0.2139... 10.0.5.5      10.0.1.1      ICMP    146 Echo (ping) reply    id=0x0700, seq=20735/65360, ttl=64 (request in 5)
  7 0.3160... 10.0.1.1      10.0.5.5      ICMP    146 Echo (ping) request  id=0x0700, seq=23551/65371, ttl=64 (reply in 8)
Frame 1: 146 bytes on wire (1168 bits), 146 bytes captured (1168 bits) on interface 0
Ethernet II, Src: 50:00:00:02:00:02 (50:00:00:02:00:02), Dst: 50:00:00:03:00:02 (50:00:00:03:00:02)
MultiProtocol Label Switching Header, Label: 203, Exp: 0, S: 1, TTL: 63
  0000 0000 0000 1100 1011 .... .... .... = MPLS Label: 203
  .... .... .... .... .... 000. .... .... = MPLS Experimental Bits: 0
  .... .... .... .... .... ...1 .... .... = MPLS Bottom Of Label Stack: 1
  .... .... .... .... .... .... 0011 1111 = MPLS TTL: 63
Internet Protocol Version 4, Src: 10.0.1.1, Dst: 10.0.5.5
Internet Control Message Protocol

→ 1 0.0000... 10.0.1.1      10.0.5.5      ICMP    146 Echo (ping) request  id=0x0700, seq=15359/65339, ttl=64 (reply in 2)
  2 0.0036... 10.0.5.5      10.0.1.1      ICMP    146 Echo (ping) reply    id=0x0700, seq=15359/65339, ttl=64 (request in 1)
  3 0.1053... 10.0.1.1      10.0.5.5      ICMP    146 Echo (ping) request  id=0x0700, seq=18175/65350, ttl=64 (no response found!)
  4 0.1086... 10.0.5.5      10.0.1.1      ICMP    146 Echo (ping) reply    id=0x0700, seq=18175/65350, ttl=64 (request in 3)
  5 0.2103... 10.0.1.1      10.0.5.5      ICMP    146 Echo (ping) request  id=0x0700, seq=20735/65360, ttl=64 (reply in 6)
  6 0.2139... 10.0.5.5      10.0.1.1      ICMP    146 Echo (ping) reply    id=0x0700, seq=20735/65360, ttl=64 (request in 5)
  7 0.3160... 10.0.1.1      10.0.5.5      ICMP    146 Echo (ping) request  id=0x0700, seq=23551/65371, ttl=64 (reply in 8)
Frame 2: 146 bytes on wire (1168 bits), 146 bytes captured (1168 bits) on interface 0
Ethernet II, Src: 50:00:00:03:00:02 (50:00:00:03:00:02), Dst: 50:00:00:02:00:02 (50:00:00:02:00:02)
MultiProtocol Label Switching Header, Label: 302, Exp: 0, S: 1, TTL: 62
  0000 0000 0001 0010 1110 .... .... .... = MPLS Label: 302
  .... .... .... .... .... 000. .... .... = MPLS Experimental Bits: 0
  .... .... .... .... .... ...1 .... .... = MPLS Bottom Of Label Stack: 1
  .... .... .... .... .... .... 0011 1110 = MPLS TTL: 62
Internet Protocol Version 4, Src: 10.0.5.5, Dst: 10.0.1.1
Internet Control Message Protocol
```

> 问题与思考

1. 在静态 MPLS LSP 配置上使用标签 3 有什么作用？
2. FEC 的分类都有哪些？目前常用的 FEC 分类是哪一种？

3.1.2 动态 MPLS LDP 协议

> 原理

标签分发协议 LDP（Label Distribution Protocol）是多协议标签交换 MPLS 的一种控制协议，相当于传统网络中的信令协议，负责转发等价类 FEC（Forwarding Equivalence Class）的分类、标签的分配以及标签交换路径 LSP（Label Switched Path）的建立和维护等。LDP 规定了标签分发过程中的各种消息以及相关处理过程。

LDP 对等体是指相互之间存在 LDP 会话、使用 LDP 来交换标签消息的两个 LSR。LDP 对等体通过它们之间的 LDP 会话获得对方的标签。

在未开启 LDP 域间 LSP 功能时，LDP 协议只有对精确匹配的路由才会将其对应的标签用于转发并往所有 LDP 邻居分发标签映射消息。开启 LDP 域间 LSP 功能后，如果路由表中存在和标签映射消息中的 FEC 最长匹配的路由，且发送标签映射消息的邻居是对应最长匹配路由的下一跳地址，则会添加对应的转发表项，并向 LDP 邻居分发标签映射消息（针对接收到的标签映射消息中的 FEC，而非从路由表中获得的最长匹配路由的 FEC 分发标签映射消息）。

➢ 任务拓扑

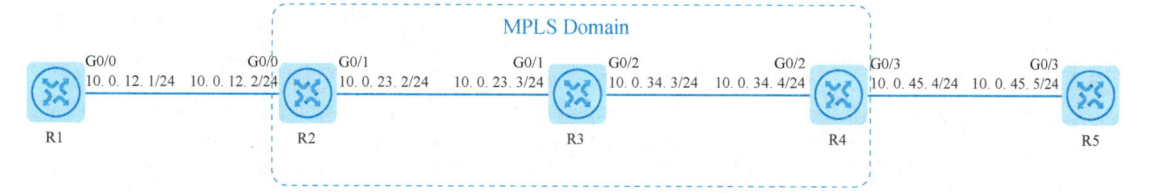

➢ 实施步骤

1. 根据任务拓扑配置各设备接口 IP 地址。

```
Ruijie>enable
Password:ruijie
Ruijie#configure terminal
Ruijie(config)#hostname   R1
R1(config)#interface   gigabitEthernet 0/0
R1(config-if-GigabitEthernet 0/0)#no   switchport
R1(config-if-GigabitEthernet 0/0)#ip address    10.0.12.1 24
R1(config-if-GigabitEthernet 0/0)#exit
R1(config)#interface   loopback 0
R1(config-if-Loopback 0)#ip address    10.0.1.1 32
R1(config-if-Loopback 0)#exit
R1(config)#

Ruijie>enable
Password:ruijie
Ruijie#configure terminal
Ruijie(config)#hostname   R2
R2(config)#interface   gigabitEthernet 0/0
R2(config-if-GigabitEthernet 0/0)#no   switchport
R2(config-if-GigabitEthernet 0/0)#ip address    10.0.12.2 24
R2(config-if-GigabitEthernet 0/0)#exit
R2(config)#interface   gigabitEthernet 0/1
R2(config-if-GigabitEthernet 0/1)#no   switchport
R2(config-if-GigabitEthernet 0/1)#ip address    10.0.23.2 24
R2(config-if-GigabitEthernet 0/1)#exit
R2(config)#interface   loopback 0
R2(config-if-Loopback 0)#ip address    10.0.2.2 32
R2(config-if-Loopback 0)#exit
R2(config)#
```

```
Ruijie>enable
Password:ruijie
Ruijie#configure terminal
Ruijie(config)#hostname   R3
R3(config)#interface   gigabitEthernet 0/1
R3(config-if-GigabitEthernet 0/1)#ip address    10.0.23.3 24
R3(config-if-GigabitEthernet 0/1)#exit
R3(config)#interface   gigabitEthernet 0/2
R3(config-if-GigabitEthernet 0/2)#no   switchport
R3(config-if-GigabitEthernet 0/2)#ip address    10.0.34.3 24
R3(config-if-GigabitEthernet 0/2)#exit
R3(config)#interface   loopback 0
R3(config-if-Loopback 0)#ip address    10.0.3.3 32
R3(config-if-Loopback 0)#exit
R3(config)#

Ruijie>enable
Password:ruijie
Ruijie#configure terminal
Ruijie(config)#hostname   R4
R4(config)#interface   gigabitEthernet 0/2
R4(config-if-GigabitEthernet 0/2)#no   switchport
R4(config-if-GigabitEthernet 0/2)#ip address    10.0.34.4 24
R4(config-if-GigabitEthernet 0/2)#exit
R4(config)#interface   gigabitEthernet 0/3
R4(config-if-GigabitEthernet 0/3)#no   switchport
R4(config-if-GigabitEthernet 0/3)#ip address    10.0.45.4 24
R4(config-if-GigabitEthernet 0/3)#exit
R4(config)#interface   loopback 0
R4(config-if-Loopback 0)#ip address    10.0.4.4 32
R4(config-if-Loopback 0)#exit
R4(config)#

Ruijie>enable
Password:ruijie
Ruijie#configure terminal
Ruijie(config)#hostname   R5
R5(config)#interface   loopback 0
R5(config-if-Loopback 0)#ip address    10.0.5.5 32
R5(config-if-Loopback 0)#exit
R5(config)#interface   gigabitEthernet 0/3
R5(config-if-GigabitEthernet 0/3)#no   switchport
R5(config-if-GigabitEthernet 0/3)#ip address    10.0.45.5 24
R5(config-if-GigabitEthernet 0/3)#exit
R5(config)#
```

2. 在 MPLS 域内配置 OSPF 协议使其 loopback 接口互通。

```
R1(config)#router   ospf   1
R1(config-router)#router-id   1.1.1.1
Change router-id and update OSPF process! [yes/no]:y
R1(config-router)#network   10.0.12.0 0.0.0.255 area   0
R1(config-router)#network   10.0.1.1 0.0.0.0 area   0
R1(config-router)#exit
```

```
R1(config)#

R2(config)#router   ospf   1
R2(config-router)#router-id   2.2.2.2
Change router-id and update OSPF process! [yes/no]:y
R2(config-router)#network   10.0.2.2 0.0.0.0 area   0
R2(config-router)#network   10.0.12.0 0.0.0.255 area 0
R2(config-router)#network   10.0.23.0 0.0.0.255 area 0
R2(config-router)#exit
R2(config)#

R3(config)#router   ospf   1
R3(config-router)#router-id   3.3.3.3
Change router-id and update OSPF process! [yes/no]:y
R3(config-router)#network   10.0.3.3 0.0.0.0 area   0
R3(config-router)#network   10.0.23.0 0.0.0.255 area   0
R3(config-router)#network   10.0.34.0 0.0.0.255 area   0
R3(config-router)#exit
R3(config)#

R4(config)#router   ospf   1
R4(config-router)#router-id   4.4.4.4
Change router-id and update OSPF process! [yes/no]:y
R4(config-router)#network   10.0.34.0 0.0.0.255 area   0
R4(config-router)#network   10.0.45.0 0.0.0.255 area   0
R4(config-router)#network   10.0.4.4 0.0.0.0 area   0
R4(config-router)#exit
R4(config)#

R5(config)#router   ospf   1
R5(config-router)#router-id   5.5.5.5
Change router-id and update OSPF process! [yes/no]:y
R5(config-router)#network   10.0.45.0 0.0.0.255 area   0
R5(config-router)#network   10.0.5.5 0.0.0.0 are 0
R5(config-router)#exit
R5(config)#
```

3. 在 R2 设备、R3 设备、R4 设备开启 MPLS IP 协议功能以及配置 LDP 协议，将 loopback 0 接口配置为 LDP 的 Router ID。

```
R2(config)#mpls   ip
R2(config)#mpls   router   ldp
R2(config-mpls-router)#ldp router-id   interface   loopback 0 force
R2(config-mpls-router)#exit
R2(config)#

R3(config)#mpls   ip
R3(config)#mpls   router   ldp
R3(config-mpls-router)#ldp router-id   interface   loopback 0 force
R3(config-mpls-router)#exit
R3(config)#

R4(config)#mpls   ip
R4(config)#mpls   router   ldp
R4(config-mpls-router)#ldp router-id   interface   loopback 0 force
```

```
R4(config-mpls-router)#exit
R4(config)#
```

4．在设备接口开启 MPLS IP 协议及标签转发功能，MPLS 域内的网络设备接口需要具备标签交换功能。

```
R2(config)#interface    gigabitEthernet 0/1
R2(config-if-GigabitEthernet 0/1)#mpls    ip
R2(config-if-GigabitEthernet 0/1)#label-switching
R2(config-if-GigabitEthernet 0/1)#exit
R2(config)#

R3(config)#interface    gigabitEthernet 0/1
R3(config-if-GigabitEthernet 0/1)#mpls    ip
R3(config-if-GigabitEthernet 0/1)#label-switching
R3(config-if-GigabitEthernet 0/1)#exit
R3(config)#interface    gigabitEthernet 0/2
R3(config-if-GigabitEthernet 0/2)#mpls    ip
R3(config-if-GigabitEthernet 0/2)#label-switching
R3(config-if-GigabitEthernet 0/2)#exit
R3(config)

R4(config)#interface    gigabitEthernet 0/2
R4(config-if-GigabitEthernet 0/2)#mpls    ip
R4(config-if-GigabitEthernet 0/2)#label-switching
R4(config-if-GigabitEthernet 0/2)#exit
R4(config)#
```

5．设置标签分配，只为/32 位主机路由分配标签。

```
R2(config)#mpls    router    ldp
R2(config-mpls-router)#advertise-labels for    host-routes
R2(config-mpls-router)#exit
R2(config)#

R3(config)#mpls    router    ldp
R3(config-mpls-router)#advertise-labels for    host-routes
R3(config-mpls-router)#exit
R3(config)#

R4(config)#mpls    router    ldp
R4(config-mpls-router)#advertise-labels for    host-routes
R4(config-mpls-router)#exit
R4(config)#
```

> 任务验证

1．检查 OSPF 协议邻居建立情况。

```
R3#show    ip ospf    neighbor

OSPF process 1, 2 Neighbors, 2 is Full:
Neighbor ID     Pri    State         BFD State    Dead Time    Address      Interface
2.2.2.2         1      Full/BDR      -            00:00:39     10.0.23.2    GigabitEthernet 0/1
4.4.4.4         1      Full/BDR      -            00:00:34     10.0.34.4    GigabitEthernet 0/2
R3#
```

2. 检查 MPLS LDP 邻居关系建立情况。

```
R3#show  mpls  ldp  neighbor
Default VRF:
    Peer LDP Ident: 10.0.2.2:0; Local LDP Ident: 10.0.3.3:0
        TCP connection: 10.0.2.2.646 - 10.0.3.3.33361
        State: OPERATIONAL; Msgs sent/recv: 121/122; UNSOLICITED
        Up time: 00:27:37
        Graceful Restart enabled; Peer reconnect time (msecs): 300000
        LDP discovery sources:
            Link Peer on GigabitEthernet 0/1, Src IP addr: 10.0.23.2
        Addresses bound to peer LDP Ident:
            10.0.12.2       10.0.23.2       10.0.2.2
    Peer LDP Ident: 10.0.4.4:0; Local LDP Ident: 10.0.3.3:0
        TCP connection: 10.0.4.4.46273 - 10.0.3.3.646
        State: OPERATIONAL; Msgs sent/recv: 116/117; UNSOLICITED
        Up time: 00:26:23
        Graceful Restart enabled; Peer reconnect time (msecs): 300000
        LDP discovery sources:
            Link Peer on GigabitEthernet 0/2, Src IP addr: 10.0.34.4
        Addresses bound to peer LDP Ident:
            10.0.34.4       10.0.45.4       10.0.4.4
```

3. 在 R3 设备检查 MPLS 标签转发表，此时的标签转发表为 R3 设备所接收的路由信息分配的标签，即所有网段的路由信息都分配标签。

```
R3#show  mpls  forwarding-table

Label Operation Code:
PH--PUSH label
PP--POP label
SW--SWAP label
SP--SWAP topmost label and push new label
DP--DROP packet
PC--POP label and continue lookup by IP or Label
PI--POP label and do ip lookup forward
PN--POP label and forward to nexthop
PM--POP label and do MAC lookup forward
PV--POP label and output to VC attach interface
IP--IP lookup forward
 s--stale
   Local    Outgoing OP FEC                    Outgoing          Nexthop
   label    label                              interface
   --       11264    PH 10.0.1.1/32             Gi0/1             10.0.23.2
   --       imp-null PH 10.0.2.2/32             Gi0/1             10.0.23.2
   --       imp-null PH 10.0.4.4/32             Gi0/2             10.0.34.4
   --       11267    PH 10.0.5.5/32             Gi0/2             10.0.34.4
   --       imp-null PH 10.0.12.0/24            Gi0/1             10.0.23.2
   --       imp-null PH 10.0.45.0/24            Gi0/2             10.0.34.4
   11264    11264    SW 10.0.1.1/32             Gi0/1             10.0.23.2
   11265    imp-null PP 10.0.2.2/32             Gi0/1             10.0.23.2
   11266    imp-null PP 10.0.4.4/32             Gi0/2             10.0.34.4
   11267    11267    SW 10.0.5.5/32             Gi0/2             10.0.34.4
   11268    imp-null PP 10.0.12.0/24            Gi0/1             10.0.23.2
   11269    imp-null PP 10.0.45.0/24            Gi0/2             10.0.34.4
R3#
```

4. 设置仅为/32 位主机路由分配标签后，标签转发信息表只接收到/32 位主机路由及为/32 位主机路由分配标签。

```
R3#show   mpls  forwarding-table

Label Operation Code:
PH--PUSH label
PP--POP label
SW--SWAP label
SP--SWAP topmost label and push new label
DP--DROP packet
PC--POP label and continue lookup by IP or Label
PI--POP label and do ip lookup forward
PN--POP label and forward to nexthop
PM--POP label and do MAC lookup forward
PV--POP label and output to VC attach interface
IP--IP lookup forward
  s--stale
  Local   Outgoing OP FEC              Outgoing       Nexthop
  label   label                        interface
  --      11264    PH 10.0.1.1/32      Gi0/1          10.0.23.2
  --      imp-null PH 10.0.2.2/32      Gi0/1          10.0.23.2
  --      imp-null PH 10.0.4.4/32      Gi0/2          10.0.34.4
  --      11267    PH 10.0.5.5/32      Gi0/2          10.0.34.4
  11264   11264    SW 10.0.1.1/32      Gi0/1          10.0.23.2
  11265   imp-null PP 10.0.2.2/32      Gi0/1          10.0.23.2
  11266   imp-null PP 10.0.4.4/32      Gi0/2          10.0.34.4
  11267   11267    SW 10.0.5.5/32      Gi0/2          10.0.34.4

R3#
```

5. 检查其中一条路由的标签信息。

```
R3#show   mpls  forwarding-table  10.0.5.5/32 detail

Label Operation Code:
PH--PUSH label
PP--POP label
SW--SWAP label
SP--SWAP topmost label and push new label
DP--DROP packet
PC--POP label and continue lookup by IP or Label
PI--POP label and do ip lookup forward
PN--POP label and forward to nexthop
PM--POP label and do MAC lookup forward
PV--POP label and output to VC attach interface
IP--IP lookup forward
  s--stale
  Local   Outgoing OP FEC              Outgoing       Nexthop
  label   label                        interface
  --      11267    PH 10.0.5.5/32      Gi0/2          10.0.34.4
          Added by Route(vrf Global), Tag Stack: { 11267 }
  11267   11267    SW 10.0.5.5/32      Gi0/2          10.0.34.4
          Added by Route(vrf Global), Tag Stack: { 11267 }

R3#
```

6. 在 R1 设备使用 loopback 0 接口测试与 R5 设备的 loopback 0 接口互通情况,能够正常通信。

```
R1#ping    10.0.5.5 source    10.0.1.1
Sending 5, 100-byte ICMP Echoes to 10.0.5.5, timeout is 2 seconds:
   < press Ctrl+C to break >
!!!!!
Success rate is 100 percent (5/5), round-trip min/avg/max = 5/6/8 ms.
R1#
```

7. 通过 Wireshark 抓包工具验证标签分配。

```
 54.327601 10.0.1.1          10.0.5.5         ICMP          146 Echo (ping) request  id=0x0300, seq=64069/17914
 64.331666 10.0.5.5          10.0.1.1         ICMP          146 Echo (ping) reply    id=0x0300, seq=64069/17914
 74.435572 10.0.1.1          10.0.5.5         ICMP          146 Echo (ping) request  id=0x0300, seq=1094/17924

> Frame 5: 146 bytes on wire (1168 bits), 146 bytes captured (1168 bits) on interface 0
> Ethernet II, Src: 50:00:00:02:00:02 (50:00:00:02:00:02), Dst: 50:00:00:03:00:02 (50:00:00:03:00:02)
v MultiProtocol Label Switching Header, Label: 11267, Exp: 0, S: 1, TTL: 63
    0000 0010 1100 0000 0011 .... .... .... = MPLS Label: 11267
    .... .... .... .... .... 000. .... .... = MPLS Experimental Bits: 0
    .... .... .... .... .... ...1 .... .... = MPLS Bottom Of Label Stack: 1
    .... .... .... .... .... .... 0011 1111 = MPLS TTL: 63
> Internet Protocol Version 4, Src: 10.0.1.1, Dst: 10.0.5.5
> Internet Control Message Protocol

No.    Time       Source            Destination        Protocol    Length  Info
 54.327601 10.0.1.1          10.0.5.5         ICMP          146 Echo (ping) request  id=0x0300, seq=64069/17914
 64.331666 10.0.5.5          10.0.1.1         ICMP          146 Echo (ping) reply    id=0x0300, seq=64069/17914
 74.435572 10.0.1.1          10.0.5.5         ICMP          146 Echo (ping) request  id=0x0300, seq=1094/17924

> Frame 6: 146 bytes on wire (1168 bits), 146 bytes captured (1168 bits) on interface 0
> Ethernet II, Src: 50:00:00:03:00:02 (50:00:00:03:00:02), Dst: 50:00:00:02:00:02 (50:00:00:02:00:02)
v MultiProtocol Label Switching Header, Label: 11264, Exp: 0, S: 1, TTL: 62
    0000 0010 1100 0000 0000 .... .... .... = MPLS Label: 11264
    .... .... .... .... .... 000. .... .... = MPLS Experimental Bits: 0
    .... .... .... .... .... ...1 .... .... = MPLS Bottom Of Label Stack: 1
    .... .... .... .... .... .... 0011 1110 = MPLS TTL: 62
> Internet Protocol Version 4, Src: 10.0.5.5, Dst: 10.0.1.1
> Internet Control Message Protocol
```

➤ 问题与思考

1. 标签分发协议 LDP 会话建立过程需经过哪几个阶段?
2. 标签分发协议 LDP 的标签行为都有哪些?

3.2　MPLS VPN 协议

3.2.1　单域 VPN 实例

➤ 原理

MPLS 作为一种标签转发技术,集成了路由转发和 ATM 信元转发的优势,在骨干网络等大型网络中得到广泛应用,然而在数据转发之余,用户数据的安全问题也需要更加关注。

传统的 VPN 一般是通过 GRE、L2TP、PPTP 等隧道协议来实现私有网络数据流在公网上的传送。BGP/MPLS IP VPN 是另外一种实现 VPN 的方式,可以说是一种介于第二层和第三层隧道协议的 VPN。LSP 就是公网上的隧道,只不过这个公网隧道是通过 MPLS 所使用的标签分发协议建立起来的。基于 MPLS 的 VPN 就是通过 LSP 将私有网络在地域上的不同分支联结起来，形成一个统一的网络。基于 MPLS 的 VPN 还支持不同 VPN 间的互通。用 MPLS 来实现 VPN 有天然的优势，如对于 VPN 用户而言，采用 MPLS VPN 可以大大简化 VPN 用户的工作量，不再需要专门的 VPN 设备，只需要用传统的路由器就可以构建 VPN；对于运营商而言，采用 MPLS VPN 很容易实现 VPN 扩展。

使用 MPLS VPN 实现的 L3VPN 主要有如下特点：

（1）VPN 隧道是在网络服务提供商 PE 上建立的，而不是在用户 CE 之间建立的。VPN 路由也在 PE 之间传递，用户不用花费精力维护 VPN 信息。

（2）直接利用现有的路由协议，VPN 隧道的建立和路由发布都变为动态实现，有利于 VPN 规模的扩大。

（3）支持地址重叠，不同的 VPN 用户可以使用相同的地址空间。

（4）在服务提供商的网络中，VPN 的业务交流使用标签交换而不是传统的路由分发。

（5）能够达到和用户租用专线一样的安全性。

MPLS VPN 需要实现的功能有：

（1）骨干网络中采用标签分发协议 LDP 建立 LSP 隧道，这个过程通常在服务商的网络中建立，拓扑稳定时已经完成了。

（2）数据转发，根据报文打上的标签和本地映射表进行报文的转发。

（3）MP-BGP 和 BGP 扩展属性，用来传递 VPN 路由和承载 VPN 属性、标签内容。

（4）对 VPN 路由的管理，建立多张 VPN 路由表，维护 VPN 路由信息。

关于 MPLS VPN 组网结构相关名词术语：

（1）CE（Customer Edge Router）：位于用户网络边缘的设备，CE 设备逻辑上属于用户的 VPN，CE 的某个接口与服务提供商的设备直接相连。CE 可以是主机也可以是路由器或者交换机，可以不支持 MPLS 功能。

（2）PE（Provider Edge Router）：是 ISP 骨干网络的边缘设备，PE 设备逻辑上属于服务提供商，PE 与 CE 直连，且一台 PE 可以连接多个不同的 CE。PE 主要负责接收 CE 端发送的 VPN 信息，向其他 PE 发送 VPN 信息，并从其他 PE 接收 VPN 信息，分发给对应的 CE。PE 要求支持 MPLS 功能。

（3）P（Provider Router）：是位于 ISP 骨干网络内的核心设备，P 设备不和 CE 设备相连，负责路由和报文快速转发。P 作为 MPLS 核心骨干网络中的设备，要求 P 支持 MPLS 功能。P 知道到达骨干网络内任何目的地的路由，而不知道到达 VPN 的路由。

（4）VRN 路由转发表（VPN Routing and Forwarding table），简称 VRF，VRF 主要是为了解决本地路由冲突问题。PE 和 CE 之间的连接都需要关联一个 VRF。在一台 PE 上可以有多个用来和 CE 端交换路由信息的 VRF。可以把每个 VRF 想象成一台"虚拟路由器"，每个虚拟路由器都和 CE 相连，负责从 CE 端接收路由信息，或者向 CE 端通告 VPN 路由信息。

（5）路由区分符（Route Distinguisher），简称 RD，RD 的引入解决了传输过程中的路由冲突问题。可以把 RD 理解成一个区分符。如果不同的 VPN 具有相同的网络地址，通过 BGP 在骨干网络中通告路由信息，BGP 决策只会从这些重叠的地址中选择最优的路由通告，会导致某些 VPN 得不到对应的路由信息，加入了 RD 值，可以为这些重叠的网络地址加上不同的区分符，BGP 决策过程中根据 VPN 信息中携带的不同区分符区分开相同的网络地址，各自的

VPN 可以得到各自的路由信息，RD 的意义只是一个区分符，区分开相同的网络地址。如果不同的 VPN 没有地址重叠问题，那么实际上没有 RD 值也可以。

VPNV4 地址结构：

| Route Distinguisher（8 个字节） | IPv4 路由 |

一般情况下，通常为一个 VPN 指定一个唯一的 RD 值，这样不同的 VPN 有不同的 RD 值，在骨干网络中传播路由信息就没有问题，通常 RD 值用 xx:xx 标识。

TYPE（2 字节）	Administrator Field	Assigned Number Field
0x0002	2 字节 AS 号	4 字节分配编号
0x0102	4 字节 IP 地址	2 字节分配编号

（6）Route-target（RT 路由标记），RT 属性本质是一种让 VRF 表达自己的路由取舍方式。RT 属性分为"导出 RT 属性（Export Route-target）"和"导入 RT 属性（Import Route-target）"。PE 从 CE 接收到 VPN 路由信息，并为这些 VPN 路由信息加上"Export Route-target"属性，向其他 PE 通告该 VPN 路由信息；PE 也根据"Import Route-target"属性决定从其他 PE 收到的 VPN 路由信息是否需要导入到 VRF。一个原则是：当 PE 接收到一条 VPN 的路由信息时，只有该条 VPN 路由信息中携带的 RT 属性至少有一个 RT 属性和本 PE 上第一个 VRF 的"Import RT"属性相等时，这条 VPN 信息才可以安装到该"VRF"中，通过上面的这种方法可以灵活控制 VPN 路由信息的分发。一条 VPN 路由信息可以携带多个 RT 值。

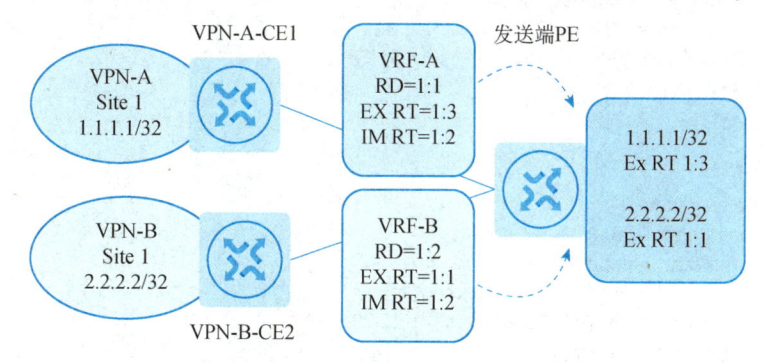

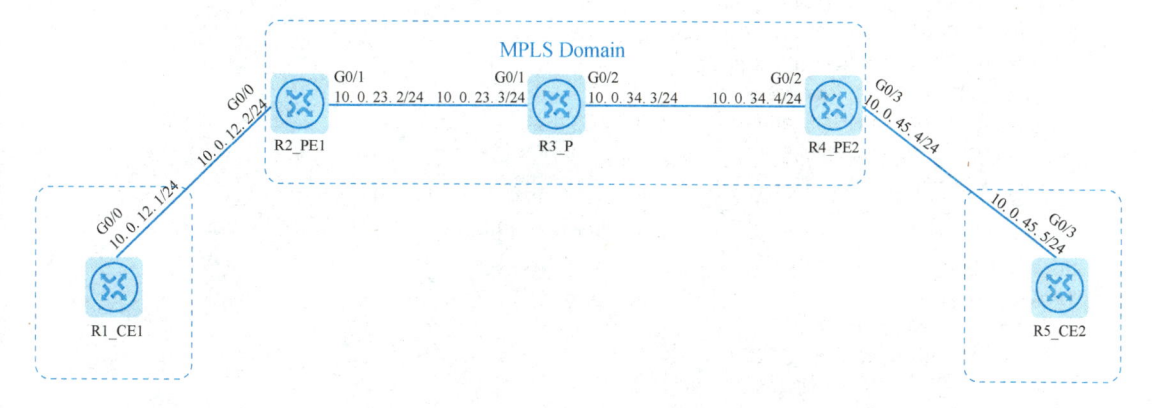

> 任务拓扑

> 实施步骤

1. 根据任务拓扑配置各设备接口 IP 地址。

```
Ruijie>enable
Password:ruijie
Ruijie#configure terminal
R1(config)#hostname    R1_CE1
R1_CE1(config)#interface    gigabitEthernet 0/0
R1_CE1(config-if-GigabitEthernet 0/0)#no    switchport
R1_CE1(config-if-GigabitEthernet 0/0)#ip address    10.0.12.1 24
R1_CE1(config-if-GigabitEthernet 0/0)#exit
R1_CE1(config)#

Ruijie>enable
Password:ruijie
Ruijie#configure terminal
Ruijie(config)#hostname    R2_PE1
R2_PE1(config)#interface    gigabitEthernet 0/0
R2_PE1(config-if-GigabitEthernet 0/0)#no    switchport
R2_PE1(config-if-GigabitEthernet 0/0)#ip address    10.0.12.2 24
R2_PE1(config-if-GigabitEthernet 0/0)#exit
R2_PE1(config)#interface    gigabitEthernet 0/1
R2_PE1(config-if-GigabitEthernet 0/1)#no    switchport
R2_PE1(config-if-GigabitEthernet 0/1)#ip address    10.0.23.2 24
R2_PE1(config-if-GigabitEthernet 0/1)#exit
R2_PE1(config)#interface    loopback 0
R2_PE1(config-if-Loopback 0)#ip address    10.0.2.2 32
R2_PE1(config-if-Loopback 0)#exit
R2_PE1(config)#

Ruijie>enable
Password:ruijie
Ruijie#configure terminal
Ruijie(config)#hostname    R3_P
R3_P(config)#interface    gigabitEthernet 0/1
R3_P(config-if-GigabitEthernet 0/1)#no    switchport
R3_P(config-if-GigabitEthernet 0/1)#ip address    10.0.23.3 24
R3_P(config-if-GigabitEthernet 0/1)#exit
R3_P(config)#interface    gigabitEthernet 0/2
R3_P(config-if-GigabitEthernet 0/2)#no    switchport
R3_P(config-if-GigabitEthernet 0/2)#ip address    10.0.34.3 24
R3_P(config-if-GigabitEthernet 0/2)#exit
R3_P(config)#interface    loopback 0
R3_P(config-if-Loopback 0)#ip address    10.0.3.3 32
R3_P(config-if-Loopback 0)#exit
R3_P(config)#

Ruijie>enable
Password:ruijie
Ruijie#configure terminal
Ruijie(config)#hostname    R4_PE2
R4_PE2(config)#interface    gigabitEthernet 0/2
R4_PE2(config-if-GigabitEthernet 0/2)#no    switchport
```

```
R4_PE2(config-if-GigabitEthernet 0/2)#ip address    10.0.34.4 24
R4_PE2(config-if-GigabitEthernet 0/2)#exit
R4_PE2(config)#interface    gigabitEthernet 0/3
R4_PE2(config-if-GigabitEthernet 0/3)#no    switchport
R4_PE2(config-if-GigabitEthernet 0/3)#ip address    10.0.45.4 24
R4_PE2(config-if-GigabitEthernet 0/3)#exit
R4_PE2(config)#interface    loopback 0
R4_PE2(config-if-Loopback 0)#ip address    10.0.4.4 32
R4_PE2(config-if-Loopback 0)#exit
R4_PE2(config)#

Ruijie>enable
Password:ruijie
Ruijie#configure terminal
Ruijie(config)#hostname    R5_CE2
R5_CE2(config)#interface    gigabitEthernet 0/3
R5_CE2(config-if-GigabitEthernet 0/3)#no    switchport
R5_CE2(config-if-GigabitEthernet 0/3)#ip address    10.0.45.5 24
R5_CE2(config-if-GigabitEthernet 0/3)#exit
R5_CE2(config)#
```

2. 在 MPLS 域内配置 OSPF 协议进程 10，各设备接口属于 Area 0，手工指定 Router ID。

```
R2_PE1(config)#router    ospf    10
R2_PE1(config-router)#router-id    2.2.2.2
Change router-id and update OSPF process! [yes/no]:y
R2_PE1(config-router)#network    10.0.23.0 0.0.0.255 area 0
R2_PE1(config-router)#network    10.0.2.2 0.0.0.0 area    0
R2_PE1(config-router)#exit
R2_PE1(config)#


R3_P(config)#router ospf    10
R3_P(config-router)#router-id    3.3.3.3
Change router-id and update OSPF process! [yes/no]:y
R3_P(config-router)#network    10.0.23.0 0.0.0.255 are 0
R3_P(config-router)#network    10.0.3.3 0.0.0.0 area    0
R3_P(config-router)#network    10.0.34.0 0.0.0.255 area    0
R3_P(config-router)#exit
R3_P(config)#


R4_PE2(config)#router    ospf    10
R4_PE2(config-router)#router-id    4.4.4.4
Change router-id and update OSPF process! [yes/no]:y
R4_PE2(config-router)#network    10.0.4.4 0.0.0.0 area    0
R4_PE2(config-router)#network    10.0.34.0 0.0.0.255 area 0
R4_PE2(config-router)#exit
R4_PE2(config)#
```

3. 在 MPLS 域内建立 LDP 协议邻居，由 LDP 协议分配公网标签。

```
R2_PE1(config)#mpls    ip
R2_PE1(config)#mpls    router    ldp
R2_PE1(config-mpls-router)#ldp router-id    interface    loopback 0 force
R2_PE1(config-mpls-router)#exiti
R2_PE1(config)#interface    gigabitEthernet 0/1
```

```
R2_PE1(config-if-GigabitEthernet 0/1)#mpls   ip
R2_PE1(config-if-GigabitEthernet 0/1)#label-switching
R2_PE1(config-if-GigabitEthernet 0/1)#exit
R2_PE1(config)#


R3_P(config)#mpls    ip
R3_P(config)#mpls    router   ldp
R3_P(config-mpls-router)#ldp router-id   interface   loopback 0 force
R3_P(config-mpls-router)#exit
R3_P(config)#interface   gigabitEthernet 0/1
R3_P(config-if-GigabitEthernet 0/1)#mpls    ip
R3_P(config-if-GigabitEthernet 0/1)#label-switching
R3_P(config-if-GigabitEthernet 0/1)#exit
R3_P(config)#interface   gigabitEthernet 0/2
R3_P(config-if-GigabitEthernet 0/2)#mpls    ip
R3_P(config-if-GigabitEthernet 0/2)#label-switching
R3_P(config-if-GigabitEthernet 0/2)#exit
R3_P(config)#


R4_PE2(config)#mpls   ip
R4_PE2(config)#mpls   router   ldp
R4_PE2(config-mpls-router)#ldp router-id   interface   loopback 0 force
R4_PE2(config-mpls-router)#exit
R4_PE2(config)#interface   gigabitEthernet 0/2
R4_PE2(config-if-GigabitEthernet 0/2)#mpls    ip
R4_PE2(config-if-GigabitEthernet 0/2)#label-switching
R4_PE2(config-if-GigabitEthernet 0/2)#exit
R4_PE2(config)#
```

4. 在 R2_PE1 设备和 R4_PE2 设备创建 VRF，设置 RD 值与 RT 值，并把 VRF 与对应接口关联。

注：接口在关联 VRF 时，该接口关于 IPv4 的配置信息会被清空，若接口配置过多则需要做备份后再操作接口关联。

```
R2_PE1(config)#ip vrf VPN1
R2_PE1(config-vrf)#rd 100:1
R2_PE1(config-vrf)#route-target both    10:1
R2_PE1(config-vrf)#exit
R2_PE1(config)#
R2_PE1(config)#interface   gigabitEthernet 0/0
R2_PE1(config-if-GigabitEthernet 0/0)#ip vrf forwarding VPN1      //关联 VRF
% Interface GigabitEthernet 0/0 IP address removed due to enabling VRF VPN1    //清空接口配置
R2_PE1(config-if-GigabitEthernet 0/0)#ip address    10.0.12.2 24
R2_PE1(config-if-GigabitEthernet 0/0)#exit
R2_PE1(config)#


R4_PE2(config)#ip vrf    VPN1
R4_PE2(config-vrf)#rd    100:2
R4_PE2(config-vrf)#route-target   both   10:1
R4_PE2(config-vrf)#exit
R4_PE2(config)#interface   gigabitEthernet 0/3
R4_PE2(config-if-GigabitEthernet 0/3)#ip vrf   forwarding   VPN1
% Interface GigabitEthernet 0/3 IP address removed due to enabling VRF VPN1
```

```
R4_PE2(config-if-GigabitEthernet 0/3)#ip address    10.0.45.4 24
R4_PE2(config-if-GigabitEthernet 0/3)#exit
R4_PE2(config)#
```

5. 在 MPLS 域内配置单播 BGP 协议，通过设备的 loopback 0 接口建立 BGP 邻居关系。

```
R2_PE1(config)#router    bgp    300
R2_PE1(config-router)#bgp    router-id    2.2.2.2
R2_PE1(config-router)#neighbor 10.0.4.4 remote-as    300
R2_PE1(config-router)#neighbor    10.0.4.4 update-source    loopback 0
R2_PE1(config-router)#exit
R2_PE1(config)#


R4_PE2(config)#router    bgp    300
R4_PE2(config-router)#bgp router-id    4.4.4.4
R4_PE2(config-router)#neighbor    10.0.2.2 remote-as    300
R4_PE2(config-router)#neighbor    10.0.2.2 update-source    loopback 0
R4_PE2(config-router)#exit
R4_PE2(config)#
```

6. 在 R2_PE1 设备和 R4_PE2 设备建立 MP-BGP 协议传递私网路由。

```
R2_PE1(config)#router    bgp    300
R2_PE1(config-router)#address-family vpnv4
R2_PE1(config-router-af)#neighbor    10.0.4.4 activate
R2_PE1(config-router-af)#exit
R2_PE1(config-router)#exit
R2_PE1(config)#


R4_PE2(config)#router bgp 300
R4_PE2(config-router)#address-family vpnv4
R4_PE2(config-router-af)#neighbor    10.0.2.2 activate
R4_PE2(config-router-af)#exit
R4_PE2(config-router)#exit
R4_PE2(config)#
```

7. 在 R1_CE1 设备与 R2_PE2 设备建立 OSPF 协议邻居传递私网路由，R1_CE1 设备创建 loopback 0 接口作测试网段并通告到 OSPF 协议中。

```
R1_CE1(config)#router    ospf    100
R1_CE1(config-router)#router-id    1.1.1.1
Change router-id and update OSPF process! [yes/no]:y
R1_CE1(config-router)#network    10.0.12.0 0.0.0.255 area    1
R1_CE1(config-router)#exit
R1_CE1(config)#


R2_PE1(config)#router    ospf    100 vrf    VPN1      //PE 设备的路由协议必须关联 VRF
R2_PE1(config-router)#router-id    22.22.22.22
Change router-id and update OSPF process! [yes/no]:y
R2_PE1(config-router)#network    10.0.12.0 0.0.0.255 area    1
R2_PE1(config-router)#exit
R2_PE1(config)#


R1_CE1(config)#interface    loopback 0
R1_CE1(config-if-Loopback 0)#ip address    192.168.1.1 24
R1_CE1(config-if-Loopback 0)#exit
R1_CE1(config)#
```

```
R1_CE1(config)#router ospf   100
R1_CE1(config-router)#network   192.168.1.0 0.0.0.255 area   1      //通告地址段
R1_CE1(config-router)#exit
R1_CE1(config)
```

8. 在 R5_CE2 设备与 R4_PE2 设备建立 IS-IS 邻居传递私网路由。

```
R4_PE2(config)#router   isis   1
R4_PE2(config-router)#vrf   VPN1
R4_PE2(config-router)#net   49.0001.0000.0000.0004.00
R4_PE2(config-router)#is-type   level-2
R4_PE2(config-router)#exit
R4_PE2(config)#interface   gigabitEthernet 0/3
R4_PE2(config-if-GigabitEthernet 0/3)#ip router   isis   1
R4_PE2(config-if-GigabitEthernet 0/3)#exit
R4_PE2(config)#

R5_CE2(config)#router   isis   1
R5_CE2(config-router)#net   49.0001.0000.0000.0005.00
R5_CE2(config-router)#is-type   level-2
R5_CE2(config-router)#exit
R5_CE2(config)#interface   gigabitEthernet 0/3
R5_CE2(config-if-GigabitEthernet 0/3)#ip router isis   1
R5_CE2(config-if-GigabitEthernet 0/3)#exit
R5_CE2(config)#

R5_CE2(config)#interface   loopback 0
R5_CE2(config-if-Loopback 0)#ip address   192.168.2.1 24
R5_CE2(config-if-Loopback 0)#ip router isis   1         //通告地址
R5_CE2(config-if-Loopback 0)#exit
R5_CE2(config)#
```

9. 在 R2_PE1 设备和 R4_PE2 设备将 VRF 路由重分布到 MP-BGP 协议中。

```
R2_PE1(config)#router   bgp   300
R2_PE1(config-router)#address-family   ipv4 vrf VPN1
R2_PE1(config-router-af)#redistribute ospf   100 match   external internal
R2_PE1(config-router-af)#exit
R2_PE1(config-router)#exit
R2_PE1(config)#


R4_PE2(config)#router   bgp   300
R4_PE2(config-router)#address-family ipv4 vrf   VPN1
R4_PE2(config-router-af)#redistribute   isis 1 level-2
R4_PE2(config-router-af)#exit
R4_PE2(config-router)#exit
R4_PE2(config)#
```

10. 在 R2_PE1 设备将 MP-BGP 路由重分布到 OSPF 协议中，在 R4_PE2 设备将 MP-BGP 路由重分布到 IS-IS 协议中。

```
R2_PE1(config)#router   ospf   100
R2_PE1(config-router)#redistribute   bgp   subnets
R2_PE1(config-router)#exit
R2_PE1(config)#
```

```
R4_PE2(config)#router   isis   1
R4_PE2(config-router)#redistribute   bgp level-2
R4_PE2(config-router)#exit
R4_PE2(config)#
```

> 任务验证

1. 检查 MPLS 域内即公网互通情况。

```
R3_P#show   ip ospf   neighbor

OSPF process 10, 2 Neighbors, 2 is Full:
Neighbor ID      Pri    State        BFD State    Dead Time    Address      Interface
2.2.2.2           1     Full/BDR     -            00:00:31     10.0.23.2
GigabitEthernet 0/1
4.4.4.4           1     Full/DR      -            00:00:40     10.0.34.4
GigabitEthernet 0/2
R3_P#


R2_PE1#show   ip route   ospf
O      10.0.3.3/32 [110/1] via 10.0.23.3, 00:09:58, GigabitEthernet 0/1
O      10.0.4.4/32 [110/2] via 10.0.23.3, 00:09:33, GigabitEthernet 0/1
O      10.0.34.0/24 [110/2] via 10.0.23.3, 00:09:58, GigabitEthernet 0/1
R2_PE1#

R2_PE1#ping   10.0.3.3
Sending 5, 100-byte ICMP Echoes to 10.0.3.3, timeout is 2 seconds:
  < press Ctrl+C to break >
!!!!!
Success rate is 100 percent (5/5), round-trip min/avg/max = 1/1/3 ms.
R2_PE1#
```

2. 在 R3_P 设备检查 LDP 协议邻居建立情况，正常查看 10.0.2.2 和 10.0.4.4 两个 LDP 邻居。

```
R3_P#show   mpls   ldp   neighbor
Default VRF:
    Peer LDP Ident: 10.0.2.2:0; Local LDP Ident: 10.0.3.3:0
        TCP connection: 10.0.2.2.646 - 10.0.3.3.35503
        State: OPERATIONAL; Msgs sent/recv: 37/39; UNSOLICITED
        Up time: 00:07:30
        Graceful Restart enabled; Peer reconnect time (msecs): 300000
        LDP discovery sources:
            Link Peer on GigabitEthernet 0/1, Src IP addr: 10.0.23.2
        Addresses bound to peer LDP Ident:
            10.0.12.2        10.0.23.2        10.0.2.2
    Peer LDP Ident: 10.0.4.4:0; Local LDP Ident: 10.0.3.3:0
        TCP connection: 10.0.4.4.37103 - 10.0.3.3.646
        State: OPERATIONAL; Msgs sent/recv: 35/37; UNSOLICITED
        Up time: 00:07:03
        Graceful Restart enabled; Peer reconnect time (msecs): 300000
        LDP discovery sources:
            Link Peer on GigabitEthernet 0/2, Src IP addr: 10.0.34.4
```

```
        Addresses bound to peer LDP Ident:
            10.0.34.4        10.0.45.4        10.0.4.4
R3_P#
```

3. 检查公网标签信息转发表。

```
R3_P#show   mpls   forwarding-table

Label Operation Code:
PH--PUSH label
PP--POP label
SW--SWAP label
SP--SWAP topmost label and push new label
DP--DROP packet
PC--POP label and continue lookup by IP or Label
PI--POP label and do ip lookup forward
PN--POP label and forward to nexthop
PM--POP label and do MAC lookup forward
PV--POP label and output to VC attach interface
IP--IP lookup forward
 s--stale
   Local    Outgoing OP FEC                  Outgoing        Nexthop
   label    label                            interface
   --       imp-null PH 10.0.2.2/32          Gi0/1           10.0.23.2
   --       imp-null PH 10.0.4.4/32          Gi0/2           10.0.34.4
   11264    imp-null PP 10.0.2.2/32          Gi0/1           10.0.23.2
   11265    imp-null PP 10.0.4.4/32          Gi0/2           10.0.34.4

R3_P#
```

4. 检查 VRF 配置，接口关联 VRF 后需要携带 VRF 参数测试地址连通性。

```
R2_PE1#show   ip vrf   detail
VRF VPN1 (VRF ID = 1); default RD 100:1
  Interfaces:
     GigabitEthernet 0/0
VRF Table ID = 1
  Export VPN route-target communities
    RT: 10:1
  Import VPN route-target communities
    RT: 10:1
  No import route-map
  No export route-map
  Alloc-label per-vrf: -/aggregate(VPN1)

R2_PE1#


R2_PE1#ping   vrf   VPN1 10.0.12.1
Sending 5, 100-byte ICMP Echoes to 10.0.12.1, timeout is 2 seconds:
  < press Ctrl+C to break >
!!!!!
Success rate is 100 percent (5/5), round-trip min/avg/max = 1/1/4 ms.
R2_PE1#



R4_PE2#show   ip vrf   detail
VRF VPN1 (VRF ID = 1); default RD 100:2
```

```
     Interfaces:
        GigabitEthernet 0/3
  VRF Table ID = 1
     Export VPN route-target communities
        RT: 10:1
     Import VPN route-target communities
        RT: 10:1
     No import route-map
     No export route-map
     Alloc-label per-vrf: -/aggregate(VPN1)


  R4_PE2#


  R4_PE2#ping   vrf   VPN1 10.0.45.5
  Sending 5, 100-byte ICMP Echoes to 10.0.45.5, timeout is 2 seconds:
     < press Ctrl+C to break >
  !!!!!
  Success rate is 100 percent (5/5), round-trip min/avg/max = 1/2/3 ms.
  R4_PE2#
```

5. 检查 MP-BGP 邻居关系建立情况。

```
  R2_PE1#show   bgp vpnv4 unicast   all   summary
  For address family: VPNv4 Unicast
  BGP router identifier 2.2.2.2, local AS number 300
  BGP table version is 1
  0 BGP AS-PATH entries
  0 BGP Community entries
  0 BGP Prefix entries (Maximum-prefix:4294967295)


  Neighbor       V     AS MsgRcvd MsgSent    TblVer  InQ OutQ Up/Down   State/PfxRcd
  10.0.4.4       4    300     23      22         1    0    0 00:16:06         0

  Total number of neighbors 1, established neighbors 1
```

6. 检查私网 R1_CE1 设备的路由传递情况。

```
  R1_CE1#show   ip ospf   neighbor     //正常建立 OSPF 邻居关系


  OSPF process 100, 1 Neighbors, 1 is Full:
  Neighbor ID      Pri    State              BFD State   Dead Time   Address         Interface
  22.22.22.22       1     Full/BDR             -         00:00:40    10.0.12.2
  GigabitEthernet 0/0
  R1_CE1#


  R2_PE1#show   ip route   vrf VPN1
  Routing Table: VPN1


  Codes:   C - Connected, L - Local, S - Static
           R - RIP, O - OSPF, B - BGP, I - IS-IS, V - Overflow route
           N1 - OSPF NSSA external type 1, N2 - OSPF NSSA external type 2
           E1 - OSPF external type 1, E2 - OSPF external type 2
           SU - IS-IS summary, L1 - IS-IS level-1, L2 - IS-IS level-2
           IA - Inter area, EV - BGP EVPN, A - Arp to host
           LA - Local aggregate route
           * - candidate default
```

```
Gateway of last resort is no set
C      10.0.12.0/24 is directly connected, GigabitEthernet 0/0
C      10.0.12.2/32 is local host.
O      192.168.1.1/32 [110/1] via 10.0.12.1, 00:00:51, GigabitEthernet 0/0
R2_PE1#


R5_CE2#show   isis   neighbors       //正常建立 IS-IS 邻居关系

Area 1:
System Id         Type    IP Address       State      Holdtime   Circuit             Interface
R4_PE2            L2      10.0.45.4        Up         23         R5_CE2.01           GigabitEthernet 0/3

R5_CE2#
```

7. 检查私网 R5_CE2 设备传递给 R4_PE2 设备的路由信息。

```
R4_PE2#show    ip route   vrf   VPN1
Routing Table: VPN1

Codes:    C - Connected, L - Local, S - Static
          R - RIP, O - OSPF, B - BGP, I - IS-IS, V - Overflow route
          N1 - OSPF NSSA external type 1, N2 - OSPF NSSA external type 2
          E1 - OSPF external type 1, E2 - OSPF external type 2
          SU - IS-IS summary, L1 - IS-IS level-1, L2 - IS-IS level-2
          IA - Inter area, EV - BGP EVPN, A - Arp to host
          LA - Local aggregate route
          * - candidate default

Gateway of last resort is no set
C       10.0.45.0/24 is directly connected, GigabitEthernet 0/3
C       10.0.45.4/32 is local host.
I L2    192.168.2.0/24 [115/1] via 10.0.45.5, 00:00:49, GigabitEthernet 0/3
R4_PE2#
```

8. 检查 MP-BGP 路由传递情况。

```
R2_PE1#show    bgp    vpnv4 unicast all
BGP table version is 1, local router ID is 2.2.2.2
Status codes: s suppressed, d damped, h history, * valid, > best, i - internal,
              S Stale, b - backup entry, m - multipath, f Filter, a additional-path
Origin codes: i - IGP, e - EGP, ? - incomplete

     Network          Next Hop         Metric       LocPrf       Weight Path
Route Distinguisher: 100:1 (Default for VRF VPN1)
 *>   10.0.12.0/24     0.0.0.0          1                        32768   ?
 *>i  10.0.45.0/24     10.0.4.4         1            100         0       ?
 *>   192.168.1.1/32   10.0.12.1        1                        32768   ?
 *>i  192.168.2.0      10.0.4.4         1            100         0       ?

Total number of prefixes 4
Route Distinguisher: 100:2
 *>i  10.0.45.0/24     10.0.4.4         1            100         0       ?
 *>i  192.168.2.0      10.0.4.4         1            100         0       ?
```

Total number of prefixes 2
R2_PE1#

R4_PE2#show   bgp   vpnv4 unicast   all
BGP table version is 1, local router ID is 4.4.4.4
Status codes: s suppressed, d damped, h history, * valid, > best, i - internal,
              S Stale, b - backup entry, m - multipath, f Filter, a additional-path
Origin codes: i - IGP, e - EGP, ? - incomplete

```
   Network          Next Hop           Metric        LocPrf      Weight Path
Route Distinguisher: 100:2 (Default for VRF VPN1)
*>i 10.0.12.0/24    10.0.2.2           1             100         0       ?
*>  10.0.45.0/24    0.0.0.0            1                         32768   ?
*>i 192.168.1.1/32  10.0.2.2           1             100         0       ?
*>  192.168.2.0     10.0.45.5          1                         32768   ?

Total number of prefixes 4
Route Distinguisher: 100:1
*>i 10.0.12.0/24    10.0.2.2           1             100         0       ?
*>i 192.168.1.1/32  10.0.2.2           1             100         0       ?
```

Total number of prefixes 2
R4_PE2#

9. 在 R1_CE1 设备和 R5_CE2 设备查看全局 IP 路由表，正常接收对方发送的私网路由。

```
R1_CE1#show   ip route

Codes:   C - Connected, L - Local, S - Static
         R - RIP, O - OSPF, B - BGP, I - IS-IS, V - Overflow route
         N1 - OSPF NSSA external type 1, N2 - OSPF NSSA external type 2
         E1 - OSPF external type 1, E2 - OSPF external type 2
         SU - IS-IS summary, L1 - IS-IS level-1, L2 - IS-IS level-2
         IA - Inter area, EV - BGP EVPN, A - Arp to host
         LA - Local aggregate route
         * - candidate default

Gateway of last resort is no set
C       10.0.12.0/24 is directly connected, GigabitEthernet 0/0
C       10.0.12.1/32 is local host.
O E2    10.0.45.0/24 [110/1] via 10.0.12.2, 00:00:28, GigabitEthernet 0/0
C       192.168.1.0/24 is directly connected, Loopback 0
C       192.168.1.1/32 is local host.
O E2    192.168.2.0/24 [110/1] via 10.0.12.2, 00:00:28, GigabitEthernet 0/0
R1_CE1#


R5_CE2#show   ip route

Codes:   C - Connected, L - Local, S - Static
         R - RIP, O - OSPF, B - BGP, I - IS-IS, V - Overflow route
         N1 - OSPF NSSA external type 1, N2 - OSPF NSSA external type 2
         E1 - OSPF external type 1, E2 - OSPF external type 2
         SU - IS-IS summary, L1 - IS-IS level-1, L2 - IS-IS level-2
         IA - Inter area, EV - BGP EVPN, A - Arp to host
```

```
                LA - Local aggregate route
                * - candidate default

Gateway of last resort is no set
I L2    10.0.12.0/24 [115/1] via 10.0.45.4, 00:00:18, GigabitEthernet 0/3
C       10.0.45.0/24 is directly connected, GigabitEthernet 0/3
C       10.0.45.5/32 is local host.
I L2    192.168.1.1/32 [115/1] via 10.0.45.4, 00:00:18, GigabitEthernet 0/3
C       192.168.2.0/24 is directly connected, Loopback 0
C       192.168.2.1/32 is local host.
R5_CE2#
```

10. 测试私网地址互通情况。

```
R1_CE1#ping    192.168.2.1 source    192.168.1.1
Sending 5, 100-byte ICMP Echoes to 192.168.2.1, timeout is 2 seconds:
  < press Ctrl+C to break >
!!!!!
Success rate is 100 percent (5/5), round-trip min/avg/max = 3/14/35 ms.
R1_CE1#

R5_CE2#ping    192.168.1.1 source    192.168.2.1
Sending 5, 100-byte ICMP Echoes to 192.168.1.1, timeout is 2 seconds:
  < press Ctrl+C to break >
!!!!!
Success rate is 100 percent (5/5), round-trip min/avg/max = 5/6/8 ms.
R5_CE2#
```

11. 通过 Wireshark 抓包工具在 R2_PE1 设备或者 R4_PE2 设备检测该路由的公网标签及私网标签。

```
10 2.8332...  192.168.1.1    192.168.2.1    ICMP    150 Echo (ping) request  id=0x0200, seq=3416/22541, ttl=63 (reply in 11)
11 2.8366...  192.168.2.1    192.168.1.1    ICMP    146 Echo (ping) reply    id=0x0200, seq=3416/22541, ttl=63 (request in 10)
12 2.9398...  192.168.1.1    192.168.2.1    ICMP    150 Echo (ping) request  id=0x0200, seq=6232/22552, ttl=63 (reply in 13)
13 2.9425...  192.168.2.1    192.168.1.1    ICMP    146 Echo (ping) reply    id=0x0200, seq=6232/22552, ttl=63 (request in 12)
14 3.0445...  192.168.1.1    192.168.2.1    ICMP    150 Echo (ping) request  id=0x0200, seq=8792/22562, ttl=63 (reply in 15)
15 3.0470...  192.168.2.1    192.168.1.1    ICMP    146 Echo (ping) reply    id=0x0200, seq=8792/22562, ttl=63 (request in 14)

Frame 10: 150 bytes on wire (1200 bits), 150 bytes captured (1200 bits) on interface 0
Ethernet II, Src: 50:00:00:02:00:02 (50:00:00:02:00:02), Dst: 50:00:00:03:00:02 (50:00:00:03:00:02)
MultiProtocol Label Switching Header, Label: 11265, Exp: 0, S: 0, TTL: 63
    0000 0010 1100 0000 0001 .... .... .... = MPLS Label: 11265
    .... .... .... .... .... 000. .... .... = MPLS Experimental Bits: 0
    .... .... .... .... .... ...0 .... .... = MPLS Bottom Of Label Stack: 0
    .... .... .... .... .... .... 0011 1111 = MPLS TTL: 63
MultiProtocol Label Switching Header, Label: 60928, Exp: 0, S: 1, TTL: 63
    0000 1110 1110 0000 0000 .... .... .... = MPLS Label: 60928
    .... .... .... .... .... 000. .... .... = MPLS Experimental Bits: 0
    .... .... .... .... .... ...1 .... .... = MPLS Bottom Of Label Stack: 1
    .... .... .... .... .... .... 0011 1111 = MPLS TTL: 63
Internet Protocol Version 4, Src: 192.168.1.1, Dst: 192.168.2.1
Internet Control Message Protocol

0000  50 00 00 03 00 02 50 00  00 02 00 02 88           P.....P......
0010  10 3f 0e e0 01 3f 45 00  00 80 36 1f 00           .?...?E..6..
0020  c1 0b c0 a8 01 01 c0 a8  02 01 08 00 5a           ............Z
0030  0d 58 ab cd ab cd ab cd  ab cd ab cd ab           .X..........
0040  ab cd ab cd ab cd ab cd  ab cd ab cd ab           ............
0050  ab cd ab cd ab cd ab cd  ab cd ab cd ab           ............
0060  ab cd ab cd ab cd ab cd  ab cd ab cd ab           ............
0070  ab cd ab cd ab cd ab cd  ab cd ab cd ab           ............
0080  ab cd ab cd ab cd ab cd  ab cd ab cd ab           ............
0090  ab cd ab cd ab                                     .....
```

➢ 问题与思考

1. MPLS VPN 中会有几层标签？标签分别都由哪种协议分配？
2. MPLS VPN 的 VRF 路由表和 IP 路由表有什么区别？

3.2.2 MPLS VPN-Hub Spoke

➢ 原理

MPLS VPN-Hub Spoke 是 MPLS VPN 组网中的一种 Hub Spoke 组网模式，VPN 内部数据不能直接交换，必须经过统一的控制中心才能进行交互，且只有该控制中心了解 VPN 内部的全部信息资源，其他属于该 VPN 的用户要获得 VPN 内部的资源必须由控制中心告知。

➢ 任务拓扑

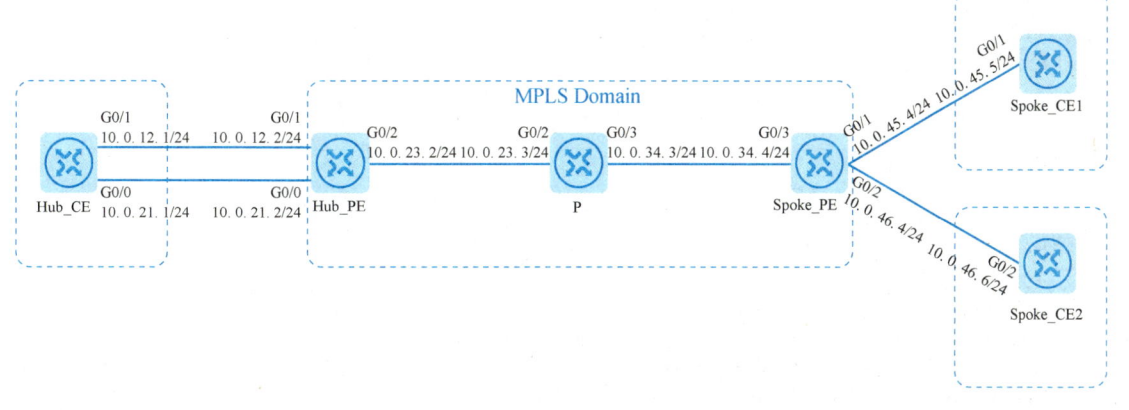

➢ 实施步骤

1. 根据任务拓扑配置各设备接口 IP 地址。（略）
2. 在 MPLS 域内 Hub_PE 设备、P 设备、Spoke_PE 设备配置 IS-IS 协议互通。

```
Hub_PE(config)#router   isis    1
Hub_PE(config-router)#net    49.0001.0000.0000.0002.00
Hub_PE(config-router)#is-type    level-2
```

```
Hub_PE(config-router)#exit
Hub_PE(config)#interface    gigabitEthernet 0/2
Hub_PE(config-if-GigabitEthernet 0/2)#ip router    isis    1
Hub_PE(config-if-GigabitEthernet 0/2)#exit
Hub_PE(config)#interface    loopback 0
Hub_PE(config-if-Loopback 0)#ip router    isis    1
Hub_PE(config-if-Loopback 0)#exit
Hub_PE(config)#

P(config)#router    isis    1
P(config-router)#net    49.0001.0000.0000.0003.00
P(config-router)#is-type    level-2
P(config-router)#exit
P(config)#interface    gigabitEthernet 0/2
P(config-if-GigabitEthernet 0/2)#ip router    isis    1
P(config-if-GigabitEthernet 0/2)#exit
P(config)#interface    gigabitEthernet 0/3
P(config-if-GigabitEthernet 0/3)#ip router    isis    1
P(config-if-GigabitEthernet 0/3)#exit
P(config)#interface    loopback 0
P(config-if-Loopback 0)#ip router    isis    1
P(config-if-Loopback 0)#exit
P(config)#

Spoke_PE(config)#router    isis    1
Spoke_PE(config-router)#net    49.0001.0000.0000.0004.00
Spoke_PE(config-router)#is-type    level-2
Spoke_PE(config-router)#exit
Spoke_PE(config)#interface    gigabitEthernet 0/3
Spoke_PE(config-if-GigabitEthernet 0/3)#ip router    isis    1
Spoke_PE(config-if-GigabitEthernet 0/3)#exit
Spoke_PE(config)#interface    loopback 0
Spoke_PE(config-if-Loopback 0)#ip address    10.0.4.4 32
Spoke_PE(config-if-Loopback 0)#ip router    isis    1
Spoke_PE(config-if-Loopback 0)#exit
Spoke_PE(config)#
```

3. 在 MPLS 域内配置 LDP 协议，由 LDP 协议分配公网标签。

```
Hub_PE(config)#mpls    ip
Hub_PE(config)#mpls    router    ldp
Hub_PE(config-mpls-router)#ldp router-id    interface    loopback 0 force
Hub_PE(config-mpls-router)#exit
Hub_PE(config)#interface    gigabitEthernet 0/2
Hub_PE(config-if-GigabitEthernet 0/2)#mpls    ip
Hub_PE(config-if-GigabitEthernet 0/2)#label-switching
Hub_PE(config-if-GigabitEthernet 0/2)#exit
Hub_PE(config)#

P(config)#mpls    ip
P(config)#mpls    router    ldp
P(config-mpls-router)#ldp router-id    interface    loopback 0 force
P(config-mpls-router)#exit
P(config)#interface    gigabitEthernet 0/2
P(config-if-GigabitEthernet 0/2)#mpls    ip
P(config-if-GigabitEthernet 0/2)#label-switching
```

```
P(config-if-GigabitEthernet 0/2)#exit
P(config)#interface    gigabitEthernet 0/3
P(config-if-GigabitEthernet 0/3)#mpls    ip
P(config-if-GigabitEthernet 0/3)#label-switching
P(config-if-GigabitEthernet 0/3)#exit
P(config)#

Spoke_PE(config)#mpls    ip
Spoke_PE(config)#mpls    router    ldp
Spoke_PE(config-mpls-router)#ldp router-id    interface    loopback 0 force
Spoke_PE(config-mpls-router)#exit
Spoke_PE(config)#interface    gigabitEthernet 0/3
Spoke_PE(config-if-GigabitEthernet 0/3)#mpls    ip
Spoke_PE(config-if-GigabitEthernet 0/3)#label-switching
Spoke_PE(config-if-GigabitEthernet 0/3)#exit
Spoke_PE(config)#
```

4. 在 Hub_PE 设备和 Spoke_PE 设备创建 VRF。在 Hub_PE 设备创建两个 VRF，一个用作接收 Spoke_PE 设备发送的路由，其 Import Target 为 10:1、20:1，另一个用作向 Spoke_PE 设备发送路由，其 Export Target 为 30:1。

```
Hub_PE(config)#ip vrf    VPN1_IN
Hub_PE(config-vrf)#rd    100:1
Hub_PE(config-vrf)#route-target    import    10:1
Hub_PE(config-vrf)#route-target    import    20:1
Hub_PE(config-vrf)#exit
Hub_PE(config)#

Hub_PE(config)#ip vrf    VPN1_OUT
Hub_PE(config-vrf)#rd 100:10
Hub_PE(config-vrf)#route-target    export    30:1
Hub_PE(config-vrf)#exit
Hub_PE(config)#

Hub_PE(config)#interface    gigabitEthernet 0/1
Hub_PE(config-if-GigabitEthernet 0/1)#ip vrf    forwarding VPN1_IN
% Interface GigabitEthernet 0/1 IP address removed due to enabling VRF VPN1_IN
Hub_PE(config-if-GigabitEthernet 0/1)#ip address    10.0.12.2 24
Hub_PE(config-if-GigabitEthernet 0/1)#exit
Hub_PE(config)#interface    gigabitEthernet 0/0
Hub_PE(config-if-GigabitEthernet 0/0)#ip vrf    forwarding    VPN1_OUT
% Interface GigabitEthernet 0/0 IP address removed due to enabling VRF VPN1_OUT
Hub_PE(config-if-GigabitEthernet 0/0)#ip address    10.0.21.2 24
Hub_PE(config-if-GigabitEthernet 0/0)#exit
Hub_PE(config)#

Spoke_PE(config)#ip vrf    VPN2
Spoke_PE(config-vrf)#rd    200:1
Spoke_PE(config-vrf)#route-target    export    10:1
Spoke_PE(config-vrf)#route-target    import    30:1
Spoke_PE(config-vrf)#exit
Spoke_PE(config)#

Spoke_PE(config)#ip vrf    VPN3
Spoke_PE(config-vrf)#rd    300:1
```

```
Spoke_PE(config-vrf)#route-target    export    20:1
Spoke_PE(config-vrf)#route-target    import    30:1
Spoke_PE(config-vrf)#exit
Spoke_PE(config)#
Spoke_PE(config)#interface    gigabitEthernet 0/1
Spoke_PE(config-if-GigabitEthernet 0/1)#ip vrf    forwarding    VPN2
% Interface GigabitEthernet 0/1 IP address removed due to enabling VRF VPN2
Spoke_PE(config-if-GigabitEthernet 0/1)#ip address    10.0.45.4 24
Spoke_PE(config-if-GigabitEthernet 0/1)#exit
Spoke_PE(config)#interface    gigabitEthernet 0/2
Spoke_PE(config-if-GigabitEthernet 0/2)#ip vrf    forwarding    VPN3
% Interface GigabitEthernet 0/2 IP address removed due to enabling VRF VPN3
Spoke_PE(config-if-GigabitEthernet 0/2)#ip address    10.0.46.4 24
Spoke_PE(config-if-GigabitEthernet 0/2)#exit
Spoke_PE(config)#
```

5. 在 Hub_PE 设备和 Spoke_PE 设备建立 IBGP 邻居关系，BGP 自治系统编号为 65312。

```
Hub_PE(config)#router    bgp    65312
Hub_PE(config-router)#bgp    router-id    2.2.2.2
Hub_PE(config-router)#neighbor    10.0.4.4 remote-as    65312
Hub_PE(config-router)#neighbor    10.0.4.4 update-source    loopback 0
Hub_PE(config-router)#exit
Hub_PE(config)#

Spoke_PE(config)#router    bgp    65312
Spoke_PE(config-router)#bgp    router-id    4.4.4.4
Spoke_PE(config-router)#neighbor    10.0.2.2 remote-as    65312
Spoke_PE(config-router)#neighbor    10.0.2.2 update-source    loopback 0
Spoke_PE(config-router)#exit
Spoke_PE(config)#
```

6. 在 Hub_PE 设备和 Spoke_PE 设备建立 MP-BGP 邻居关系。注：若不需要使用 IPv4 单播 IBGP 邻居，可通过以下命令关闭。

```
Hub_PE(config)#router    bgp    65312
Hub_PE(config-router)#address-family ipv4
Hub_PE(config-router-af)#no neighbor    10.0.4.4 activate        //关闭 IPv4 单播 IBGP 邻居
Hub_PE(config-router-af)#exit
Hub_PE(config-router)#

Spoke_PE(config)#router    bgp    65312
Spoke_PE(config-router)#address-family ipv4
Spoke_PE(config-router-af)#no neighbor    10.0.2.2 activate
Spoke_PE(config-router-af)#exit
Spoke_PE(config-router)#
Hub_PE(config)#router    bgp    65312
Hub_PE(config-router)#address-family vpnv4
Hub_PE(config-router-af)#neighbor    10.0.4.4 activate
Hub_PE(config-router-af)#exit

Spoke_PE(config)#router    bgp    65312
Spoke_PE(config-router)#address-family vpnv4
Spoke_PE(config-router-af)#neighbor    10.0.2.2 activate
Spoke_PE(config-router-af)#exit
Spoke_PE(config-router)#exit
Spoke_PE(config)#
```

7. 在 Hub_PE 设备和 HuB_CE1 设备之间运行 BGP 协议传递私网路由信息，Hub_CE 设备自治系统号为 100。

```
Hub_CE1(config)#router bgp    100
Hub_CE1(config-router)#bgp    router-id    1.1.1.1
Hub_CE1(config-router)#neighbor    10.0.12.2 remote-as    65312
Hub_CE1(config-router)#neighbor    10.0.21.2 remote-as    65312
Hub_CE1(config-router)#exit
Hub_CE1(config)#

Hub_PE(config)#router    bgp    65312
Hub_PE(config-router)#address-family ipv4 vrf VPN1_IN
Hub_PE(config-router-af)#neighbor    10.0.12.1 remote-as    100
Hub_PE(config-router-af)#exit
Hub_PE(config-router)#address-family ipv4 vrf    VPN1_OUT
Hub_PE(config-router-af)#neighbor    10.0.21.1 remote-as    100
Hub_PE(config-router-af)#exit
Hub_PE(config-router)#
```

8. 在 Spoke_PE 设备和 Spoke_CE1 设备建立 OSPF 邻居关系传递私网路由，Spoke_CE1 设备进程号为 10。

```
Spoke_PE(config)#router    ospf    10 vrf    VPN2
Spoke_PE(config-router)#router-id    4.4.4.4
Change router-id and update OSPF process! [yes/no]:y
Spoke_PE(config-router)#network    10.0.45.0 0.0.0.255 area    0
Spoke_PE(config-router)#exit
Spoke_PE(config)#

Spoke_CE1(config)#router    ospf    10
Spoke_CE1(config-router)#router-id    5.5.5.5
Change router-id and update OSPF process! [yes/no]:y
Spoke_CE1(config-router)#network    10.0.45.0 0.0.0.255 area    0
Spoke_CE1(config-router)#exit
Spoke_CE1(config)#
```

9. 在 Spoke_PE 设备和 Spoke_CE2 设备建立 IS-IS 邻居传递私网路由，Spoke_CE2 设备进程号为 100。

```
Spoke_PE(config)#router    isis    100
Spoke_PE(config-router)#vrf    VPN3
Spoke_PE(config-router)#net    49.0002.0000.0000.0014.00
Spoke_PE(config-router)#is-type    level-2
Spoke_PE(config-router)#exit
Spoke_PE(config)#interface    gigabitEthernet 0/2
Spoke_PE(config-if-GigabitEthernet 0/2)#ip router isis    100
Spoke_PE(config-if-GigabitEthernet 0/2)#exit
Spoke_PE(config)#

Spoke_CE2(config)#router    isis    100
Spoke_CE2(config-router)#net    49.0002.0000.0000.0006.00
Spoke_CE2(config-router)#is-type    level-2
Spoke_CE2(config-router)#exit
Spoke_CE2(config)#interface    gigabitEthernet 0/2
Spoke_CE2(config-if-GigabitEthernet 0/2)#ip router    isis    100
Spoke_CE2(config-if-GigabitEthernet 0/2)#exit
Spoke_CE2(config)#
```

10. 在 Hub_CE1 设备创建 loopback 0 接口，并将 loopback 0 接口通告到 BGP 协议中。

```
Hub_CE1(config)#interface    loopback 0
Hub_CE1(config-if-Loopback 0)#ip address    192.168.1.1 24
Hub_CE1(config-if-Loopback 0)#exit
Hub_CE1(config)#router   bgp    100
Hub_CE1(config-router)#network    192.168.1.0 mask    255.255.255.0
Hub_CE1(config-router)#exit
Hub_CE1(config)#
```

11. 在 Spoke_CE1 设备和 Spoke_CE2 设备创建 loopback 0 接口，并通告 loopback 0 接口。

```
Spoke_CE1(config)#interface    loopback 0
Spoke_CE1(config-if-Loopback 0)#ip address    192.168.2.1 24
Spoke_CE1(config-if-Loopback 0)#exit
Spoke_CE1(config)#router   ospf    10
Spoke_CE1(config-router)#network 192.168.2.0 0.0.0.255 area    0
Spoke_CE1(config-router)#exit
Spoke_CE1(config)#

Spoke_CE2(config)#interface    loopback 0
Spoke_CE2(config-if-Loopback 0)#ip address    192.168.3.1 24
Spoke_CE2(config-if-Loopback 0)#ip router    isis    100
Spoke_CE2(config-if-Loopback 0)#exit
Spoke_CE2(config)#
```

12. 在 Spoke_PE 设备将 OSPF 10 的路由重分布到 MP-BGP 协议中。

```
Spoke_PE(config)#router    bgp    65312
Spoke_PE(config-router)#address-family ipv4 vrf VPN2
Spoke_PE(config-router-af)#redistribute    ospf   10 match    internal    external
Spoke_PE(config-router-af)#exit
Spoke_PE(config-router)#exit
Spoke_PE(config)#
```

13. 在 Spoke_PE 设备将 IS-IS 100 的路由重分布到 MP-BGP 协议中。

```
Spoke_PE(config)#router bgp    65312
Spoke_PE(config-router)#address-family ipv4 vrf    VPN3
Spoke_PE(config-router-af)#redistribute    isis 100 level-2
Spoke_PE(config-router-af)#exit
Spoke_PE(config-router)#exit
Spoke_PE(config)#
```

14. 在 Hub_PE 设备针对 VPN1_OUT 方向设置允许接收包含本身 AS 号的 Update 报文，即打破 BGP 协议的防环规则经过 Hub 设备中转。

```
Hub_PE(config)#router    bgp    65312
Hub_PE(config-router)#address-family ipv4 vrf    VPN1_OUT
Hub_PE(config-router-af)#neighbor    10.0.21.1 allowas-in
Hub_PE(config-router-af)#exit
Hub_PE(config-router)#
Hub_PE(config)#
```

> 任务验证

1. 检查 MPLS 域内 IS-IS 协议邻居建立情况，测试 loopback 接口地址互通情况。

```
P#show   isis   neighbors

Area 1:
System Id      Type   IP Address      State    Holdtime   Circuit        Interface
Hub_PE         L2     10.0.23.2       Up       24         P.01           GigabitEthernet 0/2
Spoke_PE       L2     10.0.34.4       Up       7          Spoke_PE.01    GigabitEthernet 0/3

P#

Hub_PE#show ip route

Codes:   C - Connected, L - Local, S - Static
         R - RIP, O - OSPF, B - BGP, I - IS-IS, V - Overflow route
         N1 - OSPF NSSA external type 1, N2 - OSPF NSSA external type 2
         E1 - OSPF external type 1, E2 - OSPF external type 2
         SU - IS-IS summary, L1 - IS-IS level-1, L2 - IS-IS level-2
         IA - Inter area, EV - BGP EVPN, A - Arp to host
         LA - Local aggregate route
         * - candidate default

Gateway of last resort is no set
C       10.0.2.2/32 is local host.
I L2    10.0.3.3/32 [115/1] via 10.0.23.3, 01:05:15, GigabitEthernet 0/2
I L2    10.0.4.4/32 [115/2] via 10.0.23.3, 01:04:15, GigabitEthernet 0/2
C       10.0.23.0/24 is directly connected, GigabitEthernet 0/2
C       10.0.23.2/32 is local host.
I L2    10.0.34.0/24 [115/2] via 10.0.23.3, 01:08:07, GigabitEthernet 0/2
Hub_PE#

Hub_PE#ping   10.0.4.4
Sending 5, 100-byte ICMP Echoes to 10.0.4.4, timeout is 2 seconds:
  < press Ctrl+C to break >
!!!!!
Success rate is 100 percent (5/5), round-trip min/avg/max = 2/2/3 ms.
Hub_PE#
```

2. 检查 LDP 协议邻居建立情况。

```
P#show   mpls   ldp   neighbor
Default VRF:
    Peer LDP Ident: 10.0.2.2:0; Local LDP Ident: 10.0.3.3:0
        TCP connection: 10.0.2.2.646 - 10.0.3.3.41205
        State: OPERATIONAL; Msgs sent/recv: 65/66; UNSOLICITED
        Up time: 00:14:32
        Graceful Restart enabled; Peer reconnect time (msecs): 300000
        LDP discovery sources:
            Link Peer on GigabitEthernet 0/2, Src IP addr: 10.0.23.2
        Addresses bound to peer LDP Ident:
            10.0.23.2          10.0.2.2
    Peer LDP Ident: 10.0.4.4:0; Local LDP Ident: 10.0.3.3:0
        TCP connection: 10.0.4.4.44109 - 10.0.3.3.646
        State: OPERATIONAL; Msgs sent/recv: 15/16; UNSOLICITED
        Up time: 00:02:12
        Graceful Restart enabled; Peer reconnect time (msecs): 300000
        LDP discovery sources:
            Link Peer on GigabitEthernet 0/3, Src IP addr: 10.0.34.4
        Addresses bound to peer LDP Ident:
```

```
             10.0.34.4         10.0.4.4
P#
```

3. 检查 MPLS 公网标签信息转发表。

```
P#show   mpls   forwarding-table

Label Operation Code:
PH--PUSH label
PP--POP label
SW--SWAP label
SP--SWAP topmost label and push new label
DP--DROP packet
PC--POP label and continue lookup by IP or Label
PI--POP label and do ip lookup forward
PN--POP label and forward to nexthop
PM--POP label and do MAC lookup forward
PV--POP label and output to VC attach interface
IP--IP lookup forward
  s--stale
  Local    Outgoing OP FEC                 Outgoing         Nexthop
  label    label                           interface
  --       imp-null PH 10.0.2.2/32         Gi0/2            10.0.23.2
  --       imp-null PH 10.0.4.4/32         Gi0/3            10.0.34.4
  11264    imp-null PP 10.0.2.2/32         Gi0/2            10.0.23.2
  11265    imp-null PP 10.0.4.4/32         Gi0/3            10.0.34.4

P#
```

4. 检查 MP-BGP 邻居关系建立情况。

```
Hub_PE#show   bgp   vpnv4 unicast all   summary
For address family: VPNv4 Unicast
BGP router identifier 2.2.2.2, local AS number 65312
BGP table version is 1
0 BGP AS-PATH entries
0 BGP Community entries
0 BGP Prefix entries (Maximum-prefix:4294967295)

Neighbor      V         AS MsgRcvd MsgSent   TblVer   InQ OutQ Up/Down   State/PfxRcd
10.0.4.4      4       65312       3       2        1    0    0 00:00:05       0

Total number of neighbors 1, established neighbors 1

Hub_PE#
```

5. 检查 Hub_PE 设备和 Hub_CE 设备之间的 BGP 邻居关系建立情况。

```
Hub_CE1#show ip bgp   summary
For address family: IPv4 Unicast
BGP router identifier 1.1.1.1, local AS number 100
BGP table version is 1
0 BGP AS-PATH entries
0 BGP Community entries
0 BGP Prefix entries (Maximum-prefix:4294967295)

Neighbor      V         AS MsgRcvd MsgSent   TblVer   InQ OutQ Up/Down   State/PfxRcd
10.0.12.2     4       65312       4       3        1    0    0 00:01:14       0
```

```
10.0.21.2     4      65312     4     2     1     0     0 00:01:01     0
```

Total number of neighbors 2, established neighbors 2

Hub_CE1#

6. 检查 Spoke_PE 设备和 Spoke_CE1 设备的 OSPF 协议邻居建立情况。

```
Spoke_CE1#show   ip ospf    neighbor

OSPF process 10, 1 Neighbors, 1 is Full:
Neighbor ID   Pri   State        BFD State   Dead Time   Address      Interface
4.4.4.4       1     Full/BDR     -           00:00:40    10.0.45.4    GigabitEthernet 0/1
Spoke_CE1#
```

7. 检查 Spoke_PE 设备和 Spoke_CE2 设备的 IS-IS 协议邻居建立情况。

```
Spoke_CE2#show   isis   neighbors
Area 100:
System Id     Type   IP Address    State   Holdtime  Circuit        Interface
Spoke_PE      L2     10.0.46.4     Up      21        Spoke_CE2.01   GigabitEthernet 0/2
```

8. 检查 MP-BGP 路由。

```
Hub_PE#show   bgp  vpnv4 unicast   all
BGP table version is 1, local router ID is 2.2.2.2
Status codes: s suppressed, d damped, h history, * valid, > best, i - internal,
              S Stale, b - backup entry, m - multipath, f Filter, a additional-path
Origin codes: i - IGP, e - EGP, ? - incomplete

     Network          Next Hop         Metric      LocPrf     Weight Path
Route Distinguisher: 100:1 (Default for VRF VPN1_IN)
*>   192.168.1.0      10.0.12.1          0                    0 100 i

Total number of prefixes 1
Route Distinguisher: 100:10 (Default for VRF VPN1_OUT)
*>   192.168.1.0      10.0.21.1          0                    0 100 i

Total number of prefixes 1
Hub_PE#
```

9. 在 Hub_PE 设备查看 VPNv4 路由表，VPN1_OUT 的 VRF 下有相关的私网路由信息，此路由需要经过 Hub 端设备中转传递给 Spoke_CE1 设备或者 Spoke_CE2 设备。

```
Hub_PE#show   bgp  vpnv4 unicast   all
BGP table version is 1, local router ID is 2.2.2.2
Status codes: s suppressed, d damped, h history, * valid, > best, i - internal,
              S Stale, b - backup entry, m - multipath, f Filter, a additional-path
Origin codes: i - IGP, e - EGP, ? - incomplete

     Network          Next Hop         Metric      LocPrf     Weight Path
Route Distinguisher: 100:1 (Default for VRF VPN1_IN)
*>i  10.0.45.0/24     10.0.4.4           1         100        0      ?
*>i  10.0.46.0/24     10.0.4.4           1         100        0      ?
*>   192.168.1.0      10.0.12.1          0                    0 100  i
*>i  192.168.2.1/32   10.0.4.4           1         100        0      ?
*>i  192.168.3.0      10.0.4.4           1         100        0      ?
```

```
Total number of prefixes 5
Route Distinguisher: 100:10 (Default for VRF VPN1_OUT)
*>   10.0.45.0/24      10.0.21.1         0                0 100 65312 ?
*>   10.0.46.0/24      10.0.21.1         0                0 100 65312 ?
*>   192.168.1.0       10.0.21.1         0                0 100 i
*>   192.168.2.1/32    10.0.21.1         0                0 100 65312 ?
*>   192.168.3.0       10.0.21.1         0                0 100 65312 ?

Total number of prefixes 5
Route Distinguisher: 300:1
*>i 10.0.46.0/24       10.0.4.4          1        100      0       ?
*>i 192.168.3.0        10.0.4.4          1        100      0       ?

Total number of prefixes 2
Route Distinguisher: 200:1
*>i 10.0.45.0/24       10.0.4.4          1        100      0       ?
*>i 192.168.2.1/32     10.0.4.4          1        100      0       ?

Total number of prefixes 2
Hub_PE#
```

10. 在 Hub_CE1 设备查看全局路由表，接收 Spoke_CE1 设备和 Spoke_CE2 设备的私网路由。

```
Hub_CE1#show    ip route

Codes:    C - Connected, L - Local, S - Static
          R - RIP, O - OSPF, B - BGP, I - IS-IS, V - Overflow route
          N1 - OSPF NSSA external type 1, N2 - OSPF NSSA external type 2
          E1 - OSPF external type 1, E2 - OSPF external type 2
          SU - IS-IS summary, L1 - IS-IS level-1, L2 - IS-IS level-2
          IA - Inter area, EV - BGP EVPN, A - Arp to host
          LA - Local aggregate route
          * - candidate default

Gateway of last resort is no set
C       10.0.12.0/24 is directly connected, GigabitEthernet 0/1
C       10.0.12.1/32 is local host.
C       10.0.21.0/24 is directly connected, GigabitEthernet 0/0
C       10.0.21.1/32 is local host.
B       10.0.45.0/24 [20/0] via 10.0.12.2, 00:43:08
B       10.0.46.0/24 [20/0] via 10.0.12.2, 00:45:49
C       192.168.1.0/24 is directly connected, Loopback 0
C       192.168.1.1/32 is local host.
B       192.168.2.1/32 [20/0] via 10.0.12.2, 00:42:41
B       192.168.3.0/24 [20/0] via 10.0.12.2, 00:45:49
Hub_CE1#
```

11. 在 Spoke_CE1 设备查看全局路由表，接收 Hub_CE 设备和 Spoke_CE2 设备的私网路由。

```
Spoke_CE1#show    ip route

Codes:    C - Connected, L - Local, S - Static
          R - RIP, O - OSPF, B - BGP, I - IS-IS, V - Overflow route
          N1 - OSPF NSSA external type 1, N2 - OSPF NSSA external type 2
          E1 - OSPF external type 1, E2 - OSPF external type 2
```

```
         SU - IS-IS summary, L1 - IS-IS level-1, L2 - IS-IS level-2
         IA - Inter area, EV - BGP EVPN, A - Arp to host
         LA - Local aggregate route
         * - candidate default

Gateway of last resort is no set
C        10.0.45.0/24 is directly connected, GigabitEthernet 0/1
C        10.0.45.5/32 is local host.
O E2     10.0.46.0/24 [110/1] via 10.0.45.4, 00:09:24, GigabitEthernet 0/1
O E2     192.168.1.0/24 [110/1] via 10.0.45.4, 00:13:11, GigabitEthernet 0/1
C        192.168.2.0/24 is directly connected, Loopback 0
C        192.168.2.1/32 is local host.
O E2     192.168.3.0/24 [110/1] via 10.0.45.4, 00:09:24, GigabitEthernet 0/1
Spoke_CE1#
```

12. 在 Spoke_CE2 设备查看全局路由表，接收 Hub_CE 设备和 Spoke_CE1 设备的私网路由。

```
Spoke_CE2#show    ip route

Codes:   C - Connected, L - Local, S - Static
         R - RIP, O - OSPF, B - BGP, I - IS-IS, V - Overflow route
         N1 - OSPF NSSA external type 1, N2 - OSPF NSSA external type 2
         E1 - OSPF external type 1, E2 - OSPF external type 2
         SU - IS-IS summary, L1 - IS-IS level-1, L2 - IS-IS level-2
         IA - Inter area, EV - BGP EVPN, A - Arp to host
         LA - Local aggregate route
         * - candidate default

Gateway of last resort is no set
I L2     10.0.45.0/24 [115/1] via 10.0.46.4, 00:30:23, GigabitEthernet 0/2
C        10.0.46.0/24 is directly connected, GigabitEthernet 0/2
C        10.0.46.6/32 is local host.
I L2     192.168.1.0/24 [115/1] via 10.0.46.4, 00:34:10, GigabitEthernet 0/2
I L2     192.168.2.1/32 [115/1] via 10.0.46.4, 00:30:23, GigabitEthernet 0/2
C        192.168.3.0/24 is directly connected, Loopback 0
C        192.168.3.1/32 is local host.
Spoke_CE2#
```

13. 在 Spoke_CE1 设备和 Spoke_CE2 设备用 traceroute 检查路径，此路径为经过 Hub 端设备中转后到达 Spoke 端，即 Spoke_CE1 设备和 Spoke_CE2 设备用户通信必须经过 Hub 端执行中转。

```
Spoke_CE1#traceroute    192.168.3.1 source    192.168.2.1
  < press Ctrl+C to break >
Tracing the route to 192.168.3.1

   1        10.0.45.4        2 msec      1 msec      <1 msec
   2        10.0.34.3        4 msec      3 msec      3 msec
   3        10.0.21.2        2 msec      2 msec      3 msec
   4        10.0.12.1        4 msec      <1 msec     2 msec
   5        10.0.12.2        5 msec      3 msec      2 msec
   6        10.0.23.3        5 msec      4 msec      4 msec
   7        10.0.46.4        4 msec      6 msec      16 msec
   8        192.168.3.1      11 msec     9 msec      7 msec
Spoke_CE1#
```

```
Spoke_CE2#traceroute    192.168.2.1 source    192.168.3.1
< press Ctrl+C to break >
Tracing the route to 192.168.2.1

1          10.0.46.4         1 msec      1 msec      <1 msec
2          10.0.34.3         3 msec      3 msec      3 msec
3          10.0.21.2         3 msec      2 msec      4 msec
4          10.0.12.1         3 msec      2 msec      1 msec
5          10.0.12.2         2 msec      2 msec      2 msec
6          10.0.23.3         4 msec      4 msec      5 msec
7          10.0.45.4         6 msec      7 msec      4 msec
8          192.168.2.1       9 msec      6 msec      5 msec
Spoke_CE2#
```

> **问题与思考**

以上 MPLS VPN 的 Hub Spoke 组网模式通过物理接口实现，请使用子接口方式实现 Hub Spoke 组网模式的私网用户互访。

3.3 跨域 MPLS VPN

3.3.1 跨域 MPLS VPN-OptionA

> **原理**

跨域 MPLS VPN-OptionA 方式被称为"VRF-to-VRF"方式，也被称为"VRF 背靠背"方式，这种方式实现较简单，自治域的 ASBR 为有跨域需求的 VPN 各自建立一个 VRF，分别为这些 VRF 绑定接口，ASBR 的 VRF 利用这些接口互联交换 VPN 路由。

VRF-to-VRF 方式的特点是实现简单，直接利用 MP-IBGP 就可以实现，业务部署相对简单，但是这种配置方案要求 ASBR 为每个跨域的 VPN 配置一个接口（通常是逻辑子接口）并与之绑定，绑定接口的数量至少要和跨域的 VPN 的数量相当，并需要在 ASBR 上逐个对 VPN 进行配置，因而存在可扩展性问题。此外为每个 VPN 单独创建子接口也提高了对 ASBR 设备的要求。这种方案一般使用在跨域 VPN 数量较少的网络中。

> **任务拓扑**

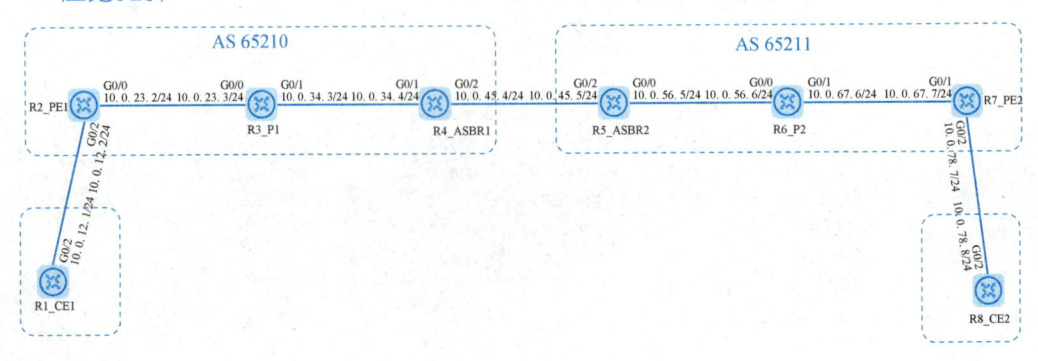

➢ 实施步骤

1. 根据任务拓扑配置各设备接口 IP 地址。(略)
2. 配置 AS 65210 内公网路由互通。

```
R2_PE1(config)#router   ospf   10
R2_PE1(config-router)#router-id   2.2.2.2
Change router-id and update OSPF process! [yes/no]:y
R2_PE1(config-router)#network   10.0.23.0 0.0.0.255 area   0
R2_PE1(config-router)#network   10.0.2.2 0.0.0.0 area   0
R2_PE1(config-router)#exit
R2_PE1(config)#


R3_P1(config)#router   ospf   10
R3_P1(config-router)#router-id   3.3.3.3
Change router-id and update OSPF process! [yes/no]:y
R3_P1(config-router)#network   10.0.3.3 0.0.0.0 area   0
R3_P1(config-router)#network   10.0.23.0 0.0.0.255 area   0
R3_P1(config-router)#network   10.0.34.0 0.0.0.255 area   0
R3_P1(config-router)#exit
R3_P1(config)#


R4_ASBR1(config)#router   ospf   10
R4_ASBR1(config-router)#router-id   4.4.4.4
Change router-id and update OSPF process! [yes/no]:y
R4_ASBR1(config-router)#network 10.0.4.4 0.0.0.0 area   0
R4_ASBR1(config-router)#network   10.0.34.0 0.0.0.255 area   0
R4_ASBR1(config-router)#exit
R4_ASBR1(config)#
```

3. 配置 AS 65211 内公网路由互通。

```
R5_ASBR2(config)#router ospf   20
R5_ASBR2(config-router)#router-id   5.5.5.5
Change router-id and update OSPF process! [yes/no]:y
R5_ASBR2(config-router)#network   10.0.5.5 0.0.0.0 area   0
R5_ASBR2(config-router)#network   10.0.56.0 0.0.0.255 area   0
R5_ASBR2(config-router)#exit
R5_ASBR2(config)#


R6_P2(config)#router   ospf   20
R6_P2(config-router)#router-id   6.6.6.6
Change router-id and update OSPF process! [yes/no]:y
R6_P2(config-router)#network   10.0.56.0 0.0.0.255 area   0
R6_P2(config-router)#network   10.0.67.0 0.0.0.255 area   0
R6_P2(config-router)#network   10.0.6.6 0.0.0.0 area   0
R6_P2(config-router)#exit
R6_P2(config)#


R7_PE2(config)#router   ospf   20
R7_PE2(config-router)#router-id   7.7.7.7
Change router-id and update OSPF process! [yes/no]:y
```

R7_PE2(config-router)#network   10.0.67.0 0.0.0.255 area   0
R7_PE2(config-router)#network   10.0.7.7 0.0.0.0 area   0
R7_PE2(config-router)#exit
R7_PE2(config)#

4. 在公网 AS 65210 和 AS 65211 内配置 LDP 协议同时接口开启标签交换功能分配公网标签。

```
R2_PE1(config)#mpls    ip
R2_PE1(config)#mpls    router   ldp
R2_PE1(config-mpls-router)#ldp router-id   interface   loopback 0 force
R2_PE1(config-mpls-router)#exit
R2_PE1(config)#interface   gigabitEthernet 0/0
R2_PE1(config-if-GigabitEthernet 0/0)#mpls    ip
R2_PE1(config-if-GigabitEthernet 0/0)#label-switching
R2_PE1(config-if-GigabitEthernet 0/0)#exit
R2_PE1(config)#

R3_P1(config)#mpls    ip
R3_P1(config)#mpls    rou ldp
R3_P1(config-mpls-router)#ldp router-id   interface   loopback 0 force
R3_P1(config-mpls-router)#exit
R3_P1(config)#interface   gigabitEthernet 0/0
R3_P1(config-if-GigabitEthernet 0/0)#mpls    ip
R3_P1(config-if-GigabitEthernet 0/0)#label-switching
R3_P1(config-if-GigabitEthernet 0/0)#exit
R3_P1(config)#interface   gigabitEthernet 0/1
R3_P1(config-if-GigabitEthernet 0/1)#mpls    ip
R3_P1(config-if-GigabitEthernet 0/1)#label-switching
R3_P1(config-if-GigabitEthernet 0/1)#exit
R3_P1(config)#

R4_ASBR1(config)#mpls    ip
R4_ASBR1(config)#mpls    router    ldp
R4_ASBR1(config-mpls-router)#ldp router-id   interface   loopback 0 force
R4_ASBR1(config-mpls-router)#exit
R4_ASBR1(config)#interface   gigabitEthernet 0/1
R4_ASBR1(config-if-GigabitEthernet 0/1)#mpls    ip
R4_ASBR1(config-if-GigabitEthernet 0/1)#label-switching
R4_ASBR1(config-if-GigabitEthernet 0/1)#exit
R4_ASBR1(config)#

R5_ASBR2(config)#mpls    ip
R5_ASBR2(config)#mpls    router    ldp
R5_ASBR2(config-mpls-router)#ldp router-id   interface   loopback   0   force
R5_ASBR2(config-mpls-router)#exit
R5_ASBR2(config)#interface   gigabitEthernet 0/0
R5_ASBR2(config-if-GigabitEthernet 0/0)#mpls ip
R5_ASBR2(config-if-GigabitEthernet 0/0)#label-switching
R5_ASBR2(config-if-GigabitEthernet 0/0)#exit
R5_ASBR2(config)#


R6_P2(config)#mpls    ip
R6_P2(config)#mpls    router    ldp
```

```
R6_P2(config-mpls-router)#ldp router-id   interface   loopback 0 force
R6_P2(config-mpls-router)#exit
R6_P2(config)#interface   gigabitEthernet 0/0
R6_P2(config-if-GigabitEthernet 0/0)#mpls   ip
R6_P2(config-if-GigabitEthernet 0/0)#label-switching
R6_P2(config-if-GigabitEthernet 0/0)#exit
R6_P2(config)#interface   gigabitEthernet 0/1
R6_P2(config-if-GigabitEthernet 0/1)#mpls   ip
R6_P2(config-if-GigabitEthernet 0/1)#label-switching
R6_P2(config-if-GigabitEthernet 0/1)#exit
R6_P2(config)#


R7_PE2(config)#mpls   ip
R7_PE2(config)#mpls   router   ldp
R7_PE2(config-mpls-router)#ldp router-id   interface   loopback 0 force
R7_PE2(config-mpls-router)#exit
R7_PE2(config)#interface   gigabitEthernet 0/1
R7_PE2(config-if-GigabitEthernet 0/1)#mpls   ip
R7_PE2(config-if-GigabitEthernet 0/1)#label-switching
R7_PE2(config-if-GigabitEthernet 0/1)#exit
R7_PE2(config)#
```

5．在公网各 AS 内部之间建立 MP-IBGP 邻居关系。

```
R2_PE1(config)#router   bgp   65210
R2_PE1(config-router)#bgp   router-id   2.2.2.2
R2_PE1(config-router)#neighbor   10.0.4.4 remote-as   65210
R2_PE1(config-router)#neighbor   10.0.4.4 update-source loopback 0
R2_PE1(config-router)#address-family vpnv4
R2_PE1(config-router-af)#neighbor   10.0.4.4 activate
R2_PE1(config-router-af)#exit
R2_PE1(config-router)#


R4_ASBR1(config)#router   bgp   65210
R4_ASBR1(config-router)#bgp   router-id   4.4.4.4
R4_ASBR1(config-router)#neighbor   10.0.2.2 remote-as   65210
R4_ASBR1(config-router)#neighbor   10.0.2.2 update-source   loopback 0
R4_ASBR1(config-router)#address-family vpnv4
R4_ASBR1(config-router-af)#neighbor   10.0.2.2 activate
R4_ASBR1(config-router-af)#exit
R4_ASBR1(config-router)#exit
R4_ASBR1(config)#


R5_ASBR2(config)#router   bgp   65211
R5_ASBR2(config-router)#bgp   router 5.5.5.5
R5_ASBR2(config-router)#neighbor   10.0.7.7 remote-as   65211
R5_ASBR2(config-router)#neighbor   10.0.7.7 update-source   loopback 0
R5_ASBR2(config-router)#address-family vpnv4
R5_ASBR2(config-router-af)#neighbor   10.0.7.7 activate
R5_ASBR2(config-router-af)#exit
R5_ASBR2(config-router)#exit
R5_ASBR2(config)#
```

```
R7_PE2(config)#router bgp    65211
R7_PE2(config-router)#bgp    router-id    7.7.7.7
R7_PE2(config-router)#neighbor    10.0.5.5 remote-as    65211
R7_PE2(config-router)#neighbor    10.0.5.5 update-source    loopback 0
R7_PE2(config-router)#address-family vpnv4
R7_PE2(config-router-af)#neighbor    10.0.5.5 activate
R7_PE2(config-router-af)#exit
R7_PE2(config-router)#exit
R7_PE2(config)#
```

6. 跨域 MPLS VPN-OptionA 配置，R4_ASBR1 设备创建 VRF，其 RD 值为 100:2，RT 值出入方向一致为 10:1，再将 R4_ASBR1 互联接口 G0/2 关联 VRF，R5_ASBR2 设备创建 VRF，RD 值为 100:3，RT 值出入方向为 20:1。

```
R4_ASBR1(config)#ip vrf    VPN1
R4_ASBR1(config-vrf)#rd    100:2
R4_ASBR1(config-vrf)#route-target    both    10:1
R4_ASBR1(config-vrf)#exit
R4_ASBR1(config)#interface    gigabitEthernet 0/2
R4_ASBR1(config-if-GigabitEthernet 0/2)#ip vrf    forwarding    VPN1
% Interface GigabitEthernet 0/2 IP address removed due to enabling VRF VPN1
R4_ASBR1(config-if-GigabitEthernet 0/2)#ip address    10.0.45.4 24
R4_ASBR1(config-if-GigabitEthernet 0/2)#exit
R4_ASBR1(config)#

R5_ASBR2(config)#ip vrf    VPN1
R5_ASBR2(config-vrf)#rd    100:3
R5_ASBR2(config-vrf)#route-target    both    20:1
R5_ASBR2(config-vrf)#exit
R5_ASBR2(config)#interface    gigabitEthernet 0/2
R5_ASBR2(config-if-GigabitEthernet 0/2)#ip vrf    forwarding    VPN1
% Interface GigabitEthernet 0/2 IP address removed due to enabling VRF VPN1
R5_ASBR2(config-if-GigabitEthernet 0/2)#ip address    10.0.45.5 24
R5_ASBR2(config-if-GigabitEthernet 0/2)#exit
R5_ASBR2(config)#
```

7. 在 R4_ASBR1 设备和 R5_ASBR2 设备之间采用 BGP 路由协议传递私网路由。

```
R4_ASBR1(config)#router    bgp    65210
R4_ASBR1(config-router)#address-family ipv4 vrf    VPN1
R4_ASBR1(config-router-af)#neighbor    10.0.45.5 remote-as    65211
R4_ASBR1(config-router-af)#exit
R4_ASBR1(config-router)#exit
R4_ASBR1(config)#

R5_ASBR2(config)#router    bgp    65211
R5_ASBR2(config-router)#address-family ipv4 vrf    VPN1
R5_ASBR2(config-router-af)#neighbor    10.0.45.4 remote-as    65210
R5_ASBR2(config-router-af)#exit
R5_ASBR2(config-router)#exit
R5_ASBR2(config)#
```

8. 在 R2_PE1 设备和 R7_PE2 设备之间创建 VRF，并且关联接口。

```
R2_PE1(config)#ip vrf    VPN1
```

```
R2_PE1(config-vrf)#rd    100:1
R2_PE1(config-vrf)#route-target    both    10:1
R2_PE1(config-vrf)#exit
R2_PE1(config)#interface    gigabitEthernet 0/2
R2_PE1(config-if-GigabitEthernet 0/2)#ip vrf    forwarding    VPN1
% Interface GigabitEthernet 0/2 IP address removed due to enabling VRF VPN1
R2_PE1(config-if-GigabitEthernet 0/2)#ip address    10.0.12.2 24
R2_PE1(config-if-GigabitEthernet 0/2)#exit
R2_PE1(config)#


R7_PE2(config)#ip vrf    VPN1
R7_PE2(config-vrf)#rd    100:4
R7_PE2(config-vrf)#route-target    both    20:1
R7_PE2(config-vrf)#exit
R7_PE2(config)#interface gigabitEthernet 0/2
R7_PE2(config-if-GigabitEthernet 0/2)#ip vrf    forwarding    VPN1
% Interface GigabitEthernet 0/2 IP address removed due to enabling VRF VPN1
R7_PE2(config-if-GigabitEthernet 0/2)#ip address    10.0.78.7 24
R7_PE2(config-if-GigabitEthernet 0/2)#exit
R7_PE2(config)#
```

9. 在 R2_PE1 设备与 R1_CE1 设备配置 IS-IS 协议传递私网路由，在 R1_CE1 设备创建 loopback 0 接口，并通告该接口。

```
R1_CE1(config)#router    isis    10
R1_CE1(config-router)#net    49.0001.0000.0000.0001.00
R1_CE1(config-router)#is-type    level-2
R1_CE1(config-router)#exit
R1_CE1(config)#interface    gigabitEthernet 0/2
R1_CE1(config-if-GigabitEthernet 0/2)#ip router    isis    10
R1_CE1(config-if-GigabitEthernet 0/2)#exit
R1_CE1(config)#
R1_CE1(config)#interface    loopback 0
R1_CE1(config-if-Loopback 0)#ip address    192.168.1.1 24
R1_CE1(config-if-Loopback 0)#ip router    isis    10
R1_CE1(config-if-Loopback 0)#exit
R1_CE1(config)#


R2_PE1(config)#router    isis    10
R2_PE1(config-router)#vrf    VPN1
R2_PE1(config-router)#net    49.0001.0000.0000.0002.00
R2_PE1(config-router)#is-type    level-2
R2_PE1(config-router)#exit
R2_PE1(config)#interface    gigabitEthernet 0/2
R2_PE1(config-if-GigabitEthernet 0/2)#ip router    isis    10
R2_PE1(config-if-GigabitEthernet 0/2)#exit
R2_PE1(config)#
```

10. 在 R7_PE2 设备和 R8_CE2 设备采用 OSPF 协议传递私网路由，进程为 100，在 R8_CE2 设备创建 loopback 0 接口并通告到 OSPF 进程中。

```
R7_PE2(config)#no router    ospf    100
R7_PE2(config)#router    ospf    100 vrf    VPN1
R7_PE2(config-router)#router-id    77.7.7.7
```

```
Change router-id and update OSPF process! [yes/no]:y
R7_PE2(config-router)#network    10.0.78.0 0.0.0.255 area    1
R7_PE2(config-router)#exit
R7_PE2(config)#


R8_CE2(config)#interface    loopback 0
R8_CE2(config-if-Loopback 0)#ip address    192.168.2.1 24
R8_CE2(config-if-Loopback 0)#exit
R8_CE2(config)#router    ospf    100
R8_CE2(config-router)#router-id    8.8.8.8
Change router-id and update OSPF process! [yes/no]:y
R8_CE2(config-router)#network    192.168.2.0 0.0.0.255 area    1
R8_CE2(config-router)#network    10.0.78.0 0.0.0.255 area    1
R8_CE2(config-router)#exit
R8_CE2(config)#
```

11. 在 R2_PE1 设备和 R7_PE2 设备将 IGP 路由重分布到 MP-BGP 协议中。

```
R2_PE1(config)#router    bgp    65210
R2_PE1(config-router)#address-family ipv4 vrf    VPN1
R2_PE1(config-router-af)#redistribute    isis 10 level-2
R2_PE1(config-router-af)#exit
R2_PE1(config-router)#


R7_PE2(config)#router    bgp    65211
R7_PE2(config-router)#address-family ipv4 vrf    VPN1
R7_PE2(config-router-af)#redistribute    ospf    100 match internal    external
R7_PE2(config-router-af)#exit
R7_PE2(config-router)#
```

12. 在 R2_PE1 设备和 R7_PE2 设备将 MP-BGP 路由重分布到 IGP 协议中。

```
R2_PE1(config)#router    isis    10
R2_PE1(config-router)#redistribute    bgp    level-2
R2_PE1(config-router)#exit
R2_PE1(config)#


R7_PE2(config)#router    ospf    100
R7_PE2(config-router)#redistribute    bgp    subnets
R7_PE2(config-router)#exit
R7_PE2(config)#
```

> 任务验证

1. 检查公网 AS 65210 内部 IGP 协议邻居建立情况。

```
R3_P1#show    ip ospf    neighbor

OSPF process 10, 2 Neighbors, 2 is Full:
    Neighbor ID      Pri      State          BFD State    Dead Time      Address         Interface
    2.2.2.2           1       Full/DR           -          00:00:30      10.0.23.2       GigabitEthernet 0/0
    4.4.4.4           1       Full/DR           -          00:00:36      10.0.34.4       GigabitEthernet 0/1
```

```
R3_P1#


R2_PE1#show    ip route

Codes:    C - Connected, L - Local, S - Static
          R - RIP, O - OSPF, B - BGP, I - IS-IS, V - Overflow route
          N1 - OSPF NSSA external type 1, N2 - OSPF NSSA external type 2
          E1 - OSPF external type 1, E2 - OSPF external type 2
          SU - IS-IS summary, L1 - IS-IS level-1, L2 - IS-IS level-2
          IA - Inter area, EV - BGP EVPN, A - Arp to host
          LA - Local aggregate route
          * - candidate default

Gateway of last resort is no set
C       10.0.2.2/32 is local host.
O       10.0.3.3/32 [110/1] via 10.0.23.3, 04:13:16, GigabitEthernet 0/0
O       10.0.4.4/32 [110/2] via 10.0.23.3, 04:12:07, GigabitEthernet 0/0
C       10.0.23.0/24 is directly connected, GigabitEthernet 0/0
C       10.0.23.2/32 is local host.
O       10.0.34.0/24 [110/2] via 10.0.23.3, 04:13:16, GigabitEthernet 0/0
R2_PE1#


R2_PE1#ping    10.0.4.4
Sending 5, 100-byte ICMP Echoes to 10.0.4.4, timeout is 2 seconds:
   < press Ctrl+C to break >
!!!!!
Success rate is 100 percent (5/5), round-trip min/avg/max = 2/2/3 ms.
R2_PE1#
```

2．检查公网 AS 65211 内部 IGP 协议邻居建立情况。

```
R6_P2#show     ip ospf neighbor

OSPF process 20, 2 Neighbors, 2 is Full:
Neighbor ID       Pri    State         BFD State    Dead Time    Address         Interface
5.5.5.5            1     Full/BDR      -            00:00:37     10.0.56.5
GigabitEthernet 0/0
7.7.7.7            1     Full/DR       -            00:00:30     10.0.67.7
GigabitEthernet 0/1
R6_P2#



R5_ASBR2#show    ip route

Codes:    C - Connected, L - Local, S - Static
          R - RIP, O - OSPF, B - BGP, I - IS-IS, V - Overflow route
          N1 - OSPF NSSA external type 1, N2 - OSPF NSSA external type 2
          E1 - OSPF external type 1, E2 - OSPF external type 2
          SU - IS-IS summary, L1 - IS-IS level-1, L2 - IS-IS level-2
          IA - Inter area, EV - BGP EVPN, A - Arp to host
          LA - Local aggregate route
          * - candidate default

Gateway of last resort is no set
```

```
C       10.0.5.5/32 is local host.
O       10.0.6.6/32 [110/1] via 10.0.56.6, 04:12:43, GigabitEthernet 0/0
O       10.0.7.7/32 [110/2] via 10.0.56.6, 04:12:21, GigabitEthernet 0/0
C       10.0.56.0/24 is directly connected, GigabitEthernet 0/0
C       10.0.56.5/32 is local host.
O       10.0.67.0/24 [110/2] via 10.0.56.6, 04:12:43, GigabitEthernet 0/0
R5_ASBR2#


R5_ASBR2#ping   10.0.7.7
Sending 5, 100-byte ICMP Echoes to 10.0.7.7, timeout is 2 seconds:
  < press Ctrl+C to break >
!!!!!
Success rate is 100 percent (5/5), round-trip min/avg/max = 1/3/5 ms.
R5_ASBR2#
```

3. 检查公网 AS 65210 的 LDP 协议邻居建立情况及公网标签转发信息表。

```
R3_P1#show   mpls   ldp   neighbor
Default VRF:
    Peer LDP Ident: 10.0.2.2:0; Local LDP Ident: 10.0.3.3:0
        TCP connection: 10.0.2.2.646 - 10.0.3.3.46379
        State: OPERATIONAL; Msgs sent/recv: 1021/1025; UNSOLICITED
        Up time: 04:13:25
        Graceful Restart enabled; Peer reconnect time (msecs): 300000
        LDP discovery sources:
           Link Peer on GigabitEthernet 0/0, Src IP addr: 10.0.23.2
        Addresses bound to peer LDP Ident:
           10.0.23.2        10.0.2.2
    Peer LDP Ident: 10.0.4.4:0; Local LDP Ident: 10.0.3.3:0
        TCP connection: 10.0.4.4.39559 - 10.0.3.3.646
        State: OPERATIONAL; Msgs sent/recv: 1020/1024; UNSOLICITED
        Up time: 04:13:04
        Graceful Restart enabled; Peer reconnect time (msecs): 300000
        LDP discovery sources:
           Link Peer on GigabitEthernet 0/1, Src IP addr: 10.0.34.4
        Addresses bound to peer LDP Ident:
           10.0.34.4        10.0.4.4
R3_P1#


R3_P1#show   mpls   forwarding-table

Label Operation Code:
PH--PUSH label
PP--POP label
SW--SWAP label
SP--SWAP topmost label and push new label
DP--DROP packet
PC--POP label and continue lookup by IP or Label
PI--POP label and do ip lookup forward
PN--POP label and forward to nexthop
PM--POP label and do MAC lookup forward
PV--POP label and output to VC attach interface
IP--IP lookup forward
 s--stale
   Local     Outgoing OP FEC                          Outgoing          Nexthop
```

label	label		interface	
--	imp-null PH 10.0.2.2/32		Gi0/0	10.0.23.2
--	imp-null PH 10.0.4.4/32		Gi0/1	10.0.34.4
11264	imp-null PP 10.0.2.2/32		Gi0/0	10.0.23.2
11265	imp-null PP 10.0.4.4/32		Gi0/1	10.0.34.4

R3_P1#

4. 检查公网 AS 65211 的 LDP 协议邻居建立情况及公网标签转发信息表。

```
R6_P2#show  mpls  ldp  neighbor
Default VRF:
    Peer LDP Ident: 10.0.5.5:0; Local LDP Ident: 10.0.6.6:0
        TCP connection: 10.0.5.5.646 - 10.0.6.6.34345
        State: OPERATIONAL; Msgs sent/recv: 1023/1027; UNSOLICITED
        Up time: 04:13:52
        Graceful Restart enabled; Peer reconnect time (msecs): 300000
        LDP discovery sources:
            Link Peer on GigabitEthernet 0/0, Src IP addr: 10.0.56.5
        Addresses bound to peer LDP Ident:
            10.0.56.5        10.0.5.5
    Peer LDP Ident: 10.0.7.7:0; Local LDP Ident: 10.0.6.6:0
        TCP connection: 10.0.7.7.35613 - 10.0.6.6.646
        State: OPERATIONAL; Msgs sent/recv: 1021/1025; UNSOLICITED
        Up time: 04:13:29
        Graceful Restart enabled; Peer reconnect time (msecs): 300000
        LDP discovery sources:
            Link Peer on GigabitEthernet 0/1, Src IP addr: 10.0.67.7
        Addresses bound to peer LDP Ident:
            10.0.7.7         10.0.67.7
R6_P2#


R6_P2#show  mpls  forwarding-table

Label Operation Code:
PH--PUSH label
PP--POP label
SW--SWAP label
SP--SWAP topmost label and push new label
DP--DROP packet
PC--POP label and continue lookup by IP or Label
PI--POP label and do ip lookup forward
PN--POP label and forward to nexthop
PM--POP label and do MAC lookup forward
PV--POP label and output to VC attach interface
IP--IP lookup forward
  s--stale
```

Local label	Outgoing label	OP	FEC	Outgoing interface	Nexthop
--	imp-null	PH	10.0.5.5/32	Gi0/0	10.0.56.5
--	imp-null	PH	10.0.7.7/32	Gi0/1	10.0.67.7
11264	imp-null	PP	10.0.5.5/32	Gi0/0	10.0.56.5
11265	imp-null	PP	10.0.7.7/32	Gi0/1	10.0.67.7

R6_P2#

5. 检查 MP-BGP 邻居关系建立情况。

```
R2_PE1#show   bgp   vpnv4 unicast   all summary
For address family: VPNv4 Unicast
BGP router identifier 2.2.2.2, local AS number 65210
BGP table version is 1
0 BGP AS-PATH entries
0 BGP Community entries
0 BGP Prefix entries (Maximum-prefix:4294967295)

Neighbor        V    AS  MsgRcvd MsgSent    TblVer   InQ OutQ Up/Down    State/PfxRcd
10.0.4.4        4   65210   28      27         1     0    0  00:24:50        0

Total number of neighbors 1, established neighbors 1


R5_ASBR2#show   bgp   vpnv4 unicast   all   summary
For address family: VPNv4 Unicast
BGP router identifier 5.5.5.5, local AS number 65211
BGP table version is 1
0 BGP AS-PATH entries
0 BGP Community entries
0 BGP Prefix entries (Maximum-prefix:4294967295)

Neighbor        V    AS  MsgRcvd MsgSent    TblVer   InQ OutQ Up/Down    State/PfxRcd
10.0.7.7        4   65211   28      27         1     0    0  00:24:48        0

Total number of neighbors 1, established neighbors 1
```

6. 检查 R4_ASBR1 设备和 R5_ASBR2 设备之间的 BGP 邻居关系建立情况。

```
R5_ASBR2#show   bgp   vpnv4 unicast   all   summary
For address family: IPv4 Unicast

BGP VRF VPN1 Route Distinguisher: 100:3
BGP table version is 1
0 BGP AS-PATH entries
0 BGP Community entries
0 BGP Prefix entries (Maximum-prefix:4294967295)

Neighbor        V    AS  MsgRcvd MsgSent    TblVer   InQ OutQ Up/Down    State/PfxRcd
10.0.45.4       4   65210    4       2         1     0    0  00:01:04        0

Total number of neighbors 1, established neighbors 1

R5_ASBR2#
```

7. 检查 R2_PE1 设备与 R1_CE1 设备的 IS-IS 邻居建立情况及私网路由传递情况。

```
R2_PE1#show   isis   neighbors

Area 10:
System Id   Type   IP Address     State   Holdtime   Circuit      Interface
R1_CE1      L2     10.0.12.1      Up      23         R2_PE1.01    GigabitEthernet 0/2

R2_PE1#
```

```
R2_PE1#show   ip route   vrf   VPN1
Routing Table: VPN1

Codes:    C - Connected, L - Local, S - Static
          R - RIP, O - OSPF, B - BGP, I - IS-IS, V - Overflow route
          N1 - OSPF NSSA external type 1, N2 - OSPF NSSA external type 2
          E1 - OSPF external type 1, E2 - OSPF external type 2
          SU - IS-IS summary, L1 - IS-IS level-1, L2 - IS-IS level-2
          IA - Inter area, EV - BGP EVPN, A - Arp to host
          LA - Local aggregate route
          * - candidate default

Gateway of last resort is no set
C        10.0.12.0/24 is directly connected, GigabitEthernet 0/2
C        10.0.12.2/32 is local host.
I L2     192.168.1.0/24 [115/1] via 10.0.12.1, 00:00:22, GigabitEthernet 0/2
R2_PE1#


R2_PE1#ping    vrf   VPN1 192.168.1.1
Sending 5, 100-byte ICMP Echoes to 192.168.1.1, timeout is 2 seconds:
  < press Ctrl+C to break >
!!!!!
Success rate is 100 percent (5/5), round-trip min/avg/max = 1/4/14 ms.
R2_PE1#
```

8. 检查 R7_PE2 设备与 R8_CE2 设备的路由传递情况。

```
R8_CE2#show   ip ospf   neighbor

OSPF process 100, 1 Neighbors, 1 is Full:
Neighbor ID          Pri    State            BFD State    Dead Time    Address          Interface
77.7.7.7             1      Full/DR          -            00:00:38     10.0.78.7        GigabitEthernet 0/2
R8_CE2#


R7_PE2#show   ip route   vrf   VPN1
Routing Table: VPN1

Codes:    C - Connected, L - Local, S - Static
          R - RIP, O - OSPF, B - BGP, I - IS-IS, V - Overflow route
          N1 - OSPF NSSA external type 1, N2 - OSPF NSSA external type 2
          E1 - OSPF external type 1, E2 - OSPF external type 2
          SU - IS-IS summary, L1 - IS-IS level-1, L2 - IS-IS level-2
          IA - Inter area, EV - BGP EVPN, A - Arp to host
          LA - Local aggregate route
          * - candidate default

Gateway of last resort is no set
C        10.0.78.0/24 is directly connected, GigabitEthernet 0/2
C        10.0.78.7/32 is local host.
O        192.168.2.1/32 [110/1] via 10.0.78.8, 00:02:16, GigabitEthernet 0/2
R7_PE2#


R7_PE2#ping    vrf   VPN1 192.168.2.1
```

Sending 5, 100-byte ICMP Echoes to 192.168.2.1, timeout is 2 seconds:
  < press Ctrl+C to break >
!!!!!
Success rate is 100 percent (5/5), round-trip min/avg/max = 1/1/2 ms.
R7_PE2#

9. 检查 MP-BGP 路由表，跨域 R2_PE1 设备和 R7_PE2 设备接收到对端传递的私网路由信息。

```
R2_PE1#show   bgp   vpnv4 unicast   all
BGP table version is 1, local router ID is 2.2.2.2
Status codes: s suppressed, d damped, h history, * valid, > best, i - internal,
              S Stale, b - backup entry, m - multipath, f Filter, a additional-path
Origin codes: i - IGP, e - EGP, ? - incomplete

     Network          Next Hop        Metric      LocPrf      Weight Path
Route Distinguisher: 100:1 (Default for VRF VPN1)
*>   10.0.12.0/24     0.0.0.0         1                       32768    ?
*>i  10.0.78.0/24     10.0.4.4        0           100         0 65211 ?
*>   192.168.1.0      10.0.12.1       1                       32768    ?
*>i  192.168.2.1/32   10.0.4.4        0           100         0 65211 ?

Total number of prefixes 4
Route Distinguisher: 100:2
*>i  10.0.78.0/24     10.0.4.4        0           100         0 65211 ?
*>i  192.168.2.1/32   10.0.4.4        0           100         0 65211 ?

Total number of prefixes 2
R2_PE1#


R7_PE2#show   bgp   vpnv4 unicast   all
BGP table version is 1, local router ID is 7.7.7.7
Status codes: s suppressed, d damped, h history, * valid, > best, i - internal,
              S Stale, b - backup entry, m - multipath, f Filter, a additional-path
Origin codes: i - IGP, e - EGP, ? - incomplete

     Network          Next Hop        Metric      LocPrf      Weight Path
Route Distinguisher: 100:4 (Default for VRF VPN1)
*>i  10.0.12.0/24     10.0.5.5        0           100         0 65210 ?
*>   10.0.78.0/24     0.0.0.0         1                       32768    ?
*>i  192.168.1.0      10.0.5.5        0           100         0 65210 ?
*>   192.168.2.1/32   10.0.78.8       1                       32768    ?

Total number of prefixes 4
Route Distinguisher: 100:3
*>i  10.0.12.0/24     10.0.5.5        0           100         0 65210 ?
*>i  192.168.1.0      10.0.5.5        0           100         0 65210 ?

Total number of prefixes 2
R7_PE2#
```

10. 检查 R1_CE1 设备的全局 IP 路由表，接收到对方的私网路由。

```
R1_CE1#show   ip route

Codes:   C - Connected, L - Local, S - Static
         R - RIP, O - OSPF, B - BGP, I - IS-IS, V - Overflow route
```

```
       N1 - OSPF NSSA external type 1, N2 - OSPF NSSA external type 2
       E1 - OSPF external type 1, E2 - OSPF external type 2
       SU - IS-IS summary, L1 - IS-IS level-1, L2 - IS-IS level-2
       IA - Inter area, EV - BGP EVPN, A - Arp to host
       LA - Local aggregate route
       * - candidate default

Gateway of last resort is no set
C      10.0.12.0/24 is directly connected, GigabitEthernet 0/2
C      10.0.12.1/32 is local host.
I L2   10.0.78.0/24 [115/1] via 10.0.12.2, 00:03:15, GigabitEthernet 0/2
C      192.168.1.0/24 is directly connected, Loopback 0
C      192.168.1.1/32 is local host.
I L2   192.168.2.1/32 [115/1] via 10.0.12.2, 00:03:15, GigabitEthernet 0/2
R1_CE1#


R8_CE2#show    ip ospf    neighbor

OSPF process 100, 1 Neighbors, 1 is Full:
Neighbor ID        Pri    State           BFD State    Dead Time    Address      Interface
77.7.7.7            1     Full/DR            -          00:00:38    10.0.78.7
GigabitEthernet 0/2
R8_CE2#
R8_CE2#show
R8_CE2#show    ip rou
R8_CE2#show    ip route

Codes:   C - Connected, L - Local, S - Static
         R - RIP, O - OSPF, B - BGP, I - IS-IS, V - Overflow route
         N1 - OSPF NSSA external type 1, N2 - OSPF NSSA external type 2
         E1 - OSPF external type 1, E2 - OSPF external type 2
         SU - IS-IS summary, L1 - IS-IS level-1, L2 - IS-IS level-2
         IA - Inter area, EV - BGP EVPN, A - Arp to host
         LA - Local aggregate route
         * - candidate default

Gateway of last resort is no set
O E2   10.0.12.0/24 [110/1] via 10.0.78.7, 00:03:26, GigabitEthernet 0/2
C      10.0.78.0/24 is directly connected, GigabitEthernet 0/2
C      10.0.78.8/32 is local host.
O E2   192.168.1.0/24 [110/1] via 10.0.78.7, 00:03:26, GigabitEthernet 0/2
C      192.168.2.0/24 is directly connected, Loopback 0
C      192.168.2.1/32 is local host.
R8_CE2#
```

11．测试私网目标网段连通情况。

```
R1_CE1#ping    192.168.2.1 source    192.168.1.1
Sending 5, 100-byte ICMP Echoes to 192.168.2.1, timeout is 2 seconds:
  < press Ctrl+C to break >
!!!!!
Success rate is 100 percent (5/5), round-trip min/avg/max = 9/12/18 ms.
R1_CE1#

R8_CE2#ping    192.168.1.1 source    192.168.2.1
```

```
Sending 5, 100-byte ICMP Echoes to 192.168.1.1, timeout is 2 seconds:
  < press Ctrl+C to break >
!!!!!
Success rate is 100 percent (5/5), round-trip min/avg/max = 8/9/12 ms.
R8_CE2#
```

12. 在 R2_PE1 设备和 R7_PE2 设备查看标签转发信息表，标签转发信息表包括公网标签信息和私网标签信息。

```
R2_PE1#show   mpls   forwarding-table

Label Operation Code:
PH--PUSH label
PP--POP label
SW--SWAP label
SP--SWAP topmost label and push new label
DP--DROP packet
PC--POP label and continue lookup by IP or Label
PI--POP label and do ip lookup forward
PN--POP label and forward to nexthop
PM--POP label and do MAC lookup forward
PV--POP label and output to VC attach interface
IP--IP lookup forward
  s--stale
  Local    Outgoing OP FEC                    Outgoing          Nexthop
  label    label                              interface
  --       imp-null PH 10.0.3.3/32            Gi0/0             10.0.23.3
  --       11265    PH 10.0.4.4/32            Gi0/0             10.0.23.3
  --       imp-null PH 10.0.34.0/24           Gi0/0             10.0.23.3
  --       60928    PH 10.0.78.0/24(V)        Gi0/0             10.0.23.3
  --       60928    PH 192.168.2.1/32(V)      Gi0/0             10.0.23.3
  11264    imp-null PP 10.0.3.3/32            Gi0/0             10.0.23.3
  11265    11265    SW 10.0.4.4/32            Gi0/0             10.0.23.3
  11266    imp-null PP 10.0.34.0/24           Gi0/0             10.0.23.3
  60928    --       PI VRF(VPN1)              --                0.0.0.0
R2_PE1#


R7_PE2#show   mpls   forwarding-table

Label Operation Code:
PH--PUSH label
PP--POP label
SW--SWAP label
SP--SWAP topmost label and push new label
DP--DROP packet
PC--POP label and continue lookup by IP or Label
PI--POP label and do ip lookup forward
PN--POP label and forward to nexthop
PM--POP label and do MAC lookup forward
PV--POP label and output to VC attach interface
IP--IP lookup forward
  s--stale
  Local    Outgoing OP FEC                    Outgoing          Nexthop
  label    label                              interface
  --       11264    PH 10.0.5.5/32            Gi0/1             10.0.67.6
```

--	imp-null	PH 10.0.6.6/32	Gi0/1	10.0.67.6
--	imp-null	PH 10.0.56.0/24	Gi0/1	10.0.67.6
--	60928	PH 10.0.12.0/24(V)	Gi0/1	10.0.67.6
--	**60928**	**PH 192.168.1.0/24(V)**	**Gi0/1**	**10.0.67.6**
11264	11264	SW 10.0.5.5/32	Gi0/1	10.0.67.6
11265	imp-null	PP 10.0.6.6/32	Gi0/1	10.0.67.6
11266	imp-null	PP 10.0.56.0/24	Gi0/1	10.0.67.6
60928	--	PI VRF(VPN1)	--	0.0.0.0

R7_PE2#

13. 通过 Wireshark 抓包工具在 PE 设备抓取报文，检查报文标签信息，报文经过 ASBR 互联链路不携带标签信息（即和 CE 和 PE 之间链路转发的报文信息相同），标签：11265 为公网标签，标签：60928 为目标 192.168.2.1/24 的私网标签。

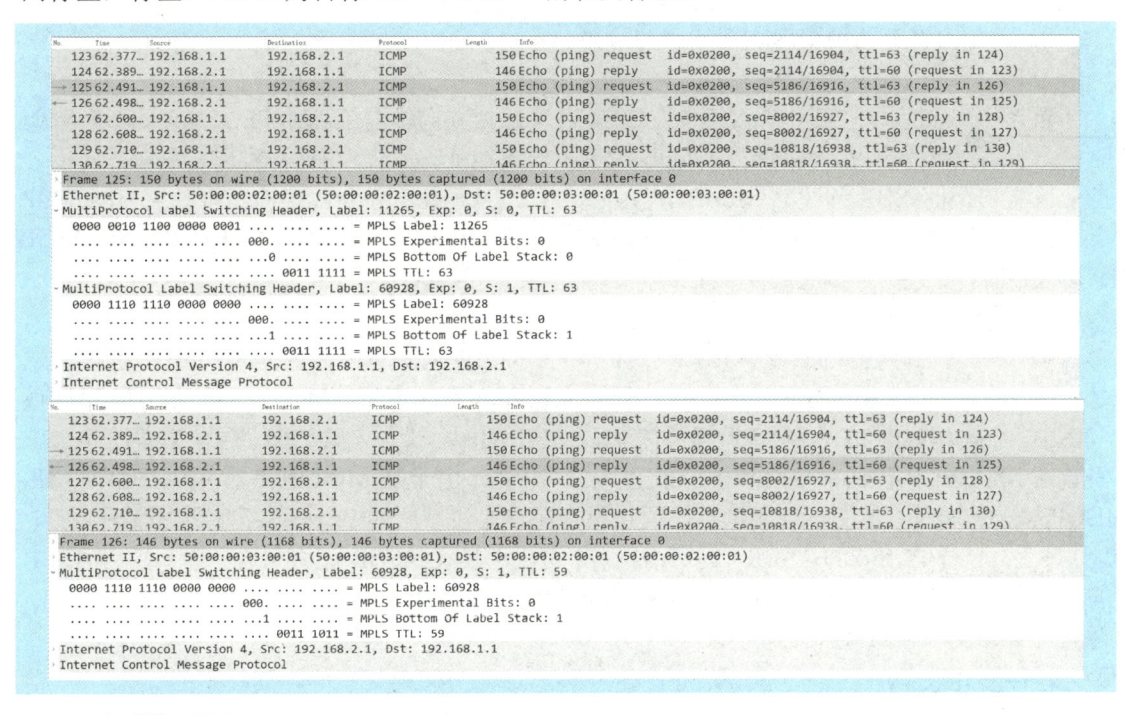

> ➤ 问题与思考

1. 根据以上配置步骤，完成多个不同客户端之间的私网互访，将 CE 端设置为 MCE 设备。
2. 在跨域 MPLS VPN OptionA 方案中的 ASBR 设备之间配置 IGP（OSPF、IS-IS、RIP）协议，实现私网互访。

3.3.2 跨域 MPLS VPN-OptionB

> ➤ 原理

OptionA 的实现方案，需要在 ASBR 上为每个 VPN 配置 VRF，并绑定接口，其原因在于能通过 MP-IBGP 携带 VPN，VPN 路由无法在 EBGP 之间直接传输，如果 VPN 路由可以直接在 EBGP 之间传输，在 ASBR 上就可以不用配置 VRF，这显然是一个更好的实现方式。所以

OptionB 方案扩展了 MP-IBGP，使得 VPN 路由可以在 ASBR 之间直接分发，称为单跳 MP-EBGP。

MP-EBGP 这种方案的优点是不需要在 ASBR 上为每个 VPN 的用户站点分配一个子接口，也不用建立跨域的 LSP，VPN 路由通过单跳 MP-EBGP 邻居直接发送。但是由于 VPN 的路由信息是通过自治系统之间的 ASBR 来保存和扩散的，所以当 VPN 路由数量比较多的时候，会对 ASBR 产生巨大的压力；由于这些 ASBR 一般都承担着公网 IP 转发的任务，这样对设备的要求就比较高。此外，由于 ASBR 之间对接收到的 VPN 路由取消了 RT 过滤功能，PE 上的 VPN 路由可能扩散至其他域的 ASBR 上，这就可能会造成 VPN 路由的泄露。因此交换 VPN 路由的各 ASBR 之间必须就这种路由交换达成信任协议，在 ASBR 之间互相信任并在 ASBR 上实施相应的路由过滤策略。OptionB 方案适用于在有较多跨域 VPN 业务的网络中部署。

OptionB 方案有两种实现方案：

ASBR 设备改变下一跳地址方案

当 ASBR 收到其他自治域 ASBR 发送的 VPN 路由要向本自治域的 PE 发送时，改变 VPN 路由下一跳地址为自己，称为"OptionB 改变下一跳地址方案"。在这种实现方案中，同一个自治域的 PE 和 ASBR 间建立 MP-IBGP 会话交互 VPN 路由；两个 ASBR 之间建立 MP-EBGP 会话交互 VPN 路由，在收到另外一个 ASBR 邻居发送的 VPN 路由，向本自治域的 MP-IBGP 邻居通告时，改变下一跳地址为自己。

ASBR 设备不改变下一跳地址方案

ASBR 在收到其他自治域 ASBR 发送的 VPN 路由要向本自治域内的 MP-IBGP 邻居发送时，不改变 VPN 路由中的下一跳地址，称为"OptionB 不改变下一跳地址方案"。在这种实现方案中，自治域内的 PE 和 ASBR 间仍然建立 MP-IBGP 会话，交互 VPN 路由，在两个 ASBR 间建立 MP-EBGP 会话，也可以直接交互 VPN 路由。由于从 MP-EBGP 收到的路由向 MP-IBGP 邻居发送时不改变下一跳地址，所以要求在该自治域中的 PE 必须存在到达该下一跳地址（即另外一个自治域的 ASBR）的路由，这个可以通过在 ASBR 上将到达另外一端的 ASBR 的路由重分布到本自治域的 IGP 协议中，从而使得另外一个自治域的 ASBR 地址变得可达，并通过 LDP 建立 LSP 路径。

➢ 任务拓扑

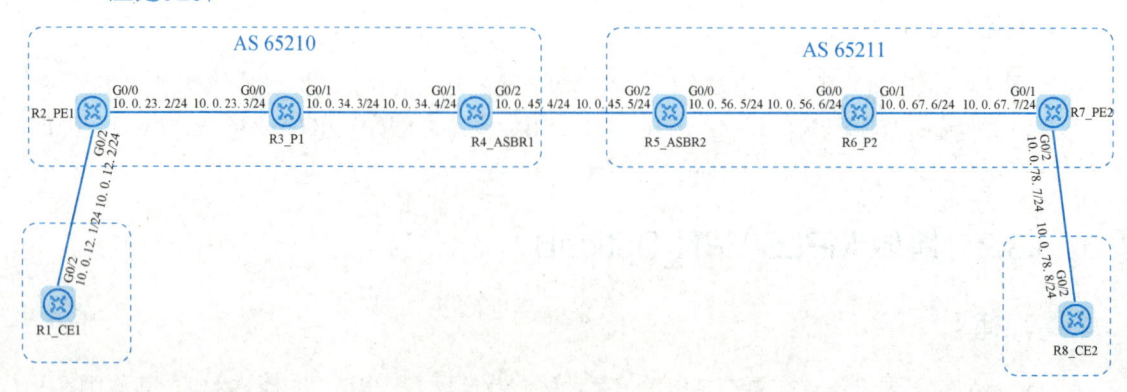

➢ 实施步骤

1. 根据任务拓扑配置各设备接口 IP 地址。（略）

2．配置 AS 65210 内公网路由互通。（参考 3.3.1）

3．配置 AS 65211 内公网路由互通。（参考 3.3.1）

4．在公网 AS 65210 和 AS 65211 内配置 LDP 协议同时接口开启标签交换功能分配公网标签。（参考 3.3.1）

5．在 R2_PE1 设备与 R7_PE2 设备创建 VRF，并且关联接口。

```
R2_PE1(config)#ip vrf VPN1
R2_PE1(config-vrf)#rd   100:1
R2_PE1(config-vrf)#route-target   both   10:100
R2_PE1(config-vrf)#exit
R2_PE1(config)#interface   gigabitEthernet 0/2
R2_PE1(config-if-GigabitEthernet 0/2)#ip vrf   forwarding   VPN1
% Interface GigabitEthernet 0/2 IP address removed due to enabling VRF VPN1
R2_PE1(config-if-GigabitEthernet 0/2)#ip address   10.0.12.2 24
R2_PE1(config-if-GigabitEthernet 0/2)#exit
R2_PE1(config)#


R7_PE2(config)#ip vrf   VPN1
R7_PE2(config-vrf)#rd 100:2
R7_PE2(config-vrf)#route-target   both   10:100
R7_PE2(config-vrf)#exit
R7_PE2(config)#interface   gigabitEthernet 0/2
R7_PE2(config-if-GigabitEthernet 0/2)#ip vrf   forwarding   VPN1
% Interface GigabitEthernet 0/2 IP address removed due to enabling VRF VPN1
R7_PE2(config-if-GigabitEthernet 0/2)#ip address   10.0.78.7 24
R7_PE2(config-if-GigabitEthernet 0/2)#exit
R7_PE2(config)#
```

6．在公网各 AS 内部之间建立 MP-IBGP 邻居关系。

```
R2_PE1(config)#router   bgp   65210
R2_PE1(config-router)#bgp   router-id   2.2.2.2
R2_PE1(config-router)#neighbor   10.0.4.4 remote-as   65210
R2_PE1(config-router)#neighbor   10.0.4.4 update-source   loopback 0
R2_PE1(config-router)#address-family vpnv4
R2_PE1(config-router-af)#neighbor   10.0.4.4 activate
R2_PE1(config-router-af)#exit
R2_PE1(config-router)#exit
R2_PE1(config)#


R4_ASBR1(config)#router   bgp   65210
R4_ASBR1(config-router)#bgp   router-id   4.4.4.4
R4_ASBR1(config-router)#neighbor   10.0.2.2 remote-as   65210
R4_ASBR1(config-router)#neighbor   10.0.2.2 update-source   loopback 0
R4_ASBR1(config-router)#address-family vpnv4
R4_ASBR1(config-router-af)#neighbor   10.0.2.2 activate
R4_ASBR1(config-router-af)#exit
R4_ASBR1(config-router)#exit
R4_ASBR1(config)#


R5_ASBR2(config)#router   bgp   65211
R5_ASBR2(config-router)#bgp   router-id   5.5.5.5
R5_ASBR2(config-router)#neighbor   10.0.7.7 remote-as   65211
```

```
R5_ASBR2(config-router)#neighbor    10.0.7.7 update-source    loopback 0
R5_ASBR2(config-router)#address-family vpnv4
R5_ASBR2(config-router-af)#neighbor    10.0.7.7 activate
R5_ASBR2(config-router-af)#exit
R5_ASBR2(config-router)#exit
R5_ASBR2(config)#


R7_PE2(config)#router    bgp    65211
R7_PE2(config-router)#bgp    router-id    7.7.7.7
R7_PE2(config-router)#neighbor    10.0.5.5 remote-as    65211
R7_PE2(config-router)#neighbor    10.0.5.5 update-source    loopback 0
R7_PE2(config-router)#address-family vpnv4
R7_PE2(config-router-af)#neighbor    10.0.5.5 activate
R7_PE2(config-router-af)#exit
R7_PE2(config-router)#exit
R7_PE2(config)#
```

7. 在 R4_ASBR1 设备和 R5_ASBR2 设备互联设备建立 MP-EBGP 邻居关系传递私网路由，本实验采用 ASBR 设备改变下一跳地址方案。

```
R4_ASBR1(config)#router bgp    65210
R4_ASBR1(config-router)#neighbor    10.0.45.5 remote-as    65211
R4_ASBR1(config-router)#address-family vpnv4
R4_ASBR1(config-router-af)#neighbor    10.0.45.5 activate
R4_ASBR1(config-router-af)#neighbor    10.0.45.5 send-community
R4_ASBR1(config-router-af)#exit
R4_ASBR1(config-router)#


R5_ASBR2(config)#router bgp    65211
R5_ASBR2(config-router)#neighbor    10.0.45.4 remote-as    65210
R5_ASBR2(config-router)#address-family vpnv4
R5_ASBR2(config-router-af)#neighbor    10.0.45.4 activate
R5_ASBR2(config-router-af)#exit
R5_ASBR2(config-router)#exit
R5_ASBR2(config)#
```

8. 在 R2_PE1 设备与 R1_CE1 设备配置 IS-IS 协议传递私网路由，在 R1_CE1 设备创建 loopback 0 接口，并通告该接口。

```
R1_CE1(config)#router isis    10
R1_CE1(config-router)#net    49.0002.0000.0000.0001.00
R1_CE1(config-router)#is-type    level-2
R1_CE1(config-router)#exit
R1_CE1(config)#interface    loopback 0
R1_CE1(config-if-Loopback 0)#ip address    192.168.1.1 24
R1_CE1(config-if-Loopback 0)#exit
R1_CE1(config)#int loopback 0
R1_CE1(config-if-Loopback 0)#ip router    isis    10
R1_CE1(config-if-Loopback 0)#exit
R1_CE1(config)#interface    gigabitEthernet 0/2
R1_CE1(config-if-GigabitEthernet 0/2)#ip router    isis    10
R1_CE1(config-if-GigabitEthernet 0/2)#exit
R1_CE1(config)#
```

```
R2_PE1(config)#router   isis   10
R2_PE1(config-router)#vrf VPN1
R2_PE1(config-router)#net   49.0002.0000.0000.0002.00
R2_PE1(config-router)#is-type   level-2
R2_PE1(config-router)#exit
R2_PE1(config)#interface   gigabitEthernet 0/2
R2_PE1(config-if-GigabitEthernet 0/2)#ip router   isis   10
R2_PE1(config-if-GigabitEthernet 0/2)#exit
R2_PE1(config)#
```

9. 在 R7_PE2 设备和 R8_CE2 设备采用 OSPF 协议传递私网路由,进程为 100,在 R8_CE2 设备创建 loopback 0 接口并通告到 OSPF 进程中。

```
R7_PE2(config)#router   ospf   100 vrf VPN1
R7_PE2(config-router)#router-id   77.7.7.7
Change router-id and update OSPF process! [yes/no]:y
R7_PE2(config-router)#network   10.0.78.0 0.0.0.255 area   0
R7_PE2(config-router)#exit
R7_PE2(config)#

R8_CE2(config)#interface   loopback 0
R8_CE2(config-if-Loopback 0)#ip   address   192.168.2.1 24
R8_CE2(config-if-Loopback 0)#exit
R8_CE2(config)#
R8_CE2(config)#router   ospf   100
R8_CE2(config-router)#router-id   8.8.8.8
Change router-id and update OSPF process! [yes/no]:y
R8_CE2(config-router)#network   10.0.78.0 0.0.0.255 area   0
R8_CE2(config-router)#network   192.168.2.0 0.0.0.255 area   0
R8_CE2(config-router)#exit
R8_CE2(config)#
```

10. R4_ASBR1 设备和 R5_ASBR2 设备的 VPNv4 地址族下关闭 RT 值过滤,BGP 默认过滤 RT 值,会导致 R4_ASBR1 设备无法接收私网路由。

```
R4_ASBR1(config)#router bgp   65210
R4_ASBR1(config-router)#address-family vpnv4
R4_ASBR1(config-router-af)#no   bgp default   route-target filter
R4_ASBR1(config-router-af)#exit
R4_ASBR1(config-router)#

R5_ASBR2(config)#router   bgp   65211
R5_ASBR2(config-router)#address-family vpnv4
R5_ASBR2(config-router-af)#no   bgp   default   route-target filter
R5_ASBR2(config-router-af)#exit
R5_ASBR2(config-router)#
```

11. R4_ASBR1 设备和 R5_ASBR2 设备互联接口需要开启标签交换功能,否则无法建立完整的标签信息。

```
R4_ASBR1(config)#interface   gigabitEthernet 0/2
R4_ASBR1(config-if-GigabitEthernet 0/2)#label-switching
R4_ASBR1(config-if-GigabitEthernet 0/2)#exit
R4_ASBR1(config)#

R5_ASBR2(config)#interface   gigabitEthernet 0/2
R5_ASBR2(config-if-GigabitEthernet 0/2)#label-switching
```

```
R5_ASBR2(config-if-GigabitEthernet 0/2)#exit
R5_ASBR2(config)#
```

12. 在 R2_PE1 设备和 R7_PE2 设备将 IGP 路由重分布到 MP-BGP 协议中。

```
R2_PE1(config)#router   bgp    65210
R2_PE1(config-router)#address-family ipv4 vrf   VPN1
R2_PE1(config-router-af)#redistribute   isis 10 level-2
R2_PE1(config-router-af)#exit
R2_PE1(config-router)#

R7_PE2(config)#router   bgp    65211
R7_PE2(config-router)#address-family ipv4 vrf   VPN1
R7_PE2(config-router-af)#redistribute   ospf   100 match   internal    external
R7_PE2(config-router-af)#exit
R7_PE2(config-router)#
```

13. 在 R4_ASBR1 设备和 R5_ASBR2 设备将下一跳地址修改。

```
R4_ASBR1(config)#router bgp   65210
R4_ASBR1(config-router)#address-family vpnv4
R4_ASBR1(config-router-af)#neighbor    10.0.2.2 next-hop-self
R4_ASBR1(config-router-af)#exit
R4_ASBR1(config-router)#

R5_ASBR2(config)#router   bgp    65211
R5_ASBR2(config-router)#address-family vpnv4
R5_ASBR2(config-router-af)#neighbor    10.0.7.7 next-hop-self
R5_ASBR2(config-router-af)#exit
R5_ASBR2(config-router)#
```

14. 在 R2_PE1 设备和 R7_PE2 设备将 MP-BGP 路由重分布到 IGP 协议中。

```
R2_PE1(config)#router   isis    10
R2_PE1(config-router)#redistribute   bgp    level-2
R2_PE1(config-router)#exit
R2_PE1(config)#

R7_PE2(config)#router    OSPF 100
R7_PE2(config-router)#redistribute   bgp    subnets
R7_PE2(config-router)#exit
R7_PE2(config)#
```

> 任务验证

1. 检查公网 AS 65210 内部 IGP 协议邻居建立情况。

```
R3_P1#show    ip ospf    neighbor

OSPF process 10, 2 Neighbors, 2 is Full:
Neighbor ID        Pri    State            BFD State   Dead Time    Address       Interface
2.2.2.2            1      Full/BDR         -           00:00:32     10.0.23.2     GigabitEthernet 0/0
4.4.4.4            1      Full/DR          -           00:00:33     10.0.34.4     GigabitEthernet 0/1
R3_P1#
```

```
R2_PE1#show    ip route

Codes:   C - Connected, L - Local, S - Static
         R - RIP, O - OSPF, B - BGP, I - IS-IS, V - Overflow route
         N1 - OSPF NSSA external type 1, N2 - OSPF NSSA external type 2
         E1 - OSPF external type 1, E2 - OSPF external type 2
         SU - IS-IS summary, L1 - IS-IS level-1, L2 - IS-IS level-2
         IA - Inter area, EV - BGP EVPN, A - Arp to host
         LA - Local aggregate route
         * - candidate default

Gateway of last resort is no set
C      10.0.2.2/32 is local host.
O      10.0.3.3/32 [110/1] via 10.0.23.3, 01:46:23, GigabitEthernet 0/0
O      10.0.4.4/32 [110/2] via 10.0.23.3, 01:46:04, GigabitEthernet 0/0
C      10.0.23.0/24 is directly connected, GigabitEthernet 0/0
C      10.0.23.2/32 is local host.
O      10.0.34.0/24 [110/2] via 10.0.23.3, 01:46:23, GigabitEthernet 0/0


R2_PE1#ping    10.0.4.4
Sending 5, 100-byte ICMP Echoes to 10.0.4.4, timeout is 2 seconds:
  < press Ctrl+C to break >
!!!!!
Success rate is 100 percent (5/5), round-trip min/avg/max = 2/2/5 ms.
R2_PE1#
```

2. 检查公网 AS 65211 内部 IGP 协议邻居建立情况。

```
R6_P2#show    ip ospf neighbor

OSPF process 20, 2 Neighbors, 2 is Full:
Neighbor ID       Pri    State        BFD State    Dead Time    Address       Interface
5.5.5.5            1     Full/BDR         -        00:00:33     10.0.56.5
GigabitEthernet 0/0
7.7.7.7            1     Full/DR          -        00:00:37     10.0.67.7
GigabitEthernet 0/1


R7_PE2#show    ip route

Codes:   C - Connected, L - Local, S - Static
         R - RIP, O - OSPF, B - BGP, I - IS-IS, V - Overflow route
         N1 - OSPF NSSA external type 1, N2 - OSPF NSSA external type 2
         E1 - OSPF external type 1, E2 - OSPF external type 2
         SU - IS-IS summary, L1 - IS-IS level-1, L2 - IS-IS level-2
         IA - Inter area, EV - BGP EVPN, A - Arp to host
         LA - Local aggregate route
         * - candidate default

Gateway of last resort is no set
O      10.0.5.5/32 [110/2] via 10.0.67.6, 01:44:28, GigabitEthernet 0/1
O      10.0.6.6/32 [110/1] via 10.0.67.6, 01:44:28, GigabitEthernet 0/1
C      10.0.7.7/32 is local host.
O      10.0.56.0/24 [110/2] via 10.0.67.6, 01:44:28, GigabitEthernet 0/1
```

```
C    10.0.67.0/24 is directly connected, GigabitEthernet 0/1
C    10.0.67.7/32 is local host.
R7_PE2#


R7_PE2#ping   10.0.5.5
Sending 5, 100-byte ICMP Echoes to 10.0.5.5, timeout is 2 seconds:
  < press Ctrl+C to break >
!!!!!
Success rate is 100 percent (5/5), round-trip min/avg/max = 1/2/4 ms.
R7_PE2#
```

3. 检查公网 AS 65210 的 LDP 协议邻居建立情况及公网标签转发信息表。

```
R3_P1#show   mpls   ldp   neighbor
Default VRF:
    Peer LDP Ident: 10.0.2.2:0; Local LDP Ident: 10.0.3.3:0
        TCP connection: 10.0.2.2.646 - 10.0.3.3.38881
        State: OPERATIONAL; Msgs sent/recv: 429/433; UNSOLICITED
        Up time: 01:45:29
        Graceful Restart enabled; Peer reconnect time (msecs): 300000
        LDP discovery sources:
            Link Peer on GigabitEthernet 0/0, Src IP addr: 10.0.23.2
        Addresses bound to peer LDP Ident:
            10.0.23.2        10.0.2.2
    Peer LDP Ident: 10.0.4.4:0; Local LDP Ident: 10.0.3.3:0
        TCP connection: 10.0.4.4.40139 - 10.0.3.3.646
        State: OPERATIONAL; Msgs sent/recv: 427/430; UNSOLICITED
        Up time: 01:45:06
        Graceful Restart enabled; Peer reconnect time (msecs): 300000
        LDP discovery sources:
            Link Peer on GigabitEthernet 0/1, Src IP addr: 10.0.34.4
        Addresses bound to peer LDP Ident:
            10.0.4.4        10.0.34.4        10.0.45.4


R3_P1#show   mpls   forwarding-table

Label Operation Code:
PH--PUSH label
PP--POP label
SW--SWAP label
SP--SWAP topmost label and push new label
DP--DROP packet
PC--POP label and continue lookup by IP or Label
PI--POP label and do ip lookup forward
PN--POP label and forward to nexthop
PM--POP label and do MAC lookup forward
PV--POP label and output to VC attach interface
IP--IP lookup forward
  s--stale
   Local    Outgoing OP FEC                Outgoing        Nexthop
   label    label                          interface
   --       imp-null PH 10.0.2.2/32        Gi0/0           10.0.23.2
   --       imp-null PH 10.0.4.4/32        Gi0/1           10.0.34.4
   11264    imp-null PP 10.0.2.2/32        Gi0/0           10.0.23.2
   11265    imp-null PP 10.0.4.4/32        Gi0/1           10.0.34.4
```

R3_P1#

4．检查公网 AS 65211 的 LDP 协议邻居建立情况及公网标签转发信息表。

```
R6_P2#show   mpls   ldp   neighbor
Default VRF:
    Peer LDP Ident: 10.0.5.5:0; Local LDP Ident: 10.0.6.6:0
        TCP connection: 10.0.5.5.646 - 10.0.6.6.36913
        State: OPERATIONAL; Msgs sent/recv: 426/429; UNSOLICITED
        Up time: 01:44:59
        Graceful Restart enabled; Peer reconnect time (msecs): 300000
        LDP discovery sources:
            Link Peer on GigabitEthernet 0/0, Src IP addr: 10.0.56.5
        Addresses bound to peer LDP Ident:
            10.0.5.5        10.0.45.5        10.0.56.5
    Peer LDP Ident: 10.0.7.7:0; Local LDP Ident: 10.0.6.6:0
        TCP connection: 10.0.7.7.44341 - 10.0.6.6.646
        State: OPERATIONAL; Msgs sent/recv: 426/430; UNSOLICITED
        Up time: 01:44:42
        Graceful Restart enabled; Peer reconnect time (msecs): 300000
        LDP discovery sources:
            Link Peer on GigabitEthernet 0/1, Src IP addr: 10.0.67.7
        Addresses bound to peer LDP Ident:
            10.0.7.7        10.0.67.7
R6_P2#show
R6_P2#show   mp
R6_P2#show   mpls   fo
R6_P2#show   mpls   forwarding-table

Label Operation Code:
PH--PUSH label
PP--POP label
SW--SWAP label
SP--SWAP topmost label and push new label
DP--DROP packet
PC--POP label and continue lookup by IP or Label
PI--POP label and do ip lookup forward
PN--POP label and forward to nexthop
PM--POP label and do MAC lookup forward
PV--POP label and output to VC attach interface
IP--IP lookup forward
 s--stale
 Local    Outgoing OP FEC              Outgoing      Nexthop
 label    label                        interface
  --      imp-null PH 10.0.5.5/32      Gi0/0         10.0.56.5
  --      imp-null PH 10.0.7.7/32      Gi0/1         10.0.67.7
 11264    imp-null PP 10.0.5.5/32      Gi0/0         10.0.56.5
 11265    imp-null PP 10.0.7.7/32      Gi0/1         10.0.67.7

R6_P2#
```

5．检查 MP-BGP 邻居关系建立情况。

```
R2_PE1#show   bgp   vpnv4 unicast   all   summary
For address family: VPNv4 Unicast
BGP router identifier 2.2.2.2, local AS number 65210
```

```
BGP table version is 2
0 BGP AS-PATH entries
0 BGP Community entries
0 BGP Prefix entries (Maximum-prefix:4294967295)

Neighbor        V    AS  MsgRcvd MsgSent   TblVer  InQ OutQ Up/Down    State/PfxRcd
10.0.4.4        4    65210    94      91        1    0    0 01:25:59        0

Total number of neighbors 1, established neighbors 1

R7_PE2#show  bgp  vpnv4 unicast  all  summary
For address family: VPNv4 Unicast
BGP router identifier 7.7.7.7, local AS number 65211
BGP table version is 2
0 BGP AS-PATH entries
0 BGP Community entries
0 BGP Prefix entries (Maximum-prefix:4294967295)

Neighbor        V    AS  MsgRcvd MsgSent   TblVer  InQ OutQ Up/Down    State/PfxRcd
10.0.5.5        4    65211    94      90        1    0    0 01:24:48        0

Total number of neighbors 1, established neighbors 1
```

6. 检查 R4_ASBR1 设备和 R5_ASBR2 设备之间的 MP-EBGP 邻居关系建立情况。

```
R4_ASBR1#show  bgp  vpnv4 unicast  all  summary
For address family: VPNv4 Unicast
BGP router identifier 4.4.4.4, local AS number 65210
BGP table version is 4
0 BGP AS-PATH entries
0 BGP Community entries
0 BGP Prefix entries (Maximum-prefix:4294967295)

Neighbor        V    AS  MsgRcvd MsgSent   TblVer  InQ OutQ Up/Down    State/PfxRcd
10.0.2.2        4    65210    93      93        4    0    0 01:26:47        0
10.0.45.5       4    65211    65      63        4    0    0 00:59:10        0

Total number of neighbors 2, established neighbors 2
```

7. 检查 R2_PE1 设备与 R1_CE1 设备的 IS-IS 邻居建立情况及私网路由传递情况。

```
R2_PE1#show  isis  neighbors

Area 10:
System Id    Type   IP Address     State   Holdtime  Circuit        Interface
R1_CE1       L2     10.0.12.1      Up      24        R2_PE1.01      GigabitEthernet 0/2

R2_PE1#show  ip route  vrf VPN1
Routing Table: VPN1

Codes:    C - Connected, L - Local, S - Static
          R - RIP, O - OSPF, B - BGP, I - IS-IS, V - Overflow route
          N1 - OSPF NSSA external type 1, N2 - OSPF NSSA external type 2
          E1 - OSPF external type 1, E2 - OSPF external type 2
          SU - IS-IS summary, L1 - IS-IS level-1, L2 - IS-IS level-2
          IA - Inter area, EV - BGP EVPN, A - Arp to host
          LA - Local aggregate route
```

```
         * - candidate default

Gateway of last resort is no set
C       10.0.12.0/24 is directly connected, GigabitEthernet 0/2
C       10.0.12.2/32 is local host.
I L2    192.168.1.0/24 [115/1] via 10.0.12.1, 01:05:30, GigabitEthernet 0/2
R2_PE1#


R2_PE1#ping    vrf VPN1 192.168.1.1
Sending 5, 100-byte ICMP Echoes to 192.168.1.1, timeout is 2 seconds:
  < press Ctrl+C to break >
!!!!!
Success rate is 100 percent (5/5), round-trip min/avg/max = 1/4/15 ms.
R2_PE1#
```

8. 检查 R7_PE2 设备与 R8_CE2 设备的路由传递。

```
R8_CE2#show    ip ospf neighbor

OSPF process 100, 1 Neighbors, 1 is Full:
Neighbor ID        Pri    State           BFD State   Dead Time   Address      Interface
77.7.7.7            1     Full/DR           -          00:00:30   10.0.78.7
GigabitEthernet 0/2
R8_CE2#


R7_PE2#show    ip route    vrf VPN1
Routing Table: VPN1

Codes:    C - Connected, L - Local, S - Static
          R - RIP, O - OSPF, B - BGP, I - IS-IS, V - Overflow route
          N1 - OSPF NSSA external type 1, N2 - OSPF NSSA external type 2
          E1 - OSPF external type 1, E2 - OSPF external type 2
          SU - IS-IS summary, L1 - IS-IS level-1, L2 - IS-IS level-2
          IA - Inter area, EV - BGP EVPN, A - Arp to host
          LA - Local aggregate route
          * - candidate default

Gateway of last resort is no set
C       10.0.78.0/24 is directly connected, GigabitEthernet 0/2
C       10.0.78.7/32 is local host.
O       192.168.2.1/32 [110/1] via 10.0.78.8, 01:07:25, GigabitEthernet 0/2
R7_PE2#


R7_PE2#ping    vrf   VPN1 192.168.2.1
Sending 5, 100-byte ICMP Echoes to 192.168.2.1, timeout is 2 seconds:
  < press Ctrl+C to break >
!!!!!
Success rate is 100 percent (5/5), round-trip min/avg/max = 1/1/2 ms.
R7_PE2#
```

9. 检查 MP-BGP 路由表，R2_PE1 设备和 R7_PE2 设备接收到对端传递的私网路由信息。

```
R2_PE1#show    bgp    vpnv4 unicast    all
BGP table version is 1, local router ID is 2.2.2.2
Status codes: s suppressed, d damped, h history, * valid, > best, i - internal,
              S Stale, b - backup entry, m - multipath, f Filter, a additional-path
Origin codes: i - IGP, e - EGP, ? - incomplete
```

```
     Network           Next Hop          Metric         LocPrf        Weight Path
Route Distinguisher: 100:1 (Default for VRF VPN1)
*>    10.0.12.0/24     0.0.0.0           1                            32768     ?
*>i   10.0.78.0/24     10.0.4.4          0              100           0 65211 ?
*>    192.168.1.0      10.0.12.1         1                            32768     ?
*>i   192.168.2.1/32   10.0.4.4          0              100           0 65211 ?

Total number of prefixes 4
Route Distinguisher: 100:2
*>i   10.0.78.0/24     10.0.4.4          0              100           0 65211 ?
*>i   192.168.2.1/32   10.0.4.4          0              100           0 65211 ?

Total number of prefixes 2
R2_PE1#


R7_PE2#show    bgp    vpnv4 unicast    all
BGP table version is 1, local router ID is 7.7.7.7
Status codes: s suppressed, d damped, h history, * valid, > best, i - internal,
              S Stale, b - backup entry, m - multipath, f Filter, a additional-path
Origin codes: i - IGP, e - EGP, ? - incomplete

     Network           Next Hop          Metric         LocPrf        Weight Path
Route Distinguisher: 100:2 (Default for VRF VPN1)
*>i   10.0.12.0/24     10.0.5.5          0              100           0 65210 ?
*>    10.0.78.0/24     0.0.0.0           1                            32768     ?
*>i   192.168.1.0      10.0.5.5          0              100           0 65210 ?
*>    192.168.2.1/32   10.0.78.8         1                            32768     ?

Total number of prefixes 4
Route Distinguisher: 100:1
*>i   10.0.12.0/24     10.0.5.5          0              100           0 65210 ?
*>i   192.168.1.0      10.0.5.5          0              100           0 65210 ?

Total number of prefixes 2
R7_PE2#
```

10. 检查 R1_CE1 设备的全局 IP 路由表，接收到对方的私网路由。

```
R1_CE1#show    ip route

Codes:   C - Connected, L - Local, S - Static
         R - RIP, O - OSPF, B - BGP, I - IS-IS, V - Overflow route
         N1 - OSPF NSSA external type 1, N2 - OSPF NSSA external type 2
         E1 - OSPF external type 1, E2 - OSPF external type 2
         SU - IS-IS summary, L1 - IS-IS level-1, L2 - IS-IS level-2
         IA - Inter area, EV - BGP EVPN, A - Arp to host
         LA - Local aggregate route
         * - candidate default

Gateway of last resort is no set
 C      10.0.12.0/24 is directly connected, GigabitEthernet 0/2
 C      10.0.12.1/32 is local host.
 I L2   10.0.78.0/24 [115/1] via 10.0.12.2, 00:57:33, GigabitEthernet 0/2
 C      192.168.1.0/24 is directly connected, Loopback 0
```

```
C       192.168.1.1/32 is local host.
I L2    192.168.2.1/32 [115/1] via 10.0.12.2, 00:57:33, GigabitEthernet 0/2
R1_CE1#


R8_CE2#show    ip route

Codes:    C - Connected, L - Local, S - Static
          R - RIP, O - OSPF, B - BGP, I - IS-IS, V - Overflow route
          N1 - OSPF NSSA external type 1, N2 - OSPF NSSA external type 2
          E1 - OSPF external type 1, E2 - OSPF external type 2
          SU - IS-IS summary, L1 - IS-IS level-1, L2 - IS-IS level-2
          IA - Inter area, EV - BGP EVPN, A - Arp to host
          LA - Local aggregate route
          * - candidate default

Gateway of last resort is no set
O E2    10.0.12.0/24 [110/1] via 10.0.78.7, 00:57:53, GigabitEthernet 0/2
C       10.0.78.0/24 is directly connected, GigabitEthernet 0/2
C       10.0.78.8/32 is local host.
O E2    192.168.1.0/24 [110/1] via 10.0.78.7, 00:57:53, GigabitEthernet 0/2
C       192.168.2.0/24 is directly connected, Loopback 0
C       192.168.2.1/32 is local host.
R8_CE2#
```

11. 测试私网目标网段连通情况。

```
R1_CE1#ping    192.168.2.1 source    192.168.1.1
Sending 5, 100-byte ICMP Echoes to 192.168.2.1, timeout is 2 seconds:
   < press Ctrl+C to break >
!!!!!
Success rate is 100 percent (5/5), round-trip min/avg/max = 9/10/13 ms.
R1_CE1#


R8_CE2#ping    192.168.1.1 source    192.168.2.1
Sending 5, 100-byte ICMP Echoes to 192.168.1.1, timeout is 2 seconds:
   < press Ctrl+C to break >
!!!!!
Success rate is 100 percent (5/5), round-trip min/avg/max = 7/9/12 ms.
R8_CE2#
```

12. 在 R2_PE1 设备和 R7_PE2 设备查看标签转发信息表，标签转发信息表包括公网标签信息和私网标签信息。

```
R2_PE1#show    mpls    forwarding-table

Label Operation Code:
PH--PUSH label
PP--POP label
SW--SWAP label
SP--SWAP topmost label and push new label
DP--DROP packet
PC--POP label and continue lookup by IP or Label
PI--POP label and do ip lookup forward
PN--POP label and forward to nexthop
PM--POP label and do MAC lookup forward
```

```
PV--POP label and output to VC attach interface
IP--IP lookup forward
  s--stale
  Local    Outgoing OP  FEC                     Outgoing       Nexthop
  label    label                                interface
  --       imp-null PH  10.0.3.3/32             Gi0/0          10.0.23.3
  --       11265    PH  10.0.4.4/32             Gi0/0          10.0.23.3
  --       imp-null PH  10.0.34.0/24            Gi0/0          10.0.23.3
  --       60930    PH  10.0.78.0/24(V)         Gi0/0          10.0.23.3
  --       60931    PH  192.168.2.1/32(V)       Gi0/0          10.0.23.3
  11264    imp-null PP  10.0.3.3/32             Gi0/0          10.0.23.3
  11265    11265    SW  10.0.4.4/32             Gi0/0          10.0.23.3
  11266    imp-null PP  10.0.34.0/24            Gi0/0          10.0.23.3
  60928    --           PI  VRF(VPN1)           --             0.0.0.0

R2_PE1#

R7_PE2#show   mpls   forwarding-table

Label Operation Code:
PH--PUSH label
PP--POP label
SW--SWAP label
SP--SWAP topmost label and push new label
DP--DROP packet
PC--POP label and continue lookup by IP or Label
PI--POP label and do ip lookup forward
PN--POP label and forward to nexthop
PM--POP label and do MAC lookup forward
PV--POP label and output to VC attach interface
IP--IP lookup forward
  s--stale
  Local    Outgoing OP  FEC                     Outgoing       Nexthop
  label    label                                interface
  --       11264    PH  10.0.5.5/32             Gi0/1          10.0.67.6
  --       imp-null PH  10.0.6.6/32             Gi0/1          10.0.67.6
  --       imp-null PH  10.0.56.0/24            Gi0/1          10.0.67.6
  --       60928    PH  10.0.12.0/24(V)         Gi0/1          10.0.67.6
  --       60929    PH  192.168.1.0/24(V)       Gi0/1          10.0.67.6
  11264    11264    SW  10.0.5.5/32             Gi0/1          10.0.67.6
  11265    imp-null PP  10.0.6.6/32             Gi0/1          10.0.67.6
  11266    imp-null PP  10.0.56.0/24            Gi0/1          10.0.67.6
  60928    --           PI  VRF(VPN1)           --             0.0.0.0

R7_PE2#
```

13. 通过 Wireshark 抓包工具在 PE 设备抓取报文，检查报文标签信息，OptionB 方案标签为连续的完整标签链路，标签：11265 为公网标签，标签：60929 为目标 192.168.2.1/24 的私网标签。

```
No.   Time          Source        Destination    Protocol  Length   Info
  34 16.720... 192.168.1.1    192.168.2.1    ICMP      150 Echo (ping) request  id=0x0100, seq=30595/33655, ttl=63 (reply in 35)
  35 16.729... 192.168.2.1    192.168.1.1    ICMP      146 Echo (ping) reply    id=0x0100, seq=30595/33655, ttl=63 (request in 34)
  36 16.836... 192.168.1.1    192.168.2.1    ICMP      150 Echo (ping) request  id=0x0100, seq=33667/33667, ttl=63 (reply in 37)
  37 16.845... 192.168.2.1    192.168.1.1    ICMP      146 Echo (ping) reply    id=0x0100, seq=33667/33667, ttl=63 (request in 36)
  38 16.948... 192.168.1.1    192.168.2.1    ICMP      150 Echo (ping) request  id=0x0100, seq=36483/33678, ttl=63 (reply in 39)
  39 16.956... 192.168.2.1    192.168.1.1    ICMP      146 Echo (ping) reply    id=0x0100, seq=36483/33678, ttl=63 (request in 38)

> Frame 34: 150 bytes on wire (1200 bits), 150 bytes captured (1200 bits) on interface 0
> Ethernet II, Src: 50:00:00:02:00:01 (50:00:00:02:00:01), Dst: 50:00:00:03:00:01 (50:00:00:03:00:01)
> MultiProtocol Label Switching Header, Label: 11265, Exp: 0, S: 0, TTL: 63
    0000 0010 1100 0000 0001 .... .... .... = MPLS Label: 11265
    .... .... .... .... .... 000. .... .... = MPLS Experimental Bits: 0
    .... .... .... .... .... ...0 .... .... = MPLS Bottom Of Label Stack: 0
    .... .... .... .... .... .... 0011 1111 = MPLS TTL: 63
> MultiProtocol Label Switching Header, Label: 60929, Exp: 0, S: 1, TTL: 63
    0000 1110 1110 0000 0001 .... .... .... = MPLS Label: 60929
    .... .... .... .... .... 000. .... .... = MPLS Experimental Bits: 0
    .... .... .... .... .... ...1 .... .... = MPLS Bottom Of Label Stack: 1
    .... .... .... .... .... .... 0011 1111 = MPLS TTL: 63
> Internet Protocol Version 4, Src: 192.168.1.1, Dst: 192.168.2.1
> Internet Control Message Protocol

No.   Time          Source        Destination    Protocol  Length   Info
  34 16.720... 192.168.1.1    192.168.2.1    ICMP      150 Echo (ping) request  id=0x0100, seq=30595/33655, ttl=63 (reply in 35)
  35 16.729... 192.168.2.1    192.168.1.1    ICMP      146 Echo (ping) reply    id=0x0100, seq=30595/33655, ttl=63 (request in 34)
  36 16.836... 192.168.1.1    192.168.2.1    ICMP      150 Echo (ping) request  id=0x0100, seq=33667/33667, ttl=63 (reply in 37)
  37 16.845... 192.168.2.1    192.168.1.1    ICMP      146 Echo (ping) reply    id=0x0100, seq=33667/33667, ttl=63 (request in 36)
  38 16.948... 192.168.1.1    192.168.2.1    ICMP      150 Echo (ping) request  id=0x0100, seq=36483/33678, ttl=63 (reply in 39)
  39 16.956... 192.168.2.1    192.168.1.1    ICMP      146 Echo (ping) reply    id=0x0100, seq=36483/33678, ttl=63 (request in 38)

> Frame 35: 146 bytes on wire (1168 bits), 146 bytes captured (1168 bits) on interface 0
> Ethernet II, Src: 50:00:00:03:00:01 (50:00:00:03:00:01), Dst: 50:00:00:02:00:01 (50:00:00:02:00:01)
> MultiProtocol Label Switching Header, Label: 60928, Exp: 0, S: 1, TTL: 59
    0000 1110 1110 0000 0000 .... .... .... = MPLS Label: 60928
    .... .... .... .... .... 000. .... .... = MPLS Experimental Bits: 0
    .... .... .... .... .... ...1 .... .... = MPLS Bottom Of Label Stack: 1
    .... .... .... .... .... .... 0011 1011 = MPLS TTL: 59
> Internet Protocol Version 4, Src: 192.168.2.1, Dst: 192.168.1.1
> Internet Control Message Protocol
```

> 问题与思考

根据以上配置步骤完成跨域 OptionB 方案的不改变下一跳地址方案的配置，并将 CE 改为 MCE。

3.3.3 跨域 MPLS VPN-OptionC（1）

> 原理

OptionA 和 OptionB 两种方案都能够满足跨域 VPN 的组网需求，这两种方案的一个共同点就是 ASBR 都需要参与 VPN 路由的维护和发布。当每个自治域内都有大量的跨域 VPN 路由需要通告时，ASBR 就可能成为阻碍网络进一步扩展的瓶颈。为了解决上述扩展性问题，提出了第三种解决方案：多跳 MP-EBGP。多跳 MP-EBGP 是指在跨域的情况下，不同自治域的 PE 之间建立多跳的 MP-EBGP 会话，直接交互 VPN 路由，这种方案就不需要 ASBR 维护和分发 VPN 路由。

多跳 MP-EBGP 的实现方式，只有 PE 保留 VPN 用户信息，ASBR 上不用保留 VPN 用户的信息，其配置相对复杂。该方案适合于需要大规模开展跨域 VPN 业务时使用。

从实现原理来说，OptionC 也有两种实现方案：
（1）只在 EBGP 邻居间启用为 IPv4 路由交换标签的功能
（2）在 EBGP 和 IBGP 邻居间都启用为 IPv4 路由交换标签的功能

为了便于 OptionC 方案的规模扩展，通常在每个自治域内都部署了路由反射器，在两个自治域的路由反射器之间建立多跳的 MP-EBGP 会话传递 VPN 路由，从部署应用角度来说，称为"路由反射器间建立多跳 MP-EBGP"方案。

方案一：只在 EBGP 邻居间启用为 IPv4 路由交换标签的功能

在这种实现方案中，需要 ASBR 设备上运行的 IGP 协议（如 OSPF、IS-IS 等）重分布到

BGP 路由中，使得自治域内的每台设备都具有到达另外一个自治域的 PE 路由。自治域之内通过 LDP 为到达另一个自治域的 PE 路由分配标签，建立 LSP，而在两个自治域之间直连的 ASBR 设备上，启用 IPv4 路由交换标签的功能，BGP 协议扮演了"MPLS 信令"的角色，为到达另一个自治域的 PE 路由分配标签，从而建立跨域的 LSP。

➢ 任务拓扑

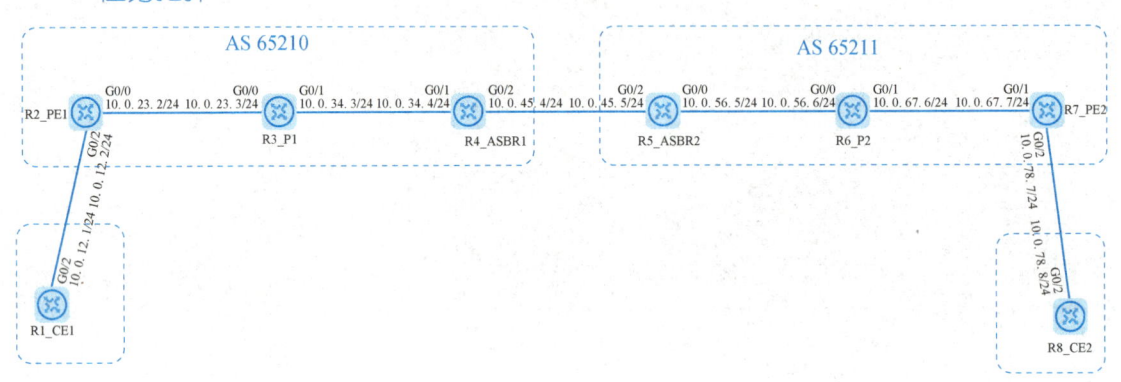

➢ 实施步骤

1. 根据任务拓扑配置各设备接口 IP 地址。（略）
2. 配置 AS 65210 内公网路由互通。（参考 3.3.1）
3. 配置 AS 65211 内公网路由互通。（参考 3.3.1）
4. 公网 AS 65210 和 AS 65211 内配置 LDP 协议同时接口开启标签交换功能分配公网标签。（参考 3.3.1）
5. 在 R2_PE1 设备和 R7-PE 设备创建 VRF，并且关联接口。

```
R2_PE1(config)#ip vrf VPN1
R2_PE1(config-vrf)#rd    100:1
R2_PE1(config-vrf)#route-target    both    10:1
R2_PE1(config-vrf)#exit
R2_PE1(config)#interface    gigabitEthernet 0/2
R2_PE1(config-if-GigabitEthernet 0/2)#ip vrf    forwarding    VPN1
% Interface GigabitEthernet 0/2 IP address removed due to enabling VRF VPN1
R2_PE1(config-if-GigabitEthernet 0/2)#ip address    10.0.12.2 24
R2_PE1(config-if-GigabitEthernet 0/2)#exit
R2_PE1(config)#


R7-PE2(config)#ip vrf    VPN1
R7_PE2(config-vrf)#rd    100:2
R7_PE2(config-vrf)#route-target    both    10:1
R7_PE2(config-vrf)#exit
R7_PE2(config)#interface    gigabitEthernet 0/2
R7_PE2(config-if-GigabitEthernet 0/2)#no    switchport
R7_PE2(config-if-GigabitEthernet 0/2)#ip vrf    forwarding    VPN1
% Interface GigabitEthernet 0/2 IP address removed due to enabling VRF VPN1
R7_PE2(config-if-GigabitEthernet 0/2)#ip address    10.0.78.7 24
R7_PE2(config-if-GigabitEthernet 0/2)#exit
R7_PE2(config)#
```

6. Option C 方案同各 AS 域内建立 IBGP 邻居关系。

```
R2_PE1(config)#router   bgp   65210
R2_PE1(config-router)#bgp   router-id   2.2.2.2
R2_PE1(config-router)#neighbor   10.0.4.4 remote-as   65210
R2_PE1(config-router)#neighbor   10.0.4.4 update-source   loopback 0
R2_PE1(config-router)#exit
R2_PE1(config)#

R4_ASBR1(config)#router bgp   65210
R4_ASBR1(config-router)#bgp   router-id   4.4.4.4
R4_ASBR1(config-router)#neighbor   10.0.2.2 remote-as   65210
R4_ASBR1(config-router)#neighbor   10.0.2.2 update-source   loopback 0
R4_ASBR1(config-router)#exit
R4_ASBR1(config)#


R5_ASBR2(config)#router   bgp   65211
R5_ASBR2(config-router)#bgp   router-id   5.5.5.5
R5_ASBR2(config-router)#neighbor   10.0.7.7 remote-as   65211
R5_ASBR2(config-router)#neighbor   10.0.7.7 update-source   loopback 0
R5_ASBR2(config-router)#exit
R5_ASBR2(config)#

R7_PE2(config)#router   bgp   65211
R7_PE2(config-router)#bgp   router-id   7.7.7.7
R7_PE2(config-router)#neighbor   10.0.5.5 remote-as   65211
R7_PE2(config-router)#neighbor   10.0.5.5 update-s loopback 0
R7_PE2(config-router)#exit
R7_PE2(config)#
```

7. 在 R4_ASBR1 设备和 R5_ASBR2 设备之间建立 EBGP 邻居关系。

```
R4_ASBR1(config)#router   bgp   65210
R4_ASBR1(config-router)#neighbor   10.0.45.5 remote-as   65211
R4_ASBR1(config-router)#exit
R4_ASBR1(config)#

R5_ASBR2(config)#router   bgp   65211
R5_ASBR2(config-router)#neighbor   10.0.45.4 remote-as   65210
R5_ASBR2(config-router)#exit
R5_ASBR2(config)#
```

8. 由于 OptionC 方案通过 PE-PE 设备之间建立 MP-EBGP 多跳邻居来传递私网路由，需要将各 AS 域内的 PE 设备的 loopback 0 接口传递到对端的 PE 设备，即各 AS 之间的 PE 设备 loopback 0 接口可以相互通信。

```
R4_ASBR1(config)#router   bgp   65210
R4_ASBR1(config-router)#network   10.0.2.2 mask   255.255.255.255
R4_ASBR1(config-router)#neighbor   10.0.2.2 next-hop-self     //修改下一跳地址
R4_ASBR1(config-router)#exit
R4_ASBR1(config)#

R5_ASBR2(config)#router   bgp   65211
R5_ASBR2(config-router)#neighbor   10.0.7.7 next-hop-self     //修改下一跳地址才能将 PE 的 Loopback 0
接口传递给对端 PE 设备
```

```
R5_ASBR2(config-router)#network    10.0.7.7 mask    255.255.255.255
```

此时完成路由传递，但由于标签值不连续，导致 PE 设备的 Loopback 0 接口无法通信

9．在 R2_PE1-R4_ASBR1-R5_ASBR2-R7_PE2 设备为 10.0.2.2/32 的地址建立一条连续的标签（即 EBGP 和 IBGP 之间都启用标签交换），设备之间的互联接口需要开启发送标签功能。

```
R4_ASBR1(config)#router    bgp 65210
R4_ASBR1(config-router)#neighbor    10.0.45.5 send-label

R4_ASBR1(config)#interface    gigabitEthernet 0/2
R4_ASBR1(config-if-GigabitEthernet 0/2)#label-switching    //ASBR 互联接口必须开启标签交换能力

R5_ASBR2(config)#interface    gigabitEthernet 0/2
R5_ASBR2(config-if-GigabitEthernet 0/2)#label-switching
R5_ASBR2(config)#ip access-list standard    1
R5_ASBR2(config-std-nacl)#permit    10.0.2.2 0.0.0.0
R5_ASBR2(config-std-nacl)#exit
R5_ASBR2(config)#mpls    router    ldp
R5_ASBR2(config-mpls-router)#advertise-labels for    bgp-routes    acl 1
R5_ASBR2(config)#router    bgp    65211
R5_ASBR2(config-router)#neighbor    10.0.7.7 send-label
R5_ASBR2(config-router)#exit
R5_ASBR2(config)#

R7_PE2(config)#router    bgp    65211
R7_PE2(config-router)#neighbor    10.0.5.5 send-label
R7_PE2(config-router)#exit
R7_PE2(config)#
```

10．在 R7_PE2-R5_ASBR2-R4_ASBR1-R2_PE1 设备为 10.0.7.7/32 的地址建立一条连续的标签。

```
R5_ASBR2(config)#router    bgp    65211
R5_ASBR2(config-router)#neighbor    10.0.45.4 send-label
R5_ASBR2(config-router)#exit
R5_ASBR2(config)#

R4_ASBR1(config)#access-list    1 permit    10.0.7.7 0.0.0.0
R4_ASBR1(config)#mpls    router    ldp
R4_ASBR1(config-mpls-router)#advertise-labels    for    bgp-routes    acl 1
R4_ASBR1(config-mpls-router)#exit
R4_ASBR1(config)#

R4_ASBR1(config)#router    bgp    65210
R4_ASBR1(config-router)#neighbor    10.0.2.2 send-label
R4_ASBR1(config-router)#exit
R4_ASBR1(config)#


R2_PE1(config)#router    bgp    65210
R2_PE1(config-router)#neighbor    10.0.4.4 send-label
R2_PE1(config-router)#exit
R2_PE1(config)#
```

此时建立了 R2_PE1-R7_PE2 的连续完整的标签信息，R2_PE1-R7_PE2 能正常通信。

11．R2_PE1-R7_PE2 设备已经完成连续标签的建立，也能通过标签信息执行通信，此时可以在 R2_PE1-R7_PE2 设备之间建立 MP-EBGP 多跳邻居关系传递私网路由。

```
R2_PE1(config)#router  bgp  65210
R2_PE1(config-router)#neighbor   10.0.7.7 remote-as  65211
R2_PE1(config-router)#neighbor   10.0.7.7 update-source   loopback 0
R2_PE1(config-router)#neighbor   10.0.7.7 ebgp-multihop      //开启 EBGP 多跳
R2_PE1(config-router)#address-family ipv4
R2_PE1(config-router-af)#no   neighbor   10.0.7.7 activate    //需要关闭 IPv4 单播 BGP 邻居建立
R2_PE1(config-router-af)#exit
R2_PE1(config-router)#address-family vpnv4
R2_PE1(config-router-af)#neighbor   10.0.7.7 activate       //开启 VPNv4 邻居
R2_PE1(config-router-af)#exit
R2_PE1(config-router)#


R7_PE2(config)#router  bgp  65211
R7_PE2(config-router)#neighbor   10.0.2.2 remote-as  65210
R7_PE2(config-router)#neighbor   10.0.2.2 update-source   loopback 0
R7_PE2(config-router)#neighbor   10.0.2.2 ebgp-multihop
R7_PE2(config-router)#address-family ipv4
R7_PE2(config-router-af)#no neighbor   10.0.2.2 activate     //关闭单播 BGP 邻居建立
R7_PE2(config-router)#address-family vpnv4
R7_PE2(config-router-af)#neighbor   10.0.2.2 activate
R7_PE2(config-router-af)#exit
R7_PE2(config-router)#
```

12．在 R2_PE1-R1_CE1 设备之间配置 OSPF 100 传递私网路由，R1_CE1 设备创建 loopback 0 接口并通告。

```
R1_CE1(config)#interface   loopback 0
R1_CE1(config-if-Loopback 0)#ip address   192.168.1.1 24
R1_CE1(config-if-Loopback 0)#exit
R1_CE1(config)#router ospf   100
R1_CE1(config-router)#router-id   1.1.1.1
Change router-id and update OSPF process! [yes/no]:y
R1_CE1(config-router)#network   10.0.12.0 0.0.0.255 area   0
R1_CE1(config-router)#network   192.168.1.0 0.0.0.255 area   0
R1_CE1(config-router)#exit
R1_CE1(config)#

R2_PE1(config)#router   ospf   100 vrf   VPN1
R2_PE1(config-router)#router-id   22.2.2.2
Change router-id and update OSPF process! [yes/no]:y
R2_PE1(config-router)#network   10.0.12.0 0.0.0.255 area   0
R2_PE1(config-router)#exit
R2_PE1(config)#
```

13．在 R7_PE2-R8_CE2 设备之间配置 IS-IS 10 传递私网路由，R8_CE2 设备创建 loopback 0 接口并通告。

```
R8_CE2(config)#router   isis   10
R8_CE2(config-router)#net   49.0002.0000.0000.0008.00
```

```
R8_CE2(config-router)#is-type    level-2
R8_CE2(config-router)#exit
R8_CE2(config)#interface    gigabitEthernet 0/2
R8_CE2(config-if-GigabitEthernet 0/2)#ip router    isis    10
R8_CE2(config-if-GigabitEthernet 0/2)#exit
R8_CE2(config)#interface    loopback 0
R8_CE2(config-if-Loopback 0)#ip address    192.168.2.1 24
R8_CE2(config-if-Loopback 0)#ip router    isis    10
R8_CE2(config-if-Loopback 0)#exit
R8_CE2(config)#


R7_PE2(config)#router    isis    10
R7_PE2(config-router)#vrf VPN1
R7_PE2(config-router)#net    49.0002.0000.0000.0007.00
R7_PE2(config-router)#is-type    level-2
R7_PE2(config-router)#exit
R7_PE2(config)#interface    gigabitEthernet 0/2
R7_PE2(config-if-GigabitEthernet 0/2)#ip router    isis    10
R7_PE2(config-if-GigabitEthernet 0/2)#exit
R7_PE2(config)#
```

14. 在 R2_PE1 设备和 R7_PE2 设备将 IGP 路由重分布到 MP-BGP 协议中。

```
R2_PE1(config)#router    bgp    65210
R2_PE1(config-router)#address-family ipv4 vrf    VPN1
R2_PE1(config-router-af)#redistribute    ospf    100 match internal    external
R2_PE1(config-router-af)#exit
R2_PE1(config-router)#


R7_PE2(config)#router    bgp    65211
R7_PE2(config-router)#address-family ipv4 vrf    VPN1
R7_PE2(config-router-af)#redistribute    isis 10 level-2
R7_PE2(config-router-af)#exit
R7_PE2(config-router)#exit
R7_PE2(config)#
```

15. 在 R2_PE1 设备和 R7_PE2 设备将 MP-BGP 路由重分布到 IGP 协议中。

```
R2_PE1(config)#router ospf    100
R2_PE1(config-router)#redistribute    bgp    subnets
R2_PE1(config-router)#exit
R2_PE1(config)#


R7_PE2(config)#router    isis    10
R7_PE2(config-router)#redistribute    bgp    level-2
R7_PE2(config-router)#exit
R7_PE2#
```

> 任务验证

1. 检查公网 AS 65210 内部 IGP 协议邻居建立情况。

```
R3_P1#show    ip ospf    neighbor
```

```
OSPF process 10, 2 Neighbors, 2 is Full:
  Neighbor ID        Pri    State         BFD State    Dead Time    Address         Interface
  2.2.2.2            1      Full/BDR      -            00:00:40     10.0.23.2
GigabitEthernet 0/0
  4.4.4.4            1      Full/DR       -            00:00:40     10.0.34.4
GigabitEthernet 0/1
  R3_P1#


R2_PE1#show    ip route

Codes:    C - Connected, L - Local, S - Static
          R - RIP, O - OSPF, B - BGP, I - IS-IS, V - Overflow route
          N1 - OSPF NSSA external type 1, N2 - OSPF NSSA external type 2
          E1 - OSPF external type 1, E2 - OSPF external type 2
          SU - IS-IS summary, L1 - IS-IS level-1, L2 - IS-IS level-2
          IA - Inter area, EV - BGP EVPN, A - Arp to host
          LA - Local aggregate route
          * - candidate default

Gateway of last resort is no set
C       10.0.2.2/32 is local host.
O       10.0.3.3/32 [110/1] via 10.0.23.3, 00:03:37, GigabitEthernet 0/0
O       10.0.4.4/32 [110/2] via 10.0.23.3, 00:01:38, GigabitEthernet 0/0
C       10.0.12.0/24 is directly connected, GigabitEthernet 0/2
C       10.0.12.2/32 is local host.
C       10.0.23.0/24 is directly connected, GigabitEthernet 0/0
C       10.0.23.2/32 is local host.
O       10.0.34.0/24 [110/2] via 10.0.23.3, 00:03:37, GigabitEthernet 0/0
R2_PE1#


R2_PE1#ping    10.0.4.4
Sending 5, 100-byte ICMP Echoes to 10.0.4.4, timeout is 2 seconds:
  < press Ctrl+C to break >
!!!!!
Success rate is 100 percent (5/5), round-trip min/avg/max = 2/2/3 ms.
R2_PE1#
```

2. 检查公网 AS 65211 内部 IGP 协议邻居建立情况。

```
R6_P2#show    ip ospf neighbor

OSPF process 20, 2 Neighbors, 2 is Full:
  Neighbor ID        Pri    State         BFD State    Dead Time    Address         Interface
  5.5.5.5            1      Full/BDR      -            00:00:31     10.0.56.5
GigabitEthernet 0/0
  7.7.7.7            1      Full/DR       -            00:00:32     10.0.67.7
GigabitEthernet 0/1
  R6_P2#



R7_PE2#show    ip route

Codes:    C - Connected, L - Local, S - Static
          R - RIP, O - OSPF, B - BGP, I - IS-IS, V - Overflow route
          N1 - OSPF NSSA external type 1, N2 - OSPF NSSA external type 2
          E1 - OSPF external type 1, E2 - OSPF external type 2
```

```
             SU - IS-IS summary, L1 - IS-IS level-1, L2 - IS-IS level-2
             IA - Inter area, EV - BGP EVPN, A - Arp to host
             LA - Local aggregate route
             * - candidate default

Gateway of last resort is no set
O       10.0.5.5/32 [110/2] via 10.0.67.6, 00:02:40, GigabitEthernet 0/1
O       10.0.6.6/32 [110/1] via 10.0.67.6, 00:02:40, GigabitEthernet 0/1
C       10.0.7.7/32 is local host.
O       10.0.56.0/24 [110/2] via 10.0.67.6, 00:02:40, GigabitEthernet 0/1
C       10.0.67.0/24 is directly connected, GigabitEthernet 0/1
C       10.0.67.7/32 is local host.
C       10.0.78.0/24 is directly connected, GigabitEthernet 0/2
C       10.0.78.7/32 is local host.
R7_PE2#


R7_PE2#ping 10.0.5.5
Sending 5, 100-byte ICMP Echoes to 10.0.5.5, timeout is 2 seconds:
  < press Ctrl+C to break >
!!!!!
Success rate is 100 percent (5/5), round-trip min/avg/max = 2/2/3 ms.
R7_PE2#
```

3．检查公网 AS 65210 的 LDP 协议邻居建立情况及公网标签转发信息表。

```
R3_P1#show   mpls   ldp   neighbor
Default VRF:
    Peer LDP Ident: 10.0.2.2:0; Local LDP Ident: 10.0.3.3:0
        TCP connection: 10.0.2.2.646 - 10.0.3.3.35103
        State: OPERATIONAL; Msgs sent/recv: 20/22; UNSOLICITED
        Up time: 00:03:27
        Graceful Restart enabled; Peer reconnect time (msecs): 300000
        LDP discovery sources:
           Link Peer on GigabitEthernet 0/0, Src IP addr: 10.0.23.2
        Addresses bound to peer LDP Ident:
           10.0.12.2       10.0.23.2       10.0.2.2
    Peer LDP Ident: 10.0.4.4:0; Local LDP Ident: 10.0.3.3:0
        TCP connection: 10.0.4.4.35271 - 10.0.3.3.646
        State: OPERATIONAL; Msgs sent/recv: 17/19; UNSOLICITED
        Up time: 00:02:41
        Graceful Restart enabled; Peer reconnect time (msecs): 300000
        LDP discovery sources:
           Link Peer on GigabitEthernet 0/1, Src IP addr: 10.0.34.4
        Addresses bound to peer LDP Ident:
           10.0.4.4        10.0.34.4       10.0.45.4


R3_P1#show   mpls   forwarding-table

Label Operation Code:
PH--PUSH label
PP--POP label
SW--SWAP label
SP--SWAP topmost label and push new label
DP--DROP packet
```

```
PC--POP label and continue lookup by IP or Label
PI--POP label and do ip lookup forward
PN--POP label and forward to nexthop
PM--POP label and do MAC lookup forward
PV--POP label and output to VC attach interface
IP--IP lookup forward
  s--stale
  Local    Outgoing OP FEC                         Outgoing        Nexthop
  label    label                                   interface
  --       imp-null PH 10.0.2.2/32                 Gi0/0           10.0.23.2
  --       imp-null PH 10.0.4.4/32                 Gi0/1           10.0.34.4
  11264    imp-null PP 10.0.2.2/32                 Gi0/0           10.0.23.2
  11265    imp-null PP 10.0.4.4/32                 Gi0/1           10.0.34.4

R3_P1#
```

4. 检查公网 AS 65211 的 LDP 协议邻居建立情况及公网标签转发信息表。

```
R6_P2#show   mpls   ldp   neighbor
Default VRF:
    Peer LDP Ident: 10.0.5.5:0; Local LDP Ident: 10.0.6.6:0
        TCP connection: 10.0.5.5.646 - 10.0.6.6.38887
        State: OPERATIONAL; Msgs sent/recv: 17/19; UNSOLICITED
        Up time: 00:02:44
        Graceful Restart enabled; Peer reconnect time (msecs): 300000
        LDP discovery sources:
            Link Peer on GigabitEthernet 0/0, Src IP addr: 10.0.56.5
        Addresses bound to peer LDP Ident:
            10.0.45.5        10.0.56.5        10.0.5.5
    Peer LDP Ident: 10.0.7.7:0; Local LDP Ident: 10.0.6.6:0
        TCP connection: 10.0.7.7.35301 - 10.0.6.6.646
        State: OPERATIONAL; Msgs sent/recv: 16/18; UNSOLICITED
        Up time: 00:02:20
        Graceful Restart enabled; Peer reconnect time (msecs): 300000
        LDP discovery sources:
            Link Peer on GigabitEthernet 0/1, Src IP addr: 10.0.67.7
        Addresses bound to peer LDP Ident:
            10.0.7.7         10.0.67.7        10.0.78.7


R6_P2#show   mpls   forwarding-table

Label Operation Code:
PH--PUSH label
PP--POP label
SW--SWAP label
SP--SWAP topmost label and push new label
DP--DROP packet
PC--POP label and continue lookup by IP or Label
PI--POP label and do ip lookup forward
PN--POP label and forward to nexthop
PM--POP label and do MAC lookup forward
PV--POP label and output to VC attach interface
IP--IP lookup forward
  s--stale
  Local    Outgoing OP FEC                         Outgoing        Nexthop
  label    label                                   interface
```

```
   --          imp-null PH 10.0.5.5/32              Gi0/0              10.0.56.5
   --          imp-null PH 10.0.7.7/32              Gi0/1              10.0.67.7
   11264       imp-null PP 10.0.5.5/32              Gi0/0              10.0.56.5
   11265       imp-null PP 10.0.7.7/32              Gi0/1              10.0.67.7

R6_P2#
```

5．检查各 AS 域内 IBGP 邻居关系建立情况。

```
R2_PE1#show ip bgp   summary
For address family: IPv4 Unicast
BGP router identifier 2.2.2.2, local AS number 65210
BGP table version is 1
0 BGP AS-PATH entries
0 BGP Community entries
0 BGP Prefix entries (Maximum-prefix:4294967295)

Neighbor        V     AS MsgRcvd MsgSent    TblVer    InQ OutQ Up/Down    State/PfxRcd
10.0.4.4        4     65210    5       4         1       0    0 00:02:23      0

Total number of neighbors 1, established neighbors 1

R2_PE1#


R5_ASBR2#show   ip bgp   summary
For address family: IPv4 Unicast
BGP router identifier 5.5.5.5, local AS number 65211
BGP table version is 1
0 BGP AS-PATH entries
0 BGP Community entries
0 BGP Prefix entries (Maximum-prefix:4294967295)

Neighbor        V     AS MsgRcvd MsgSent    TblVer    InQ OutQ Up/Down    State/PfxRcd
10.0.7.7        4     65211    4       3         1       0    0 00:01:31      0

Total number of neighbors 1, established neighbors 1

R5_ASBR2#
```

6．检查 R4_ASBR1 设备和 R5_ASBR2 设备之间的 EBGP 邻居关系建立情况。

```
R4_ASBR1#show   ip bgp   summary
For address family: IPv4 Unicast
BGP router identifier 4.4.4.4, local AS number 65210
BGP table version is 7
0 BGP AS-PATH entries
0 BGP Community entries
0 BGP Prefix entries (Maximum-prefix:4294967295)

Neighbor        V     AS MsgRcvd MsgSent    TblVer    InQ OutQ Up/Down    State/PfxRcd
10.0.2.2        4     65210   39      39        7       0    0 00:35:20      0
10.0.45.5       4     65211   49      47        6       0    0 00:43:53      0

Total number of neighbors 2, established neighbors 2

R4_ASBR1#
```

7. 检查 R2_PE1-R7_PE2 的 loopback 0 接口路由通告，在 BGP 路由表显示为有效且最优路由，由于在各 AS 域内地址携带标签信息，此时 R2_PE1 设备和 R7_PE2 设备的 loopback 0 接口标签不连续，导致无法相互通信。

```
R2_PE1#show ip bgp
BGP table version is 4, local router ID is 2.2.2.2
Status codes: s suppressed, d damped, h history, * valid, > best, i - internal,
              S Stale, b - backup entry, m - multipath, f Filter, a additional-path
Origin codes: i - IGP, e - EGP, ? - incomplete

    Network            Next Hop           Metric      LocPrf      Weight Path
*>i 10.0.2.2/32        10.0.4.4           2           100         0      i
*>i 10.0.7.7/32        10.0.4.4           2           100         0 65211 i

Total number of prefixes 2
R2_PE1#


R7_PE2#show   ip   bgp
BGP table version is 4, local router ID is 7.7.7.7
Status codes: s suppressed, d damped, h history, * valid, > best, i - internal,
              S Stale, b - backup entry, m - multipath, f Filter, a additional-path
Origin codes: i - IGP, e - EGP, ? - incomplete

    Network            Next Hop           Metric      LocPrf      Weight Path
*>i 10.0.2.2/32        10.0.5.5           2           100         0 65210 i
*>i 10.0.7.7/32        10.0.5.5           2           100         0      i

Total number of prefixes 2
R7_PE2#


R2_PE1#ping   10.0.7.7 source   10.0.2.2      //无法相互通信
Sending 5, 100-byte ICMP Echoes to 10.0.7.7, timeout is 2 seconds:
  < press Ctrl+C to break >
.....
Success rate is 0 percent (0/5).
R2_PE1#
```

8. 检查 R2_PE1 设备和 R7_PE2 设备之间 10.0.2.2/32 地址的连续标签建立情况。

```
R3_P1#show   mpls   forwarding-table

Label Operation Code:
PH--PUSH label
PP--POP label
SW--SWAP label
SP--SWAP topmost label and push new label
DP--DROP packet
PC--POP label and continue lookup by IP or Label
PI--POP label and do ip lookup forward
PN--POP label and forward to nexthop
PM--POP label and do MAC lookup forward
PV--POP label and output to VC attach interface
IP--IP lookup forward
  s--stale
    Local    Outgoing OP FEC                              Outgoing         Nexthop
```

label	label		interface	
--	imp-null PH 10.0.2.2/32		Gi0/0	10.0.23.2
--	imp-null PH 10.0.4.4/32		Gi0/1	10.0.34.4
11264	**imp-null PP 10.0.2.2/32**		**Gi0/0**	**10.0.23.2**
11265	imp-null PP 10.0.4.4/32		Gi0/1	10.0.34.4

R3_P1#
R4_ASBR1#show   mpls   forwarding-table

Label Operation Code:
PH--PUSH label
PP--POP label
SW--SWAP label
SP--SWAP topmost label and push new label
DP--DROP packet
PC--POP label and continue lookup by IP or Label
PI--POP label and do ip lookup forward
PN--POP label and forward to nexthop
PM--POP label and do MAC lookup forward
PV--POP label and output to VC attach interface
IP--IP lookup forward
  s--stale

Local label	Outgoing label	OP FEC	Outgoing interface	Nexthop
--	11264	PH 10.0.2.2/32	Gi0/1	10.0.34.3
--	imp-null	PH 10.0.3.3/32	Gi0/1	10.0.34.3
--	60928	PH 10.0.7.7/32	Gi0/2	10.0.45.5
--	imp-null	PH 10.0.23.0/24	Gi0/1	10.0.34.3
--	imp-null	PH 10.0.45.5/32	Gi0/2	10.0.45.5
11264	**11264**	**SW 10.0.2.2/32**	**Gi0/1**	**10.0.34.3**
11265	imp-null	PP 10.0.3.3/32	Gi0/1	10.0.34.3
11266	imp-null	PP 10.0.23.0/24	Gi0/1	10.0.34.3
11267	60928	SW 10.0.7.7/32	Gi0/2	10.0.45.5
11268	imp-null	PP 10.0.45.5/32	Gi0/2	10.0.45.5
60928	11264	SW 10.0.2.2/32	Gi0/1	10.0.34.3
60929	60928	SW 10.0.7.7/32	Gi0/2	10.0.45.5

R4_ASBR1#

R5_ASBR2#show   mpls   forwarding-table

Label Operation Code:
PH--PUSH label
PP--POP label
SW--SWAP label
SP--SWAP topmost label and push new label
DP--DROP packet
PC--POP label and continue lookup by IP or Label
PI--POP label and do ip lookup forward
PN--POP label and forward to nexthop
PM--POP label and do MAC lookup forward
PV--POP label and output to VC attach interface
IP--IP lookup forward
  s--stale
  Local   Outgoing OP FEC   Outgoing   Nexthop

label	label		interface	
--	60928	PH 10.0.2.2/32	Gi0/2	10.0.45.4
--	imp-null	PH 10.0.6.6/32	Gi0/0	10.0.56.6
--	11265	PH 10.0.7.7/32	Gi0/0	10.0.56.6
--	imp-null	PH 10.0.45.4/32	Gi0/2	10.0.45.4
--	imp-null	PH 10.0.67.0/24	Gi0/0	10.0.56.6
11264	imp-null	PP 10.0.6.6/32	Gi0/0	10.0.56.6
11265	11265	SW 10.0.7.7/32	Gi0/0	10.0.56.6
11266	imp-null	PP 10.0.67.0/24	Gi0/0	10.0.56.6
11267	imp-null	PP 10.0.45.4/32	Gi0/2	10.0.45.4
11268	**60928**	**SW 10.0.2.2/32**	**Gi0/2**	**10.0.45.4**
60928	11265	SW 10.0.7.7/32	Gi0/0	10.0.56.6
60929	60928	SW 10.0.2.2/32	Gi0/2	10.0.45.4

R5_ASBR2#

R7_PE2#show mpls forwarding-table

Label Operation Code:
PH--PUSH label
PP--POP label
SW--SWAP label
SP--SWAP topmost label and push new label
DP--DROP packet
PC--POP label and continue lookup by IP or Label
PI--POP label and do ip lookup forward
PN--POP label and forward to nexthop
PM--POP label and do MAC lookup forward
PV--POP label and output to VC attach interface
IP--IP lookup forward
 s--stale

Local label	Outgoing label	OP FEC	Outgoing interface	Nexthop
--	60929	PH 10.0.2.2/32	Gi0/1	10.0.67.6
--	11264	PH 10.0.5.5/32	Gi0/1	10.0.67.6
--	imp-null	PH 10.0.6.6/32	Gi0/1	10.0.67.6
--	imp-null	PH 10.0.56.0/24	Gi0/1	10.0.67.6
11264	11264	SW 10.0.5.5/32	Gi0/1	10.0.67.6
11265	imp-null	PP 10.0.6.6/32	Gi0/1	10.0.67.6
11266	imp-null	PP 10.0.56.0/24	Gi0/1	10.0.67.6

R7_PE2#

9. 检查 R2_PE1 设备和 R7_PE2 设备之间 10.0.7.7/32 地址的连续标签建立情况。

R6_P2#show mpls forwarding-table

Label Operation Code:
PH--PUSH label
PP--POP label
SW--SWAP label
SP--SWAP topmost label and push new label
DP--DROP packet
PC--POP label and continue lookup by IP or Label

PI--POP label and do ip lookup forward
PN--POP label and forward to nexthop
PM--POP label and do MAC lookup forward
PV--POP label and output to VC attach interface
IP--IP lookup forward
　s--stale
　　Local　　Outgoing OP FEC　　　　　　　　Outgoing　　　　Nexthop
　　label　　label　　　　　　　　　　　　　　interface
　　--　　　imp-null PH 10.0.5.5/32　　　　　Gi0/0　　　　　10.0.56.5
　　--　　　imp-null PH 10.0.7.7/32　　　　　Gi0/1　　　　　10.0.67.7
　　11264　 imp-null PP 10.0.5.5/32　　　　　Gi0/0　　　　　10.0.56.5
　　11265　 **imp-null PP 10.0.7.7/32**　　　　　**Gi0/1**　　　　　**10.0.67.7**

R6_P2#

R5_ASBR2#show　mpls　forwarding-table

Label Operation Code:
PH--PUSH label
PP--POP label
SW--SWAP label
SP--SWAP topmost label and push new label
DP--DROP packet
PC--POP label and continue lookup by IP or Label
PI--POP label and do ip lookup forward
PN--POP label and forward to nexthop
PM--POP label and do MAC lookup forward
PV--POP label and output to VC attach interface
IP--IP lookup forward
　s--stale
　　Local　　Outgoing OP FEC　　　　　　　　Outgoing　　　　Nexthop
　　label　　label　　　　　　　　　　　　　　interface
　　--　　　60928　　PH 10.0.2.2/32　　　　　Gi0/2　　　　　10.0.45.4
　　--　　　imp-null PH 10.0.6.6/32　　　　　Gi0/0　　　　　10.0.56.6
　　--　　　11265　　PH 10.0.7.7/32　　　　　Gi0/0　　　　　10.0.56.6
　　--　　　imp-null PH 10.0.45.4/32　　　　 Gi0/2　　　　　10.0.45.4
　　--　　　imp-null PH 10.0.67.0/24　　　　 Gi0/0　　　　　10.0.56.6
　　11264　 imp-null PP 10.0.6.6/32　　　　　Gi0/0　　　　　10.0.56.6
　　11265　 11265　　SW 10.0.7.7/32　　　　　Gi0/0　　　　　10.0.56.6
　　11266　 imp-null PP 10.0.67.0/24　　　　 Gi0/0　　　　　10.0.56.6
　　11267　 imp-null PP 10.0.45.4/32　　　　 Gi0/2　　　　　10.0.45.4
　　11268　 60928　　SW 10.0.2.2/32　　　　　Gi0/2　　　　　10.0.45.4
　　60928　 **11265**　　**SW 10.0.7.7/32**　　　　　**Gi0/0**　　　　　**10.0.56.6**
　　60929　 60928　　SW 10.0.2.2/32　　　　　Gi0/2　　　　　10.0.45.4

R5_ASBR2#

R4_ASBR1#show　mpls　forwarding-table

Label Operation Code:
PH--PUSH label
PP--POP label
SW--SWAP label
SP--SWAP topmost label and push new label
DP--DROP packet
PC--POP label and continue lookup by IP or Label

PI--POP label and do ip lookup forward
PN--POP label and forward to nexthop
PM--POP label and do MAC lookup forward
PV--POP label and output to VC attach interface
IP--IP lookup forward
  s--stale

Local label	Outgoing label	OP	FEC	Outgoing interface	Nexthop
--	11264	PH	10.0.2.2/32	Gi0/1	10.0.34.3
--	imp-null	PH	10.0.3.3/32	Gi0/1	10.0.34.3
--	60928	PH	10.0.7.7/32	Gi0/2	10.0.45.5
--	imp-null	PH	10.0.23.0/24	Gi0/1	10.0.34.3
--	imp-null	PH	10.0.45.5/32	Gi0/2	10.0.45.5
11264	11264	SW	10.0.2.2/32	Gi0/1	10.0.34.3
11265	imp-null	PP	10.0.3.3/32	Gi0/1	10.0.34.3
11266	imp-null	PP	10.0.23.0/24	Gi0/1	10.0.34.3
11267	60928	SW	10.0.7.7/32	Gi0/2	10.0.45.5
11268	imp-null	PP	10.0.45.5/32	Gi0/2	10.0.45.5
60928	11264	SW	10.0.2.2/32	Gi0/1	10.0.34.3
60929	**60928**	**SW**	**10.0.7.7/32**	**Gi0/2**	**10.0.45.5**

R4_ASBR1#

R2_PE1#show mpls forwarding-table

Label Operation Code:
PH--PUSH label
PP--POP label
SW--SWAP label
SP--SWAP topmost label and push new label
DP--DROP packet
PC--POP label and continue lookup by IP or Label
PI--POP label and do ip lookup forward
PN--POP label and forward to nexthop
PM--POP label and do MAC lookup forward
PV--POP label and output to VC attach interface
IP--IP lookup forward
  s--stale

Local label	Outgoing label	OP	FEC	Outgoing interface	Nexthop
--	imp-null	PH	10.0.3.3/32	Gi0/0	10.0.23.3
--	11265	PH	10.0.4.4/32	Gi0/0	10.0.23.3
--	**60929**	**PH**	**10.0.7.7/32**	**Gi0/0**	**10.0.23.3**
--	imp-null	PH	10.0.34.0/24	Gi0/0	10.0.23.3
11264	imp-null	PP	10.0.3.3/32	Gi0/0	10.0.23.3
11265	imp-null	PP	10.0.34.0/24	Gi0/0	10.0.23.3
11266	11265	SW	10.0.4.4/32	Gi0/0	10.0.23.3

R2_PE1#

10. 测试 R2_PE1-R7_PE2 设备的 loopback 0 接口互通情况。

R2_PE1#ping   10.0.7.7 source   10.0.2.2
Sending 5, 100-byte ICMP Echoes to 10.0.7.7, timeout is 2 seconds:

```
         < press Ctrl+C to break >
!!!!!
Success rate is 100 percent (5/5), round-trip min/avg/max = 6/8/10 ms.
R2_PE1#


R7_PE2#ping   10.0.2.2 source   10.0.7.7
Sending 5, 100-byte ICMP Echoes to 10.0.2.2, timeout is 2 seconds:
   < press Ctrl+C to break >
!!!!!
Success rate is 100 percent (5/5), round-trip min/avg/max = 8/9/10 ms.
R7_PE2#
```

11. 检查 MP-EBGP 多跳邻居关系建立情况。

```
R2_PE1#show   bgp vpnv4 unicast   all   summary
For address family: VPNv4 Unicast
BGP router identifier 2.2.2.2, local AS number 65210
BGP table version is 1
2 BGP AS-PATH entries
0 BGP Community entries
0 BGP Prefix entries (Maximum-prefix:4294967295)


Neighbor        V    AS MsgRcvd MsgSent    TblVer   InQ OutQ Up/Down    State/PfxRcd
10.0.7.7        4    65211      4          3        1   0    0 00:01:44            0

Total number of neighbors 1, established neighbors 1
```

12. 检查 R2_PE1-R1_CE1 设备之间的私网路由传递情况。

```
R1_CE1#show   ip ospf   neighbor

OSPF process 100, 1 Neighbors, 1 is Full:
Neighbor ID     Pri   State            BFD State   Dead Time   Address        Interface
22.2.2.2        1     Full/DR          -           00:00:39    10.0.12.2      GigabitEthernet 0/2
R1_CE1#


R2_PE1#show   ip route   vrf   VPN1
Routing Table: VPN1

Codes:    C - Connected, L - Local, S - Static
          R - RIP, O - OSPF, B - BGP, I - IS-IS, V - Overflow route
          N1 - OSPF NSSA external type 1, N2 - OSPF NSSA external type 2
          E1 - OSPF external type 1, E2 - OSPF external type 2
          SU - IS-IS summary, L1 - IS-IS level-1, L2 - IS-IS level-2
          IA - Inter area, EV - BGP EVPN, A - Arp to host
          LA - Local aggregate route
          * - candidate default

Gateway of last resort is no set
C     10.0.12.0/24 is directly connected, GigabitEthernet 0/2
C     10.0.12.2/32 is local host.
O     192.168.1.1/32 [110/1] via 10.0.12.1, 00:03:08, GigabitEthernet 0/2
R2_PE1#


R2_PE1#ping    vrf VPN1 192.168.1.1
Sending 5, 100-byte ICMP Echoes to 192.168.1.1, timeout is 2 seconds:
```

```
    < press Ctrl+C to break >
!!!!!
Success rate is 100 percent (5/5), round-trip min/avg/max = 1/1/3 ms.
R2_PE1#
```

13. 检查 R7_PE2-R8_CE2 设备之间的私网路由传递情况。

```
R8_CE2#show   isis   neighbors

Area 10:
System Id     Type  IP Address      State   Holdtime  Circuit        Interface
R7_PE2        L2    10.0.78.7       Up      22        R8_CE2.01      GigabitEthernet 0/2

R8_CE2#

R7_PE2#show   ip route   vrf VPN1
Routing Table: VPN1

Codes:   C - Connected, L - Local, S - Static
         R - RIP, O - OSPF, B - BGP, I - IS-IS, V - Overflow route
         N1 - OSPF NSSA external type 1, N2 - OSPF NSSA external type 2
         E1 - OSPF external type 1, E2 - OSPF external type 2
         SU - IS-IS summary, L1 - IS-IS level-1, L2 - IS-IS level-2
         IA - Inter area, EV - BGP EVPN, A - Arp to host
         LA - Local aggregate route
         * - candidate default

Gateway of last resort is no set
C       10.0.78.0/24 is directly connected, GigabitEthernet 0/2
C       10.0.78.7/32 is local host.
I L2    192.168.2.0/24 [115/1] via 10.0.78.8, 00:02:45, GigabitEthernet 0/2

R7_PE2#ping   vrf   VPN1 192.168.2.1
Sending 5, 100-byte ICMP Echoes to 192.168.2.1, timeout is 2 seconds:
   < press Ctrl+C to break >
!!!!!
Success rate is 100 percent (5/5), round-trip min/avg/max = 1/4/14 ms.
R7_PE2#
```

14. 检查 MP-BGP 路由表。

```
R2_PE1#show   bgp   vpnv4 unicast   all
BGP table version is 1, local router ID is 2.2.2.2
Status codes: s suppressed, d damped, h history, * valid, > best, i - internal,
              S Stale, b - backup entry, m - multipath, f Filter, a additional-path
Origin codes: i - IGP, e - EGP, ? - incomplete

     Network           Next Hop          Metric     LocPrf     Weight Path
Route Distinguisher: 100:1 (Default for VRF VPN1)
 *>  10.0.12.0/24      0.0.0.0           1                     32768      ?
 *>  10.0.78.0/24      10.0.7.7          1                     0 65211 ?
 *>  192.168.1.1/32    10.0.12.1         1                     32768      ?
 *>  192.168.2.0       10.0.7.7          1                     0 65211 ?

Total number of prefixes 4
Route Distinguisher: 100:2
 *>  10.0.78.0/24      10.0.7.7          1                     0 65211 ?
```

```
    *>   192.168.2.0        10.0.7.7              1                             0 65211 ?

Total number of prefixes 2
R2_PE1#
```

```
R7_PE2#show bgp   vpnv4 unicast   all
BGP table version is 1, local router ID is 7.7.7.7
Status codes: s suppressed, d damped, h history, * valid, > best, i - internal,
              S Stale, b - backup entry, m - multipath, f Filter, a additional-path
Origin codes: i - IGP, e - EGP, ? - incomplete

     Network          Next Hop           Metric        LocPrf     Weight Path
Route Distinguisher: 100:2 (Default for VRF VPN1)
    *>   10.0.12.0/24      10.0.2.2              1                             0 65210 ?
    *>   10.0.78.0/24      0.0.0.0               1                             32768   ?
    *>   192.168.1.1/32    10.0.2.2              1                             0 65210 ?
    *>   192.168.2.0       10.0.78.8             1                             32768   ?

Total number of prefixes 4
Route Distinguisher: 100:1
    *>   10.0.12.0/24      10.0.2.2              1                             0 65210 ?
    *>   192.168.1.1/32    10.0.2.2              1                             0 65210 ?

Total number of prefixes 2
R7_PE2#
```

15. 检查 R1_CE1 设备和 R8_CE2 设备的全局 IP 路由表，接收到对方的私网路由。

```
R1_CE1#show   ip route

Codes:    C - Connected, L - Local, S - Static
          R - RIP, O - OSPF, B - BGP, I - IS-IS, V - Overflow route
          N1 - OSPF NSSA external type 1, N2 - OSPF NSSA external type 2
          E1 - OSPF external type 1, E2 - OSPF external type 2
          SU - IS-IS summary, L1 - IS-IS level-1, L2 - IS-IS level-2
          IA - Inter area, EV - BGP EVPN, A - Arp to host
          LA - Local aggregate route
          * - candidate default

Gateway of last resort is no set
C       10.0.12.0/24 is directly connected, GigabitEthernet 0/2
C       10.0.12.1/32 is local host.
O E2    10.0.78.0/24 [110/1] via 10.0.12.2, 00:03:07, GigabitEthernet 0/2
C       192.168.1.0/24 is directly connected, Loopback 0
C       192.168.1.1/32 is local host.
O E2    192.168.2.0/24 [110/1] via 10.0.12.2, 00:03:07, GigabitEthernet 0/2
R1_CE1#
```

```
R8_CE2#show   ip route

Codes:    C - Connected, L - Local, S - Static
          R - RIP, O - OSPF, B - BGP, I - IS-IS, V - Overflow route
          N1 - OSPF NSSA external type 1, N2 - OSPF NSSA external type 2
          E1 - OSPF external type 1, E2 - OSPF external type 2
```

```
              SU - IS-IS summary, L1 - IS-IS level-1, L2 - IS-IS level-2
              IA - Inter area, EV - BGP EVPN, A - Arp to host
              LA - Local aggregate route
              * - candidate default

Gateway of last resort is no set
I L2    10.0.12.0/24 [115/1] via 10.0.78.7, 00:03:10, GigabitEthernet 0/2
C       10.0.78.0/24 is directly connected, GigabitEthernet 0/2
C       10.0.78.8/32 is local host.
I L2    192.168.1.1/32 [115/1] via 10.0.78.7, 00:03:10, GigabitEthernet 0/2
C       192.168.2.0/24 is directly connected, Loopback 0
C       192.168.2.1/32 is local host.
R8_CE2#
```

16. 测试私网目标网段连通情况。

```
R1_CE1#ping    192.168.2.1 source    192.168.1.1
Sending 5, 100-byte ICMP Echoes to 192.168.2.1, timeout is 2 seconds:
   < press Ctrl+C to break >
!!!!!
Success rate is 100 percent (5/5), round-trip min/avg/max = 10/14/24 ms.
R1_CE1#


R8_CE2#ping    192.168.1.1 source    192.168.2.1
Sending 5, 100-byte ICMP Echoes to 192.168.1.1, timeout is 2 seconds:
   < press Ctrl+C to break >
!!!!!
Success rate is 100 percent (5/5), round-trip min/avg/max = 10/12/13 ms.
R8_CE2#
```

17. 在 PE 设备查看标签转发信息表，标签转发信息表包括公网标签信息和私网标签信息。

> 问题与思考

根据以上配置步骤，完成以下拓扑不同 VPN 客户端之间的相互通信。

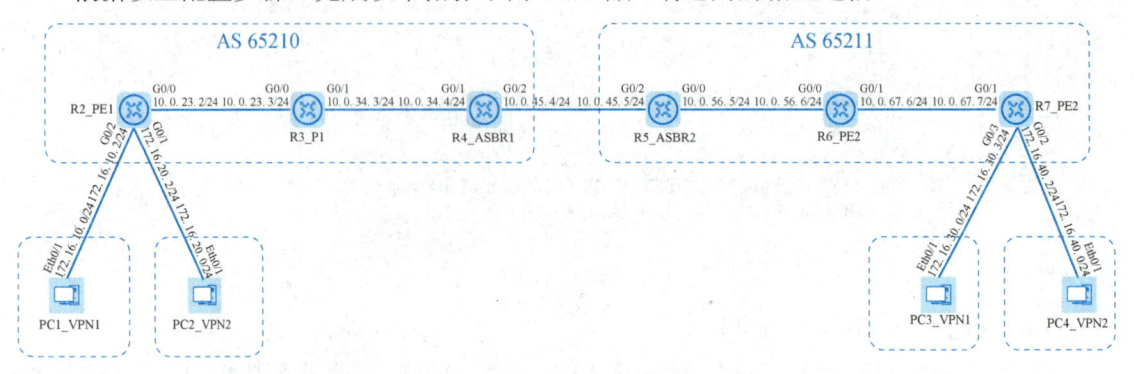

3.3.4 跨域 MPLS VPN-OptionC（2）

> 原理

方案二：在 EBGP 和 IBGP 邻居间都启用为 IPv4 交换标签的功能

方案一（只在 EBGP 邻居间启用为 IPv4 路由交换标签的功能的方案）要求其他自治域内的 IGP 和 LDP 协议要维护另外一个自治域的 PE 路由，即跨域 PE 的路由要扩散给另外一个自治域中的每台设备。实际应用中出于自治域的安全考虑，通常不希望到达另外一个自治域的 PE 路由扩散到本自治域中的每个设备，只需要 BGP 协议本身拥有这些路由，从而对自治域内的 IGP 和 LDP 透明，这可以通过在 EBGP 和 IBGP 邻居间开启为 IPv4 路由交换标签的能力的方案实现。

这种实现方案和前面的一种实现方案的主要区别是 ASBR 设备上的 IGP 协议不需要重分布 BGP 路由，LDP 也无须为 BGP 路由分配标签，仍然只负责本自治域的 LSP 建立，但是在 IBGP 邻居和 EBGP 邻居之间都要启用为 IPv4 路由交换标签的功能来完成跨域 LSP 的建立，并且要求 PE 设备支持连续压入三层标签。

> 任务拓扑

> 实施步骤

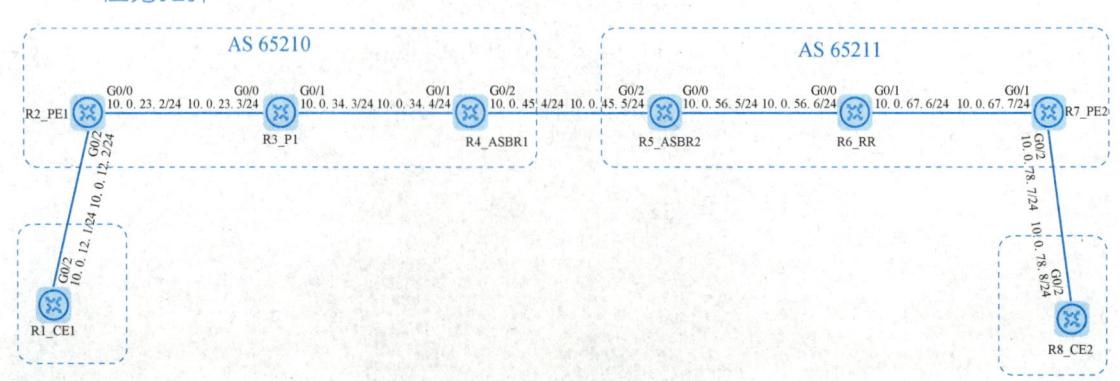

1. 根据任务拓扑配置各设备接口 IP 地址。（略）

2. 配置 AS 65210 内公网路由互通。(参考 3.3.1)
3. 配置 AS 65211 内公网路由互通。

```
R5_ASBR2(config)#router isis  1
R5_ASBR2(config-router)#net   49.0001.0000.0000.0005.00
R5_ASBR2(config-router)#is-type   level-2
R5_ASBR2(config-router)#exit
R5_ASBR2(config)#interface   gigabitEthernet 0/0
R5_ASBR2(config-if-GigabitEthernet 0/0)#ip router   isis  1
R5_ASBR2(config-if-GigabitEthernet 0/0)#exit
R5_ASBR2(config)#interface   Loopback 0
R5_ASBR2(config-if-Loopback 0)#ip router   isis  1
R5_ASBR2(config-if-Loopback 0)#exit
R5_ASBR2(config)#

R6_RR(config)#router  isis  1
R6_RR(config-router)#net   49.0001.0000.0000.0006.00
R6_RR(config-router)#is-type   level-2
R6_RR(config-router)#exit
R6_RR(config)#interface   gigabitEthernet 0/0
R6_RR(config-if-GigabitEthernet 0/0)#ip router   isis  1
R6_RR(config-if-GigabitEthernet 0/0)#exit
R6_RR(config)#interface   loopback 0
R6_RR(config-if-Loopback 0)#ip router   isis  1
R6_RR(config-if-Loopback 0)#exit
R6_RR(config)#interface   gigabitEthernet 0/1
R6_RR(config-if-GigabitEthernet 0/1)#ip router   isis  1
R6_RR(config-if-GigabitEthernet 0/1)#exit
R6_RR(config)#

R7_PE2(config)#router   isis  1
R7_PE2(config-router)#net   49.0001.0000.0000.0007.00
R7_PE2(config-router)#is-type   level-2
R7_PE2(config-router)#exit
R7_PE2(config)#interface   gigabitEthernet 0/1
R7_PE2(config-if-GigabitEthernet 0/1)#ip router   isis  1
R7_PE2(config-if-GigabitEthernet 0/1)#exit
R7_PE2(config)#interface   loopback 0
R7_PE2(config-if-Loopback 0)#ip router   isis  1
R7_PE2(config-if-Loopback 0)#exit
R7_PE2(config)#
```

4. 在公网 AS 65210 和 AS 65211 内配置 LDP 协议同时接口开启标签交换功能分配公网标签。(参考 3.3.1)

5. 在 R2_PE1 设备和 R7_PE2 设备创建 VRF,并且关联接口。(参考 3.3.3)

6. 在 R4_ASBR1 设备和 R5_ASBR2 设备之间建立 EBGP 邻居关系。(参考 3.3.3)

7. 在 R4_ASBR1 设备将 AS 65210 域内 R2_PE1 设备的 loopback 0 接口通过 EBGP 方式通告 R5_ASBR2 设备,R5_ASBR2 设备将此路由重分布到 AS 65211 域内路由器中。

```
R4_ASBR1(config)#router bgp   65210
R4_ASBR1(config-router)#network   10.0.2.2 mask   255.255.255.255
R4_ASBR1(config-router)#exit
R4_ASBR1(config)#
```

```
R5_ASBR2(config)#router isis   1
R5_ASBR2(config-router)#redistribute   bgp   level-2
R5_ASBR2(config-router)#exit
R5_ASBR2(config)#
```

8. 同上方式在 R5_ASBR2 设备将 RR 设备的 loopback 0 接口通告到 R4_ASBR1 设备，R4_ASBR1 设备将此路由重分布到 AS 65210 域内路由器中。

```
R5_ASBR2(config)#router bgp   65211
R5_ASBR2(config-router)#network    10.0.6.6 mask   255.255.255.255
R5_ASBR2(config-router)#exit
R5_ASBR2(config)#

R4_ASBR1(config)#router    ospf    10
R4_ASBR1(config-router)#redistribute    bgp    subnets
R4_ASBR1(config-router)#exit
R4_ASBR1(config)#
此时的 Loopback 0 接口都正常传递到对端设备
```

9. 在 R4_ASBR1-R5_ASBR2 设备为 10.0.2.2/32 和 10.0.6.6/32 的地址建立一条连续的标签（即 EBGP 启用标签交换），在 MPLS LDP 协议下为 BGP 路由分配标签，设备之间需要开启发送标签的功能。

```
R4_ASBR1(config)#mpls    router    ldp
R4_ASBR1(config-mpls-router)#advertise-labels    for    bgp-routes
R4_ASBR1(config-mpls-router)#exit
R4_ASBR1(config)#router    bgp    65210
R4_ASBR1(config-router)#neighbor    10.0.45.5 send-label    //向 EBGP 邻居开启发送标签功能
R4_ASBR1(config-router)#exit
R4_ASBR1(config)#

R4_ASBR1(config)#interface    gigabitEthernet 0/2
R4_ASBR1(config-if-GigabitEthernet 0/2)#label-switching        //接口需要开启标签交换能力
R4_ASBR1(config-if-GigabitEthernet 0/2)#exit
R4_ASBR1(config)#


R5_ASBR2(config)#mpls    router    ldp
R5_ASBR2(config-mpls-router)#advertise-labels for    bgp-routes
R5_ASBR2(config-mpls-router)#exit

R5_ASBR2(config)#router    bgp    65211
R5_ASBR2(config-router)#neighbor    10.0.45.4 send-label
R5_ASBR2(config-router)#exit
R5_ASBR2(config)#interface    gigabitEthernet 0/2
R5_ASBR2(config-if-GigabitEthernet 0/2)#label-switching
R5_ASBR2(config-if-GigabitEthernet 0/2)#exit
R5_ASBR2(config)#
```

10. R2_PE1-R6_RR 设备建立 MP-EBGP 多跳邻居，同时 R6_RR 设备与 R7_PE2 设备需要建立 MP-IBGP 传递私网路由（即 R7_PE2 设备是 R6_RR 设备的客户机）；无需使用 IPv4 单播 BGP 传递路由时可将 IPv4 单播 BGP 邻居关闭。

```
R2_PE1(config)#router    bgp    65210
R2_PE1(config-router)#bgp    router-id    2.2.2.2
```

```
R2_PE1(config-router)#neighbor    10.0.6.6 remote-as    65211
R2_PE1(config-router)#neighbor    10.0.6.6 update-source    loopback 0
R2_PE1(config-router)#neighbor    10.0.6.6 ebgp-multihop
R2_PE1(config-router)#address-family ipv4
R2_PE1(config-router-af)#no neighbor    10.0.6.6 activate        //关闭单播 BGP 邻居建立
R2_PE1(config-router-af)#exit
R2_PE1(config-router)#address-family vpnv4
R2_PE1(config-router-af)#neighbor    10.0.6.6 activate
R2_PE1(config-router-af)#exit
R2_PE1(config-router)#exit
R2_PE1(config)#

R6_RR(config)#router   bgp    65211
R6_RR(config-router)#bgp    router-id    6.6.6.6
R6_RR(config-router)#neighbor    10.0.2.2 remote-as    65210
R6_RR(config-router)#neighbor    10.0.2.2 update-source    loopback 0
R6_RR(config-router)#neighbor    10.0.2.2 ebgp-multihop
R6_RR(config-router)#address-family ipv4
R6_RR(config-router-af)#no neighbor    10.0.2.2 activate
R6_RR(config-router-af)#exit
R6_RR(config-router)#address-family vpnv4
R6_RR(config-router-af)#neighbor    10.0.2.2 activate
R6_RR(config-router-af)#exit

R6_RR(config-router)#neighbor    10.0.7.7 remote-as    65211
R6_RR(config-router)#neighbor    10.0.7.7 update-source    loopback 0
R6_RR(config-router)#address-family ipv4
R6_RR(config-router-af)#no neighbor    10.0.7.7 activate
R6_RR(config-router-af)#exit
R6_RR(config-router)#address-family vpnv4
R6_RR(config-router-af)#neighbor    10.0.7.7 activate
R6_RR(config-router-af)#neighbor    10.0.7.7 route-reflector-client    //指定 PE2 为客户机
R6_RR(config-router-af)#exit
R6_RR(config-router)#exit
R6_RR(config)

R7_PE2(config)#router    bgp    65211
R7_PE2(config-router)#bgp    router-id    7.7.7.7
R7_PE2(config-router)#neighbor    10.0.6.6 remote-as    65211
R7_PE2(config-router)#neighbor    10.0.6.6 update-source    loopback 0
R7_PE2(config-router-af)#no neighbor    10.0.6.6 activate
R7_PE2(config-router-af)#exit
R7_PE2(config-router)#address-family vpnv4
R7_PE2(config-router-af)#neighbor    10.0.6.6 activate
R7_PE2(config-router-af)#exit
R7_PE2(config-router)#exit
R7_PE2(config)#
```

11. 在 R2_PE1-R1_CE1 设备之间配置 IS-IS 10 传递私网路由，R1_CE1 设备创建 loopback 0 接口并通告。

```
R1_CE1(config)#router    isis    10
R1_CE1(config-router)#net    49.0002.0000.0000.0001.00
R1_CE1(config-router)#is-type    level-2
R1_CE1(config-router)#exit
R1_CE1(config)#interface loopback 0
```

```
R1_CE1(config-if-Loopback 0)#ip address   192.168.1.1 24
R1_CE1(config-if-Loopback 0)#ip router   isis   10
R1_CE1(config-if-Loopback 0)#exit
R1_CE1(config)#interface   gigabitEthernet 0/2
R1_CE1(config-if-GigabitEthernet 0/2)#ip router   isis   10
R1_CE1(config-if-GigabitEthernet 0/2)#exit
R1_CE1(config)#

R2_PE1(config)#router   isis   10
R2_PE1(config-router)#vrf   VPN1
R2_PE1(config-router)#net   49.0002.0000.0000.0002.00
R2_PE1(config-router)#is-type   level-2
R2_PE1(config-router)#exit
R2_PE1(config)#interface   gigabitEthernet 0/2
R2_PE1(config-if-GigabitEthernet 0/2)#ip router   isis   10
R2_PE1(config-if-GigabitEthernet 0/2)#exit
R2_PE1(config)#
```

12．在 R7_PE2-R8_CE2 设备之间配置 OSPF 100 传递私网路由，R8_CE2 设备创建 loopback 0 接口并通告。

```
R7_PE2(config)#router   ospf   100 vrf   VPN1
R7_PE2(config-router)#router-id   7.7.7.7
Change router-id and update OSPF process! [yes/no]:y
R7_PE2(config-router)#network   10.0.78.0 0.0.0.255 area   0
R7_PE2(config-router)#exit
R7_PE2(config)#

R8_CE2(config)#interface   loopback 0
R8_CE2(config-if-Loopback 0)#ip address   192.168.2.1 24
R8_CE2(config-if-Loopback 0)#exit
R8_CE2(config)#router   ospf   100
R8_CE2(config-router)#router-id   8.8.8.8
Change router-id and update OSPF process! [yes/no]:y
R8_CE2(config-router)#network   192.168.2.0 0.0.0.255 area   0
R8_CE2(config-router)#network   10.0.78.0 0.0.0.255 area   0
R8_CE2(config-router)#exit
R8_CE2(config)#
```

13．在 R2_PE1 设备和 R7_PE2 设备将 IGP 路由重分布到 MP-BGP 协议中。

```
R2_PE1(config)#router   bgp   65210
R2_PE1(config-router)#address-family ipv4 vrf VPN1
R2_PE1(config-router-af)#redistribute   isis   10 level-2
R2_PE1(config-router-af)#exit
R2_PE1(config-router)#

R7_PE2(config)#router   bgp   65211
R7_PE2(config-router)#address-family ipv4 vrf   VPN1
R7_PE2(config-router-af)#redistribute   ospf   100   match   internal   external
R7_PE2(config-router-af)#exit
R7_PE2(config-router)
```

14．在 R2_PE1 设备和 R7_PE2 设备将 MP-BGP 路由重分布到 IGP 协议中。

```
R2_PE1(config)#router   isis   10
R2_PE1(config-router)#redistribute   bgp   level-2
```

```
R2_PE1(config-router)#exit
R2_PE1(config)#


R7_PE2(config)#router   ospf   100
R7_PE2(config-router)#redistribute   bgp   subnets
R7_PE2(config-router)#exit
R7_PE2(config)#
```

> 任务验证

1. 检查公网 AS 65210 内部 IGP 协议邻居建立情况。

```
R3_P1#show   ip ospf   neighbor

OSPF process 10, 2 Neighbors, 2 is Full:
Neighbor ID      Pri    State        BFD State    Dead Time    Address      Interface
2.2.2.2           1     Full/BDR         -        00:00:31     10.0.23.2    GigabitEthernet 0/0
4.4.4.4           1     Full/DR          -        00:00:33     10.0.34.4    GigabitEthernet 0/1
R3_P1#
```

```
R2_PE1#show   ip route

Codes:  C - Connected, L - Local, S - Static
        R - RIP, O - OSPF, B - BGP, I - IS-IS, V - Overflow route
        N1 - OSPF NSSA external type 1, N2 - OSPF NSSA external type 2
        E1 - OSPF external type 1, E2 - OSPF external type 2
        SU - IS-IS summary, L1 - IS-IS level-1, L2 - IS-IS level-2
        IA - Inter area, EV - BGP EVPN, A - Arp to host
        LA - Local aggregate route
        * - candidate default

Gateway of last resort is no set
C    10.0.2.2/32 is local host.
O    10.0.3.3/32 [110/1] via 10.0.23.3, 00:04:57, GigabitEthernet 0/0
O    10.0.4.4/32 [110/2] via 10.0.23.3, 00:00:03, GigabitEthernet 0/0
C    10.0.12.0/24 is directly connected, GigabitEthernet 0/2
C    10.0.12.2/32 is local host.
C    10.0.23.0/24 is directly connected, GigabitEthernet 0/0
C    10.0.23.2/32 is local host.
O    10.0.34.0/24 [110/2] via 10.0.23.3, 00:04:57, GigabitEthernet 0/0
R2_PE1#

R2_PE1#ping   10.0.4.4
Sending 5, 100-byte ICMP Echoes to 10.0.4.4, timeout is 2 seconds:
  < press Ctrl+C to break >
!!!!!
Success rate is 100 percent (5/5), round-trip min/avg/max = 3/4/6 ms.
R2_PE1#
```

2. 检查公网 AS 65211 内部 IGP 协议邻居建立情况。

```
R6_RR#show   isis   neighbors
```

```
Area 1:
System Id          Type   IP Address      State    Holdtime   Circuit          Interface
R5_ASBR2           L2     10.0.56.5       Up       23         R6_RR.01         GigabitEthernet 0/0
R7_PE2             L2     10.0.67.7       Up       8          R7_PE2.01        GigabitEthernet 0/1

R6_RR#


R7_PE2#show ip route

Codes:   C - Connected, L - Local, S - Static
         R - RIP, O - OSPF, B - BGP, I - IS-IS, V - Overflow route
         N1 - OSPF NSSA external type 1, N2 - OSPF NSSA external type 2
         E1 - OSPF external type 1, E2 - OSPF external type 2
         SU - IS-IS summary, L1 - IS-IS level-1, L2 - IS-IS level-2
         IA - Inter area, EV - BGP EVPN, A - Arp to host
         LA - Local aggregate route
         * - candidate default

Gateway of last resort is no set
I L2   10.0.5.5/32 [115/2] via 10.0.67.6, 00:02:11, GigabitEthernet 0/1
I L2   10.0.6.6/32 [115/1] via 10.0.67.6, 00:02:20, GigabitEthernet 0/1
C      10.0.7.7/32 is local host.
I L2   10.0.56.0/24 [115/2] via 10.0.67.6, 00:02:20, GigabitEthernet 0/1
C      10.0.67.0/24 is directly connected, GigabitEthernet 0/1
C      10.0.67.7/32 is local host.
C      10.0.78.0/24 is directly connected, GigabitEthernet 0/2
C      10.0.78.7/32 is local host.
R7_PE2#

R7_PE2#ping    10.0.5.5
Sending 5, 100-byte ICMP Echoes to 10.0.5.5, timeout is 2 seconds:
   < press Ctrl+C to break >
!!!!!
Success rate is 100 percent (5/5), round-trip min/avg/max = 2/9/39 ms.
R7_PE2#
```

3. 检查公网 AS 65210 的 LDP 协议邻居建立情况及公网标签转发信息表。

```
R3_P1#show   mpls   ldp   neighbor
Default VRF:
    Peer LDP Ident: 10.0.2.2:0; Local LDP Ident: 10.0.3.3:0
        TCP connection: 10.0.2.2.646 - 10.0.3.3.40501
        State: OPERATIONAL; Msgs sent/recv: 28/30; UNSOLICITED
        Up time: 00:05:16
        Graceful Restart enabled; Peer reconnect time (msecs): 300000
        LDP discovery sources:
           Link Peer on GigabitEthernet 0/0, Src IP addr: 10.0.23.2
        Addresses bound to peer LDP Ident:
           10.0.12.2        10.0.23.2        10.0.2.2
    Peer LDP Ident: 10.0.4.4:0; Local LDP Ident: 10.0.3.3:0
        TCP connection: 10.0.4.4.35259 - 10.0.3.3.646
        State: OPERATIONAL; Msgs sent/recv: 26/28; UNSOLICITED
        Up time: 00:04:58
```

```
        Graceful Restart enabled; Peer reconnect time (msecs): 300000
        LDP discovery sources:
          Link Peer on GigabitEthernet 0/1, Src IP addr: 10.0.34.4
        Addresses bound to peer LDP Ident:
          10.0.4.4        10.0.34.4       10.0.45.4
R3_P1#


R3_P1#show   mpls   forwarding-table

Label Operation Code:
PH--PUSH label
PP--POP label
SW--SWAP label
SP--SWAP topmost label and push new label
DP--DROP packet
PC--POP label and continue lookup by IP or Label
PI--POP label and do ip lookup forward
PN--POP label and forward to nexthop
PM--POP label and do MAC lookup forward
PV--POP label and output to VC attach interface
IP--IP lookup forward
 s--stale
```

Local label	Outgoing label	OP	FEC	Outgoing interface	Nexthop
--	imp-null	PH	10.0.2.2/32	Gi0/0	10.0.23.2
--	imp-null	PH	10.0.4.4/32	Gi0/1	10.0.34.4
11264	imp-null	PP	10.0.2.2/32	Gi0/0	10.0.23.2
11265	imp-null	PP	10.0.4.4/32	Gi0/1	10.0.34.4

R3_P1#

4. 检查公网 AS 65211 的 LDP 协议邻居建立情况及公网标签转发信息表。

```
R6_RR#show   mpls   ldp   neighbor
Default VRF:
    Peer LDP Ident: 10.0.5.5:0; Local LDP Ident: 10.0.6.6:0
        TCP connection: 10.0.5.5.646 - 10.0.6.6.43937
        State: OPERATIONAL; Msgs sent/recv: 26/28; UNSOLICITED
        Up time: 00:04:47
        Graceful Restart enabled; Peer reconnect time (msecs): 300000
        LDP discovery sources:
          Link Peer on GigabitEthernet 0/0, Src IP addr: 10.0.56.5
        Addresses bound to peer LDP Ident:
          10.0.45.5       10.0.56.5       10.0.5.5
    Peer LDP Ident: 10.0.7.7:0; Local LDP Ident: 10.0.6.6:0
        TCP connection: 10.0.7.7.46091 - 10.0.6.6.646
        State: OPERATIONAL; Msgs sent/recv: 25/27; UNSOLICITED
        Up time: 00:04:31
        Graceful Restart enabled; Peer reconnect time (msecs): 300000
        LDP discovery sources:
          Link Peer on GigabitEthernet 0/1, Src IP addr: 10.0.67.7
        Addresses bound to peer LDP Ident:
          10.0.67.7       10.0.78.7       10.0.7.7
R6_RR#
```

```
R6_RR#show    mpls    forwarding-table

Label Operation Code:
PH--PUSH label
PP--POP label
SW--SWAP label
SP--SWAP topmost label and push new label
DP--DROP packet
PC--POP label and continue lookup by IP or Label
PI--POP label and do ip lookup forward
PN--POP label and forward to nexthop
PM--POP label and do MAC lookup forward
PV--POP label and output to VC attach interface
IP--IP lookup forward
  s--stale
  Local    Outgoing OP FEC              Outgoing         Nexthop
  label    label                        interface
  --       imp-null PH 10.0.5.5/32      Gi0/0            10.0.56.5
  --       imp-null PH 10.0.7.7/32      Gi0/1            10.0.67.7
  11264    imp-null PP 10.0.5.5/32      Gi0/0            10.0.56.5
  11265    imp-null PP 10.0.7.7/32      Gi0/1            10.0.67.7

R6_RR#
```

5. 在 AS 65211 域内查看 R2_PE1 设备的路由信息是否正常传递。

```
R7_PE2#show    ip route

Codes:   C - Connected, L - Local, S - Static
         R - RIP, O - OSPF, B - BGP, I - IS-IS, V - Overflow route
         N1 - OSPF NSSA external type 1, N2 - OSPF NSSA external type 2
         E1 - OSPF external type 1, E2 - OSPF external type 2
         SU - IS-IS summary, L1 - IS-IS level-1, L2 - IS-IS level-2
         IA - Inter area, EV - BGP EVPN, A - Arp to host
         LA - Local aggregate route
         * - candidate default

Gateway of last resort is no set
I L2     10.0.2.2/32 [115/2] via 10.0.67.6, 00:01:15, GigabitEthernet 0/1
I L2     10.0.5.5/32 [115/2] via 10.0.67.6, 00:14:50, GigabitEthernet 0/1
I L2     10.0.6.6/32 [115/1] via 10.0.67.6, 00:14:59, GigabitEthernet 0/1
C        10.0.7.7/32 is local host.
I L2     10.0.56.0/24 [115/2] via 10.0.67.6, 00:14:59, GigabitEthernet 0/1
C        10.0.67.0/24 is directly connected, GigabitEthernet 0/1
C        10.0.67.7/32 is local host.
R7_PE2#
```

6. 在 AS 65210 域内查看 R6_RR 设备的路由信息是否正常传递。

```
R2_PE1#show    ip route

Codes:   C - Connected, L - Local, S - Static
         R - RIP, O - OSPF, B - BGP, I - IS-IS, V - Overflow route
         N1 - OSPF NSSA external type 1, N2 - OSPF NSSA external type 2
         E1 - OSPF external type 1, E2 - OSPF external type 2
         SU - IS-IS summary, L1 - IS-IS level-1, L2 - IS-IS level-2
         IA - Inter area, EV - BGP EVPN, A - Arp to host
```

```
            LA - Local aggregate route
            * - candidate default

Gateway of last resort is no set
C       10.0.2.2/32 is local host.
O       10.0.3.3/32 [110/1] via 10.0.23.3, 00:20:01, GigabitEthernet 0/0
O       10.0.4.4/32 [110/2] via 10.0.23.3, 00:15:07, GigabitEthernet 0/0
O E2    10.0.6.6/32 [110/1] via 10.0.23.3, 00:01:17, GigabitEthernet 0/0
C       10.0.23.0/24 is directly connected, GigabitEthernet 0/0
C       10.0.23.2/32 is local host.
O       10.0.34.0/24 [110/2] via 10.0.23.3, 00:20:01, GigabitEthernet 0/0
R2_PE1#
```

7. 在 ASBR 设备开启标签交换后，查看是否建立完整连续的标签信息，根据各设备查询参数显示标签是否为连续一条标签路径（对 R6_RR 设备的标签信息亦是如此）。

```
R3_P1#show   mpls   forwarding-table

Label Operation Code:
PH--PUSH label
PP--POP label
SW--SWAP label
SP--SWAP topmost label and push new label
DP--DROP packet
PC--POP label and continue lookup by IP or Label
PI--POP label and do ip lookup forward
PN--POP label and forward to nexthop
PM--POP label and do MAC lookup forward
PV--POP label and output to VC attach interface
IP--IP lookup forward
  s--stale
  Local    Outgoing OP FEC                    Outgoing         Nexthop
  label    label                              interface
  --       imp-null PH 10.0.2.2/32            Gi0/0            10.0.23.2
  --       imp-null PH 10.0.4.4/32            Gi0/1            10.0.34.4
  --       11267    PH 10.0.6.6/32            Gi0/1            10.0.34.4
  11264    imp-null PP 10.0.2.2/32            Gi0/0            10.0.23.2
  11265    imp-null PP 10.0.4.4/32            Gi0/1            10.0.34.4
  11266    11267    SW 10.0.6.6/32            Gi0/1            10.0.34.4

R3_P1#

R4_ASBR1#show   mpls   forwarding-table

Label Operation Code:
PH--PUSH label
PP--POP label
SW--SWAP label
SP--SWAP topmost label and push new label
DP--DROP packet
PC--POP label and continue lookup by IP or Label
PI--POP label and do ip lookup forward
PN--POP label and forward to nexthop
PM--POP label and do MAC lookup forward
PV--POP label and output to VC attach interface
IP--IP lookup forward
```

s--stale

Local label	Outgoing label	OP	FEC	Outgoing interface	Nexthop
--	11264	PH	10.0.2.2/32	Gi0/1	10.0.34.3
--	imp-null	PH	10.0.3.3/32	Gi0/1	10.0.34.3
--	60928	PH	10.0.6.6/32	Gi0/2	10.0.45.5
--	imp-null	PH	10.0.23.0/24	Gi0/1	10.0.34.3
--	imp-null	PH	10.0.45.5/32	Gi0/2	10.0.45.5
11264	11264	SW	10.0.2.2/32	Gi0/1	10.0.34.3
11265	imp-null	PP	10.0.3.3/32	Gi0/1	10.0.34.3
11266	imp-null	PP	10.0.23.0/24	Gi0/1	10.0.34.3
11267	60928	SW	10.0.6.6/32	Gi0/2	10.0.45.5
11268	imp-null	PP	10.0.45.5/32	Gi0/2	10.0.45.5
60928	**11264**	**SW**	**10.0.2.2/32**	**Gi0/1**	**10.0.34.3**
60929	60928	SW	10.0.6.6/32	Gi0/2	10.0.45.5

R4_ASBR1#

R5_ASBR2#show mpls forwarding-table

Label Operation Code:
PH--PUSH label
PP--POP label
SW--SWAP label
SP--SWAP topmost label and push new label
DP--DROP packet
PC--POP label and continue lookup by IP or Label
PI--POP label and do ip lookup forward
PN--POP label and forward to nexthop
PM--POP label and do MAC lookup forward
PV--POP label and output to VC attach interface
IP--IP lookup forward

s--stale

Local label	Outgoing label	OP	FEC	Outgoing interface	Nexthop
--	60928	PH	10.0.2.2/32	Gi0/2	10.0.45.4
--	imp-null	PH	10.0.6.6/32	Gi0/0	10.0.56.6
--	11265	PH	10.0.7.7/32	Gi0/0	10.0.56.6
--	imp-null	PH	10.0.45.4/32	Gi0/2	10.0.45.4
--	imp-null	PH	10.0.67.0/24	Gi0/0	10.0.56.6
11264	imp-null	PP	10.0.6.6/32	Gi0/0	10.0.56.6
11265	11265	SW	10.0.7.7/32	Gi0/0	10.0.56.6
11266	imp-null	PP	10.0.67.0/24	Gi0/0	10.0.56.6
11267	60928	SW	10.0.2.2/32	Gi0/2	10.0.45.4
11268	imp-null	PP	10.0.45.4/32	Gi0/2	10.0.45.4
60928	imp-null	PP	10.0.6.6/32	Gi0/0	10.0.56.6
60929	**60928**	**SW**	**10.0.2.2/32**	**Gi0/2**	**10.0.45.4**

R5_ASBR2#

R6_RR#show mpls forwarding-table

Label Operation Code:
PH--PUSH label

PP--POP label
SW--SWAP label
SP--SWAP topmost label and push new label
DP--DROP packet
PC--POP label and continue lookup by IP or Label
PI--POP label and do ip lookup forward
PN--POP label and forward to nexthop
PM--POP label and do MAC lookup forward
PV--POP label and output to VC attach interface
IP--IP lookup forward
s--stale

Local label	Outgoing label	OP FEC	Outgoing interface	Nexthop
--	11267	PH 10.0.2.2/32	Gi0/0	10.0.56.5
--	imp-null	PH 10.0.5.5/32	Gi0/0	10.0.56.5
--	imp-null	PH 10.0.7.7/32	Gi0/1	10.0.67.7
11264	imp-null	PP 10.0.5.5/32	Gi0/0	10.0.56.5
11265	imp-null	PP 10.0.7.7/32	Gi0/1	10.0.67.7
11266	**11267**	**SW 10.0.2.2/32**	**Gi0/0**	**10.0.56.5**

R6_RR#

8. 通过 Wireshark 抓包工具检测标签，携带一层标签信息通信。

```
R2_PE1#ping    10.0.6.6 source    10.0.2.2
Sending 5, 100-byte ICMP Echoes to 10.0.6.6, timeout is 2 seconds:
  < press Ctrl+C to break >
!!!!!
Success rate is 100 percent (5/5), round-trip min/avg/max = 5/6/8 ms.
R2_PE1#
```

No.	Time	Source	Destination	Protocol	Length	Info
→	8 3.1164...	10.0.2.2	10.0.6.6	ICMP	146	Echo (ping) request id=0x0300, seq=28333/44398, ttl=64 (reply in 9)
	9 3.1191...	10.0.6.6	10.0.2.2	ICMP	146	Echo (ping) reply id=0x0300, seq=28333/44398, ttl=64 (request in 8)
	10 3.2236...	10.0.2.2	10.0.6.6	ICMP	146	Echo (ping) request id=0x0300, seq=31149/44409, ttl=64 (reply in 11)
	11 3.2258...	10.0.6.6	10.0.2.2	ICMP	146	Echo (ping) reply id=0x0300, seq=31149/44409, ttl=64 (request in 10)
	12 3.3299...	10.0.2.2	10.0.6.6	ICMP	146	Echo (ping) request id=0x0300, seq=33965/44420, ttl=64 (reply in 13)

```
Frame 8: 146 bytes on wire (1168 bits), 146 bytes captured (1168 bits) on interface 0
Ethernet II, Src: 50:00:00:04:00:03 (50:00:00:04:00:03), Dst: 50:00:00:05:00:03 (50:00:00:05:00:03)
MultiProtocol Label Switching Header, Label: 60928, Exp: 0, S: 1, TTL: 62
   0000 1110 1110 0000 0000 .... .... .... = MPLS Label: 60928
   .... .... .... .... .... 000. .... .... = MPLS Experimental Bits: 0
   .... .... .... .... .... ...1 .... .... = MPLS Bottom Of Label Stack: 1
   .... .... .... .... .... .... 0011 1110 = MPLS TTL: 62
Internet Protocol Version 4, Src: 10.0.2.2, Dst: 10.0.6.6
Internet Control Message Protocol
```

9. 检查 MP-EBGP 多跳邻居关系建立及 MP-IBGP 邻居关系建立情况。

```
R2_PE1#show    bgp vpnv4 unicast    all    summary
For address family: VPNv4 Unicast
BGP router identifier 2.2.2.2, local AS number 65210
BGP table version is 1
0 BGP AS-PATH entries
0 BGP Community entries
0 BGP Prefix entries (Maximum-prefix:4294967295)

Neighbor        V    AS MsgRcvd MsgSent    TblVer  InQ OutQ Up/Down    State/PfxRcd
10.0.6.6        4 65211      13      11         1    0    0 00:09:37          0

Total number of neighbors 1, established neighbors 1
```

R2_PE1#

R6_RR#show   bgp   vpnv4 unicast   all   summary
For address family: VPNv4 Unicast
BGP router identifier 6.6.6.6, local AS number 65211
BGP table version is 1
0 BGP AS-PATH entries
0 BGP Community entries
0 BGP Prefix entries (Maximum-prefix:4294967295)

Neighbor	V	AS	MsgRcvd	MsgSent	TblVer	InQ	OutQ	Up/Down	State/PfxRcd
10.0.2.2	4	65210	13	12	1	0	0	00:10:11	0
10.0.7.7	4	65211	8	7	1	0	0	00:05:37	0

Total number of neighbors 2, established neighbors 2

R6_RR#

10. 检查 R2_PE1-R1_CE1 设备之间的私网路由传递情况。

R2_PE1#show   ip route   vrf   VPN1
Routing Table: VPN1

Codes: C - Connected, L - Local, S - Static
         R - RIP, O - OSPF, B - BGP, I - IS-IS, V - Overflow route
         N1 - OSPF NSSA external type 1, N2 - OSPF NSSA external type 2
         E1 - OSPF external type 1, E2 - OSPF external type 2
         SU - IS-IS summary, L1 - IS-IS level-1, L2 - IS-IS level-2
         IA - Inter area, EV - BGP EVPN, A - Arp to host
         LA - Local aggregate route
         * - candidate default

Gateway of last resort is no set
C   10.0.12.0/24 is directly connected, GigabitEthernet 0/2
C   10.0.12.2/32 is local host.
I L2   192.168.1.0/24 [115/1] via 10.0.12.1, 00:05:44, GigabitEthernet 0/2
R2_PE1#

R2_PE1#ping   vrf   VPN1 192.168.1.1
Sending 5, 100-byte ICMP Echoes to 192.168.1.1, timeout is 2 seconds:
  < press Ctrl+C to break >
!!!!!
Success rate is 100 percent (5/5), round-trip min/avg/max = 2/4/14 ms.
R2_PE1#

11. 检查 R7_PE2-R8_CE2 设备之间的私网路由传递情况。

R7_PE2#show   ip route vrf   VPN1
Routing Table: VPN1

Codes: C - Connected, L - Local, S - Static
         R - RIP, O - OSPF, B - BGP, I - IS-IS, V - Overflow route
         N1 - OSPF NSSA external type 1, N2 - OSPF NSSA external type 2
         E1 - OSPF external type 1, E2 - OSPF external type 2
         SU - IS-IS summary, L1 - IS-IS level-1, L2 - IS-IS level-2
         IA - Inter area, EV - BGP EVPN, A - Arp to host

```
              LA - Local aggregate route
              * - candidate default

Gateway of last resort is no set
C        10.0.78.0/24 is directly connected, GigabitEthernet 0/2
C        10.0.78.7/32 is local host.
O        192.168.2.1/32 [110/1] via 10.0.78.8, 00:04:04, GigabitEthernet 0/2
R7_PE2#


R7_PE2#ping   vrf   VPN1 192.168.2.1
Sending 5, 100-byte ICMP Echoes to 192.168.2.1, timeout is 2 seconds:
   < press Ctrl+C to break >
!!!!!
Success rate is 100 percent (5/5), round-trip min/avg/max = 1/2/3 ms.
R7_PE2#
```

12. 检查 MP-BGP 路由表，R6_RR 设备将接收到的 MP-BGP 路由反射给客户端和非客户端。

```
R2_PE1#show   bgp   vpnv4 unicast   all
BGP table version is 1, local router ID is 2.2.2.2
Status codes: s suppressed, d damped, h history, * valid, > best, i - internal,
              S Stale, b - backup entry, m - multipath, f Filter, a additional-path
Origin codes: i - IGP, e - EGP, ? - incomplete

    Network          Next Hop           Metric      LocPrf      Weight Path
Route Distinguisher: 10:1 (Default for VRF VPN1)
 *>  10.0.12.0/24     0.0.0.0            1                      32768    ?
 *>  10.0.78.0/24     10.0.6.6           0                        0 65211 ?
 *>  192.168.1.0      10.0.12.1          1                      32768    ?
 *>  192.168.2.1/32   10.0.6.6           0                        0 65211 ?

Total number of prefixes 4
Route Distinguisher: 20:1
 *>  10.0.78.0/24     10.0.6.6           0                        0 65211 ?
 *>  192.168.2.1/32   10.0.6.6           0                        0 65211 ?

Total number of prefixes 2
R2_PE1#



R6_RR#show   bgp   vpnv4 unicast   all
BGP table version is 5, local router ID is 6.6.6.6
Status codes: s suppressed, d damped, h history, * valid, > best, i - internal,
              S Stale, b - backup entry, m - multipath, f Filter, a additional-path
Origin codes: i - IGP, e - EGP, ? - incomplete

    Network          Next Hop           Metric      LocPrf      Weight Path
Route Distinguisher: 10:1
 *>  10.0.12.0/24     10.0.2.2           1                        0 65210 ?
 *>  192.168.1.0      10.0.2.2           1                        0 65210 ?

Total number of prefixes 2
Route Distinguisher: 20:1
 *>i 10.0.78.0/24     10.0.7.7           1           100          0       ?
 *>i 192.168.2.1/32   10.0.7.7           1           100          0       ?
```

```
Total number of prefixes 2
R6_RR#


R7_PE2#show    bgp    vpnv4 unicast    all
BGP table version is 1, local router ID is 7.7.7.7
Status codes: s suppressed, d damped, h history, * valid, > best, i - internal,
              S Stale, b - backup entry, m - multipath, f Filter, a additional-path
Origin codes: i - IGP, e - EGP, ? - incomplete

     Network          Next Hop           Metric        LocPrf      Weight Path
Route Distinguisher: 20:1 (Default for VRF VPN1)
*>i 10.0.12.0/24      10.0.2.2           1             100         0 65210 ?
*>  10.0.78.0/24      0.0.0.0            1                         32768    ?
*>i 192.168.1.0       10.0.2.2           1             100         0 65210 ?
*>  192.168.2.1/32    10.0.78.8          1                         32768    ?

Total number of prefixes 4
Route Distinguisher: 10:1
*>i 10.0.12.0/24      10.0.2.2           1             100         0 65210 ?
*>i 192.168.1.0       10.0.2.2           1             100         0 65210 ?

Total number of prefixes 2
R7_PE2#
```

13. 检查 R1_CE1 设备和 R8_CE2 设备的全局 IP 路由表，接收到对方的私网路由。

```
R1_CE1#show ip route

Codes:    C - Connected, L - Local, S - Static
          R - RIP, O - OSPF, B - BGP, I - IS-IS, V - Overflow route
          N1 - OSPF NSSA external type 1, N2 - OSPF NSSA external type 2
          E1 - OSPF external type 1, E2 - OSPF external type 2
          SU - IS-IS summary, L1 - IS-IS level-1, L2 - IS-IS level-2
          IA - Inter area, EV - BGP EVPN, A - Arp to host
          LA - Local aggregate route
          * - candidate default

Gateway of last resort is no set
C        10.0.12.0/24 is directly connected, GigabitEthernet 0/2
C        10.0.12.1/32 is local host.
I L2     10.0.78.0/24 [115/1] via 10.0.12.2, 00:03:44, GigabitEthernet 0/2
C        192.168.1.0/24 is directly connected, Loopback 0
C        192.168.1.1/32 is local host.
I L2     192.168.2.1/32 [115/1] via 10.0.12.2, 00:03:44, GigabitEthernet 0/2


R8_CE2#show ip route

Codes:    C - Connected, L - Local, S - Static
          R - RIP, O - OSPF, B - BGP, I - IS-IS, V - Overflow route
          N1 - OSPF NSSA external type 1, N2 - OSPF NSSA external type 2
          E1 - OSPF external type 1, E2 - OSPF external type 2
          SU - IS-IS summary, L1 - IS-IS level-1, L2 - IS-IS level-2
          IA - Inter area, EV - BGP EVPN, A - Arp to host
          LA - Local aggregate route
```

```
                * - candidate default

Gateway of last resort is no set
O E2    10.0.12.0/24 [110/1] via 10.0.78.7, 00:02:21, GigabitEthernet 0/2
C       10.0.78.0/24 is directly connected, GigabitEthernet 0/2
C       10.0.78.8/32 is local host.
O E2    192.168.1.0/24 [110/1] via 10.0.78.7, 00:05:50, GigabitEthernet 0/2
C       192.168.2.0/24 is directly connected, Loopback 0
C       192.168.2.1/32 is local host.
R8_CE2#
```

14. 测试私网目标网段连通情况。

```
R1_CE1#ping    192.168.2.1 source    192.168.1.1
Sending 5, 100-byte ICMP Echoes to 192.168.2.1, timeout is 2 seconds:
   < press Ctrl+C to break >
!!!!!


R8_CE2#ping    192.168.1.1 source    192.168.2.1
Sending 5, 100-byte ICMP Echoes to 192.168.1.1, timeout is 2 seconds:
   < press Ctrl+C to break >
!!!!!
Success rate is 100 percent (5/5), round-trip min/avg/max = 10/12/19 ms.
R8_CE2#
```

15. 在 PE 设备查看标签转发信息表，标签转发信息表包括公网标签信息和私网标签信息。

> 问题与思考

根据以上配置步骤，完成以下拓扑不同 VPN 客户端之间的相互通信。

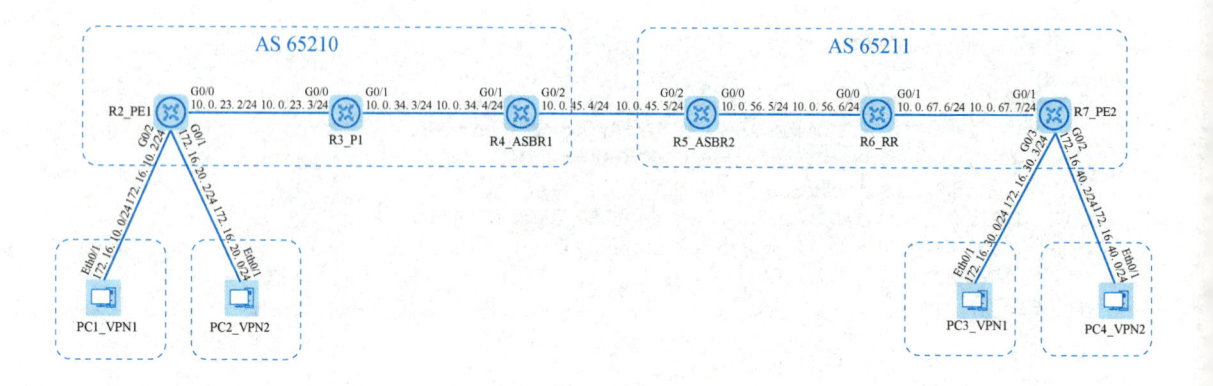

第 4 章　GRE Over IPSec VPN 协议

4.1　GRE 隧道

> 原理

1. GRE 概述

GRE（Generic Routing Encapsulation）即通用路由封装协议，GRE VPN 能够对某些网络层协议（如 IP 和 IPX）的数据报文进行封装，使这些被封装的数据报文能够在另一个网络层协议（如 IP）中传输，从而解决报文跨越异种网络进行传输的问题。

异种报文传输的通道称为 Tunnel，即"隧道"，它是一个虚拟的点对点连接，提供了一条通路使封装的数据报文能够在这个通路上传输，请求在一个 Tunnel 的两端分别对数据报文进行封装和解封装。

2. GRE 优缺点

优点：支持多种协议和组播：支持 IP 协议、IPX 协议、AppleTalk 协议等；支持组播，例如 GRE 封装 OSPF Hello 报文建立邻居关系；支持点到点 VPN 或者点到多点 VPN。

缺点：没有加密机制，需要配置其他协议以达到加密效果，如 GRE Over IPSec；缺乏标准协议监控 GRE 隧道，无法随时监控隧道是否建立成功，例如没有 keepalive 这种检测死连接的机制。

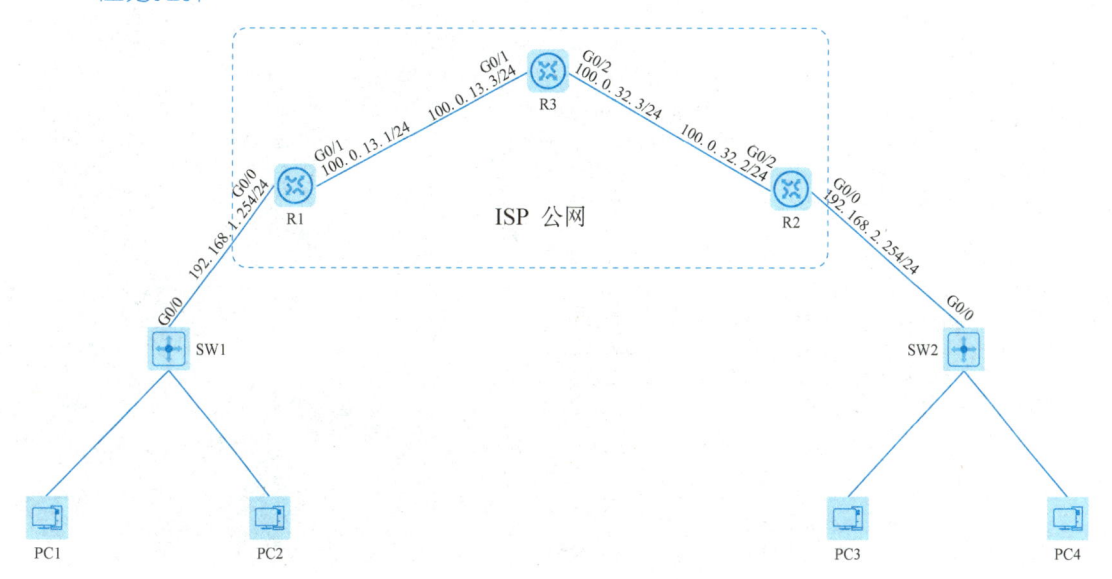

> 任务拓扑

> 实施步骤

1. 根据任务拓扑配置各设备接口 IP 地址（略）、PC 终端地址。

```
VPCS> set pcname PC1

PC1> ip 192.168.1.1 24 192.168.1.254
Checking for duplicate address...
PC1 : 192.168.1.1 255.255.255.0 gateway 192.168.1.254

PC1>

VPCS> set pcname PC2

PC2> ip 192.168.1.2 24 192.168.1.254
Checking for duplicate address...
PC2 : 192.168.1.2 255.255.255.0 gateway 192.168.1.254

PC2>

VPCS> set pcname PC3

PC3> ip 192.168.2.1 24 192.168.2.254
Checking for duplicate address...
PC3 : 192.168.2.1 255.255.255.0 gateway 192.168.2.254

PC3>

VPCS> set pcname PC4

PC4> ip 192.168.2.2 24 192.168.2.254
Checking for duplicate address...
PC4 : 192.168.2.2 255.255.255.0 gateway 192.168.2.254

PC4>
```

2. 在公网配置 OSPF 路由协议互通。

```
R1(config)#router   ospf   10
R1(config-router)#router-id   1.1.1.1
Change router-id and update OSPF process! [yes/no]:y
R1(config-router)#network    100.0.13.0 0.0.0.255 area    0
R1(config-router)#exit
R1(config)#

R3(config)#router ospf 10
R3(config-router)#router 3.3.3.3
Change router-id and update OSPF process! [yes/no]:y
R3(config-router)#network    100.0.13.0 0.0.0.255 area    0
R3(config-router)#network    100.0.32.0 0.0.0.255 area    0
R3(config-router)#exit
R3(config)#

R2(config)#route ospf    10
R2(config-router)#router-id    2.2.2.2
Change router-id and update OSPF process! [yes/no]:y
```

```
R2(config-router)#network    100.0.32.0 0.0.0.255 area    0
R2(config-router)#exit
R2(config)#
```

3. 在 R1 设备与 R2 设备配置 GRE 隧道，给 GRE 隧道接口配置 IP 地址。

```
R1(config)#interface    tunnel 1
R1(config-if-Tunnel 1)#tunnel    mode    gre    ip
R1(config-if-Tunnel 1)#tunnel    source    100.0.13.1
R1(config-if-Tunnel 1)#tunnel    destination    100.0.32.2
R1(config-if-Tunnel 1)#ip address    172.16.10.1 24
R1(config-if-Tunnel 1)#exit
R1(config)#

R2(config)#interface    tunnel    1
R2(config-if-Tunnel 1)#tunnel    mode    gre    ip
R2(config-if-Tunnel 1)#tunnel    source    100.0.32.2
R2(config-if-Tunnel 1)#tunnel    destination    100.0.13.1
R2(config-if-Tunnel 1)#ip address    172.16.10.2 24
R2(config-if-Tunnel 1)#exit
R2(config)#
```

4. 配置静态路由，选择 GRE 隧道接口作为出接口。

```
R1(config)#ip route 192.168.2.0 255.255.255.0 tunnel    1

R2(config)#ip route    192.168.1.0 255.255.255.0 tunnel    1
```

➢ 任务验证

1. 检查公网路由协议互通情况。

```
R3#show    ip ospf    neighbor

OSPF process 10, 2 Neighbors, 2 is Full:
Neighbor ID    Pri    State    BFD State    Dead Time    Address    Interface
1.1.1.1    1    Full/BDR    -    00:00:37    100.0.13.1    GigabitEthernet 0/1
2.2.2.2    1    Full/BDR    -    00:00:34    100.0.32.2    GigabitEthernet 0/2
R3#

R1#ping    100.0.32.2
Sending 5, 100-byte ICMP Echoes to 100.0.32.2, timeout is 2 seconds:
  < press Ctrl+C to break >
!!!!!
Success rate is 100 percent (5/5), round-trip min/avg/max = 2/3/4 ms.
R1#
```

2. 检查 GRE 隧道接口建立情况。

```
R1#show    interfaces    tunnel 1
Index(dec):11 (hex):b
Tunnel 1 is UP    , line protocol is UP
   Hardware is Tunnel
   Interface address is: 172.16.10.1/24
   Interface IPv6 address is:
      No IPv6 address
   MTU 1476 bytes, BW 9 Kbit
```

        Encapsulation protocol is Tunnel, loopback not set
        Keepalive interval is 10 sec ,retries 0.
        Carrier delay is 2 sec
    Tunnel attributes:
        Tunnel source 100.0.13.1, destination 100.0.32.2, routable
        Tunnel TOS/Traffic Class not set ,Tunnel TTL 254
        Tunnel config nested limit is 4, current nested number is 0
        Tunnel protocol/transport is gre ip
        Tunnel transport VPN is no set
            Key disabled, Sequencing disabled
            Checksumming of packets disabled
        RX packets
         Drop reason(Down: 0, Checksum error: 0, sequence error: 0, routing: 0)
        TX packets
         Drop reason(Too big: 0, Payload Type error: 0, Nested-limit: 0)
         Rxload is 1/255, Txload is 1/255
         Input peak rate: 604 bits/sec, at 2023-07-14 07:41:01
         Output peak rate: 604 bits/sec, at 2023-07-14 07:41:01
           10 seconds input rate 0 bits/sec, 0 packets/sec
           10 seconds output rate 0 bits/sec, 0 packets/sec
             13 packets input, 1536 bytes, 0 no buffer, 0 dropped
             Received 0 broadcasts, 0 runts, 0 giants
             0 input errors, 0 CRC, 0 frame, 0 overrun, 0 abort
             15 packets output, 1840 bytes, 0 underruns, 0 no buffer, 0 dropped
             0 output errors, 0 collisions, 0 interface resets
R1#

R2#show   interfaces   tunnel 1
Index(dec):11 (hex):b
Tunnel 1 is UP   , line protocol is UP
    Hardware is Tunnel
    Interface address is: 172.16.10.2/24
    Interface IPv6 address is:
        No IPv6 address
    MTU 1476 bytes, BW 9 Kbit
        Encapsulation protocol is Tunnel, loopback not set
        Keepalive interval is 10 sec ,retries 0.
        Carrier delay is 2 sec
    Tunnel attributes:
        Tunnel source 100.0.32.2, destination 100.0.13.1, routable
        Tunnel TOS/Traffic Class not set ,Tunnel TTL 254
        Tunnel config nested limit is 4, current nested number is 0
        Tunnel protocol/transport is gre ip
        Tunnel transport VPN is no set
            Key disabled, Sequencing disabled
            Checksumming of packets disabled
        RX packets
         Drop reason(Down: 0, Checksum error: 0, sequence error: 0, routing: 0)
        TX packets
         Drop reason(Too big: 0, Payload Type error: 0, Nested-limit: 0)
         Rxload is 1/255, Txload is 1/255
         Input peak rate: 474 bits/sec, at 2023-07-14 07:41:04
         Output peak rate: 474 bits/sec, at 2023-07-14 07:41:04
           10 seconds input rate 0 bits/sec, 0 packets/sec
           10 seconds output rate 0 bits/sec, 0 packets/sec
             13 packets input, 1536 bytes, 0 no buffer, 0 dropped

```
   Received 0 broadcasts, 0 runts, 0 giants
   0 input errors, 0 CRC, 0 frame, 0 overrun, 0 abort
   13 packets output, 1536 bytes, 0 underruns, 0 no buffer, 0 dropped
   0 output errors, 0 collisions, 0 interface resets
R2#
```

3. 总部的 PC 和分部 PC 测试连通，针对总部和分部的用户通信选择 GRE 隧道接口封装转发。

```
PC1> ping 192.168.2.1

84 bytes from 192.168.2.1 icmp_seq=1 ttl=62 time=20.625 ms
84 bytes from 192.168.2.1 icmp_seq=2 ttl=62 time=7.703 ms
84 bytes from 192.168.2.1 icmp_seq=3 ttl=62 time=7.855 ms
84 bytes from 192.168.2.1 icmp_seq=4 ttl=62 time=10.291 ms
84 bytes from 192.168.2.1 icmp_seq=5 ttl=62 time=7.857 ms
```

```
 2 2.713915      192.168.1.1    192.168.2.1    ICMP    122 Echo (ping) request  id=0xf285, seq=1/256, ttl=63 (reply in 3)
 3 2.719095      192.168.1.1    192.168.1.1    ICMP    122 Echo (ping) reply    id=0xf285, seq=1/256, ttl=63 (request in 2)
 4 3.722038      192.168.1.1    192.168.2.1    ICMP    122 Echo (ping) request  id=0xf385, seq=2/512, ttl=63 (reply in 5)
 5 3.726202      192.168.2.1    192.168.1.1    ICMP    122 Echo (ping) reply    id=0xf385, seq=2/512, ttl=63 (request in 4)
 6 4.738876      192.168.1.1    192.168.2.1    ICMP    122 Echo (ping) request  id=0xf485, seq=3/768, ttl=63 (reply in 7)
▶ Frame 2: 122 bytes on wire (976 bits), 122 bytes captured (976 bits) on interface 0
▶ Ethernet II, Src: 50:00:00:02:00:02 (50:00:00:02:00:02), Dst: 50:00:00:03:00:02 (50:00:00:03:00:02)
▶ Internet Protocol Version 4, Src: 100.0.13.1, Dst: 100.0.32.2
▼ Generic Routing Encapsulation (IP)
  ▶ Flags and Version: 0x0000
    Protocol Type: IP (0x0800)
▼ Internet Protocol Version 4, Src: 192.168.1.1, Dst: 192.168.2.1
    0100 .... = Version: 4
    .... 0101 = Header Length: 20 bytes (5)
  ▶ Differentiated Services Field: 0x00 (DSCP: CS0, ECN: Not-ECT)
    Total Length: 84
    Identification: 0x85f2 (34290)
  ▶ Flags: 0x00
    Fragment offset: 0
    Time to live: 63
    Protocol: ICMP (1)
    Header checksum: 0x7164 [validation disabled]
    [Header checksum status: Unverified]
    Source: 192.168.1.1
    Destination: 192.168.2.1
    [Source GeoIP: Unknown]
    [Destination GeoIP: Unknown]
▶ Internet Control Message Protocol
```

```
PC1> trace 192.168.2.1
trace to 192.168.2.1, 8 hops max, press Ctrl+C to stop
 1   192.168.1.254    2.876 ms    1.466 ms    1.971 ms
 2   192.168.1.254    1.249 ms    0.630 ms    4294967.218 ms
 3   172.16.10.2      3.948 ms    2.097 ms    0.482 ms
 4   *192.168.2.1     1.478 ms (ICMP type:3, code:3, Destination port unreachable)

PC1>
```

➢ 问题与思考

1. GRE 报文封装格式是什么？
2. GRE 隧道如何执行端到端验证？

4.2　GRE Over IPSec VPN 隧道

➢ 原理

1. IPSec VPN 概述

IPSec（Internet Protocol Security）VPN 是指采用 IPSec 协议来实现远程接入的一种 VPN

技术。它能够在公网上为两个私有网络提供安全通信通道，通过加密通道来保证传输数据的安全。

IPSec 是一种开放标准的框架结构，特定的通信方之间在 IP 层通过加密和数据摘要（Hash）等手段，来保证数据包在 Internet 网上传输时的私密性、完整性和真实性。

IPSec 只能在 IP 层工作，要求乘客协议和承载协议都是 IP 协议。

IPSec 安全协议描述了如何利用加密和数据摘要（Hash）来保护数据安全，主要包括 AH 和 ESP。

AH（Authentication Header）

只能进行数据摘要（Hash），不能实现数据加密，能够保证数据的完整性和真实性。

ESP（Encapsulating Security Payload）

能够进行数据加密和数据摘要（Hash），能够保证数据的机密性、完整性和真实性。

IPSec 支持两种封装模式：传输模式和隧道模式

传输模式：不改变原有的 IP 包头，通常用于主机与主机之间。

隧道模式：增加新的 IP 包头，通常用于私网与私网之间。

2．SA 安全关联

SA（Security Association，安全关联）是指两个 IPSec 实体之间经过协商建立起来的一种协定，内容包括：采用何种 IPSec 协议（AH 还是 ESP）、运行模式（传输模式还是隧道模式）、验证算法、加密算法、加密密钥、密钥生存期、抗重放窗口、计数器等，从而决定了保护什么、如何保护，以及谁来保护。可以说 SA 是构成 IPSec 的基础。

SA 是单向的，入方向（inbound）SA 负责处理接收到的数据包，出方向（outbound）SA 负责处理要发送的数据包。因此每个通信方必须要有两个 SA，一个入方向 SA，一个出方向 SA，这两个 SA 构成了一个 SA 束（SA Bundle）。

3．GRE Over IPSec 封装过程

GRE（Generic Routing Encapsulation）是一种通用的路由封装协议，用于在网络中封装和传输不同协议的数据包。IPSec（Internet Protocol Security）是一种网络层安全协议，用于对 IP 数据包进行加密和认证。

当使用 GRE Over IPSec 时，GRE 被用作在 IPSec 隧道中传输其他协议的封装协议。这种配置允许在 IPSec 加密和认证的安全隧道中传输 GRE 封装的数据包。

GRE Over IPSec 封装过程

在源主机上，待发送的数据包首先被封装为一个 GRE 的数据包，GRE 数据包被进一步封装为一个 IPSec 数据包。IPSec 对 IPSec 数据包进行加密和认证，产生一个加密的 IP 数据包，加密的 IP 数据包被封装为一个新的 IP 数据包，并添加 IPSec 头部。最终的 IP 数据包通过公网发送到目标主机以加密的形式传输。

在目标主机上，接收到的数据包将按照相反的顺序进行解封装和解密，接收到的数据包首先被解析为一个 IPSec 数据包，IPSec 对 IPSec 数据包进行解密和认证，还原为加密前的 IP 数据包，加密前的 IP 数据包被解析为一个 GRE 数据包，GRE 数据包中的原始数据被提取出来并交给目标主机解析出来。这样 GRE Over IPSec 实现了在 IPSec 隧道中传输 GRE 封装的数据包，同时提供了加密和认证的安全性保护，确保数据在公网上的安全传输。

> 任务拓扑

> 实施步骤

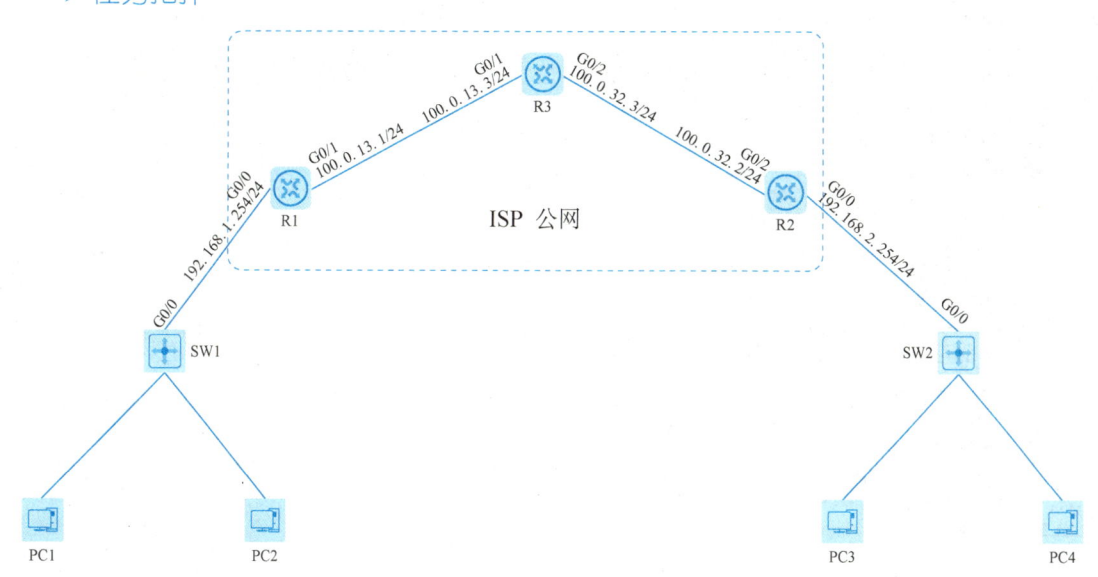

1. 根据任务拓扑配置各设备接口 IP 地址。（略）
2. 在公网配置 OSPF 路由协议互通。

```
R1(config)#router    ospf    10
R1(config-router)#router-id    1.1.1.1
Change router-id and update OSPF process! [yes/no]:y
R1(config-router)#network    100.0.13.0 0.0.0.255 area    0
R1(config-router)#exit
R1(config)#

R3(config)#router ospf 10
R3(config-router)#router 3.3.3.3
Change router-id and update OSPF process! [yes/no]:y
R3(config-router)#network    100.0.13.0 0.0.0.255 area    0
R3(config-router)#network    100.0.32.0 0.0.0.255 area    0
R3(config-router)#exit
R3(config)#

R2(config)#route ospf    10
R2(config-router)#router-id    2.2.2.2
Change router-id and update OSPF process! [yes/no]:y
R2(config-router)#network    100.0.32.0 0.0.0.255 area    0
R2(config-router)#exit
R2(config)#
```

3. 在 R1 设备与 R2 设备配置 GRE 隧道，给 GRE 隧道接口配置 IP 地址。

```
R1(config)#interface    tunnel 1
R1(config-if-Tunnel 1)#tunnel    source    100.0.13.1
R1(config-if-Tunnel 1)#tunnel    destination    100.0.32.2
R1(config-if-Tunnel 1)#ip address    172.16.10.1 24
R1(config-if-Tunnel 1)#exit
R1(config)#
```

```
R2(config)#interface  tunnel   1
R2(config-if-Tunnel 1)#tunnel   source   100.0.32.2
R2(config-if-Tunnel 1)#tunnel   destination   100.0.13.1
R2(config-if-Tunnel 1)#ip address   172.16.10.2 24
R2(config-if-Tunnel 1)#exit
R2(config)#
```

4. 在 R1 设备与 R2 设备配置 IPSec 隧道。

```
R1(config)#crypto   isakmp   enable
R1(config)#crypto   isakmp policy   1
R1(isakmp-policy)#encryption 3des
R1(isakmp-policy)#authentication   pre-share
R1(isakmp-policy)#hash md5
R1(isakmp-policy)#group 2
R1(isakmp-policy)#exit
R1(config)#crypto   isakmp   key   0 ruijie address 100.0.32.2
R1(config)#crypto   ipsec transform-set myset esp-3des esp-md5-hmac
R1(cfg-crypto-trans)#mode transport
R1(cfg-crypto-trans)#exit
R1(config)#ip access-list extended 199
R1(config-ext-nacl)#permit   gre   host   100.0.13.1   host   100.0.32.2
R1(config-ext-nacl)#exit
R1(config)#crypto   map   mymap 1 ipsec-isakmp
R1(config-crypto-map)#set   peer 100.0.32.2
R1(config-crypto-map)#set transform-set myset
R1(config-crypto-map)#match address   199
R1(config-crypto-map)#exit
R1(config)#

R1(config)#interface   gigabitEthernet 0/1
R1(config-if-GigabitEthernet 0/1)#crypto   map   mymap
R1(config-if-GigabitEthernet 0/1)#exit
R1(config)#


R2(config)#crypto   isakmp enable
R2(config)#crypto   isakmp policy   1
R2(isakmp-policy)#encryption 3des
R2(isakmp-policy)#authentication   pre-share
R2(isakmp-policy)#hash md5
R2(isakmp-policy)#group 2
R2(isakmp-policy)#exit
R2(config)#crypto   isakmp key 0 ruijie address 100.0.13.1
R2(config)#crypto   ipsec transform-set myset esp-3des esp-md5-hmac
R2(cfg-crypto-trans)#mode transport
R2(cfg-crypto-trans)#exit
R2(config)#ip access-list extended 199
R2(config-ext-nacl)#permit   gre host   100.0.32.2 host   100.0.13.1
R2(config-ext-nacl)#exit
R2(config)#crypto   map mymap 1 ipsec-isakmp
R2(config-crypto-map)#set peer 100.0.13.1
R2(config-crypto-map)#set transform-set myset
R2(config-crypto-map)#match   address   199
R2(config-crypto-map)#exit
R2(config)#
```

```
R2(config)#interface    gigabitEthernet 0/2
R2(config-if-GigabitEthernet 0/2)#crypto    map mymap
R2(config-if-GigabitEthernet 0/2)#exit
R2(config)#
```

5. 配置静态路由，选择 GRE 隧道接口作为出接口。

```
R1(config)#ip route 192.168.2.0 255.255.255.0 tunnel    1

R2(config)#ip route   192.168.1.0 255.255.255.0 tunnel    1
```

➢ 任务验证

1. 检查公网路由协议互通情况。

```
R3#show    ip ospf    neighbor

OSPF process 10, 2 Neighbors, 2 is Full:
Neighbor ID     Pri    State       BFD State    Dead Time    Address        Interface
1.1.1.1         1      Full/BDR    -            00:00:37     100.0.13.1     GigabitEthernet 0/1
2.2.2.2         1      Full/BDR    -            00:00:34     100.0.32.2     GigabitEthernet 0/2
R3#


R1#ping    100.0.32.2
Sending 5, 100-byte ICMP Echoes to 100.0.32.2, timeout is 2 seconds:
  < press Ctrl+C to break >
!!!!!
Success rate is 100 percent (5/5), round-trip min/avg/max = 2/3/4 ms.
R1#
```

2. 检查 GRE 隧道接口建立情况。

```
R1#show    interfaces    tunnel 1
Index(dec):11 (hex):b
Tunnel 1 is UP    , line protocol is UP
    Hardware is Tunnel
    Interface address is: 172.16.10.1/24
    Interface IPv6 address is:
       No IPv6 address
    MTU 1476 bytes, BW 9 Kbit
    Encapsulation protocol is Tunnel, loopback not set
    Keepalive interval is 10 sec ,retries 0.
    Carrier delay is 2 sec
Tunnel attributes:
    Tunnel source 100.0.13.1, destination 100.0.32.2, routable
    Tunnel TOS/Traffic Class not set ,Tunnel TTL 254
    Tunnel config nested limit is 4, current nested number is 0
    Tunnel protocol/transport is gre ip
    Tunnel transport VPN is no set
       Key disabled, Sequencing disabled
       Checksumming of packets disabled
  RX packets
    Drop reason(Down: 0, Checksum error: 0, sequence error: 0, routing: 0)
  TX packets
    Drop reason(Too big: 0, Payload Type error: 0, Nested-limit: 0)
    Rxload is 1/255, Txload is 1/255
```

```
    Input peak rate: 604 bits/sec, at 2023-07-14 07:41:01
    Output peak rate: 604 bits/sec, at 2023-07-14 07:41:01
      10 seconds input rate 0 bits/sec, 0 packets/sec
      10 seconds output rate 0 bits/sec, 0 packets/sec
      13 packets input, 1536 bytes, 0 no buffer, 0 dropped
      Received 0 broadcasts, 0 runts, 0 giants
      0 input errors, 0 CRC, 0 frame, 0 overrun, 0 abort
      15 packets output, 1840 bytes, 0 underruns, 0 no buffer, 0 dropped
      0 output errors, 0 collisions, 0 interface resets
R1#

R2#show    interfaces    tunnel 1
Index(dec):11 (hex):b
Tunnel 1 is UP    , line protocol is UP
    Hardware is Tunnel
    Interface address is: 172.16.10.2/24
    Interface IPv6 address is:
       No IPv6 address
    MTU 1476 bytes, BW 9 Kbit
    Encapsulation protocol is Tunnel, loopback not set
    Keepalive interval is 10 sec ,retries 0.
    Carrier delay is 2 sec
  Tunnel attributes:
    Tunnel source 100.0.32.2, destination 100.0.13.1, routable
    Tunnel TOS/Traffic Class not set ,Tunnel TTL 254
    Tunnel config nested limit is 4, current nested number is 0
    Tunnel protocol/transport is gre ip
    Tunnel transport VPN is no set
       Key disabled, Sequencing disabled
       Checksumming of packets disabled
  RX packets
    Drop reason(Down: 0, Checksum error: 0, sequence error: 0, routing: 0)
  TX packets
    Drop reason(Too big: 0, Payload Type error: 0, Nested-limit: 0)
    Rxload is 1/255, Txload is 1/255
    Input peak rate: 474 bits/sec, at 2023-07-14 07:41:04
    Output peak rate: 474 bits/sec, at 2023-07-14 07:41:04
      10 seconds input rate 0 bits/sec, 0 packets/sec
      10 seconds output rate 0 bits/sec, 0 packets/sec
      13 packets input, 1536 bytes, 0 no buffer, 0 dropped
      Received 0 broadcasts, 0 runts, 0 giants
      0 input errors, 0 CRC, 0 frame, 0 overrun, 0 abort
      13 packets output, 1536 bytes, 0 underruns, 0 no buffer, 0 dropped
      0 output errors, 0 collisions, 0 interface resets
R2#
```

3．检查 IPSec 建立情况。

```
R1#show    crypto ipsec    transform-set
transform set myset: { esp-md5-hmac,esp-3des,}
          will negotiate = {Transport,}
R1#

R1#show    crypto    isakmp sa
destination       source          state             conn-id         lifetime(second)
100.0.32.2      100.0.13.1        IKE_IDLE            1                84257
```

```
R1#

R1#show crypto    ipsec sa
      Crypto map tag:mymap
local ipv4 addr 100.0.13.1
media mtu 1500


         ==================================
         sub_map type:static, seqno:1, id=1
local    ident (addr/mask/prot/port): (100.0.13.1/0.0.0.0/47/0))
remote   ident (addr/mask/prot/port): (100.0.32.2/0.0.0.0/47/0))
PERMIT
#pkts encaps: 13, #pkts encrypt: 13, #pkts digest 13
#pkts decaps: 13, #pkts decrypt: 13, #pkts verify 13
#send errors 0, #recv errors 0
pkts encaps errors:
      #negoitate pkt drop: 0, #sab useless: 0, encap data fail: 0, compute hash fail: 0
pkts decypto errors:
      #check reply wind fail: 0, #compute hash fail: 0, verify hash fail: 0
#pkts detect send req: 0, recv reply: 0, recv req: 0, send reply: 0


          Inbound esp sas:
                spi:0x9c95cc97 (2627062935)
                  transform: esp-3des esp-md5-hmac
                  in use settings={Transport Encaps,}
                  crypto map mymap 1
                  sa timing: remaining key lifetime (k/sec): (4606996/1467)
                  IV size: 0 bytes
                  Replay detection support:Y


          Outbound esp sas:
                spi:0x574084f3 (1463846131)
                  transform: esp-3des esp-md5-hmac
                  in use settings={Transport Encaps,}
                  crypto map mymap 1
                  sa timing: remaining key lifetime (k/sec): (4606996/1467)
                  IV size: 0 bytes
                  Replay detection support:Y




R2#show   crypto   isakmp    policy
Protection suite of priority 1
      encryption algorithm:    Three key triple DES.
      hash algorithm:          Message Digest 5
      authentication method:   Pre-Shared Key
      Diffie-Hellman group:    #2 (1024 bit)
      lifetime:                86400 seconds
Default protection suite
      encryption algorithm:    DES - Data Encryption Standard (56 bit keys).
      hash algorithm:          Secure Hash Standard
      authentication method:   Pre-Shared Key
      Diffie-Hellman group:    #1 (768 bit)
      lifetime:                86400 seconds
R2#
```

4. 总部的 PC 和分部 PC 测试连通，针对总部和分部的用户通信选择隧道接口，同时对数据进行加密处理。

```
PC1> ping 192.168.2.1

84 bytes from 192.168.2.1 icmp_seq=1 ttl=62 time=20.625 ms
84 bytes from 192.168.2.1 icmp_seq=2 ttl=62 time=7.703 ms
84 bytes from 192.168.2.1 icmp_seq=3 ttl=62 time=7.855 ms
84 bytes from 192.168.2.1 icmp_seq=4 ttl=62 time=10.291 ms
84 bytes from 192.168.2.1 icmp_seq=5 ttl=62 time=7.857 ms
```

No.	Time	Source	Destination	Protocol	Length	Info
817	2693.360326	100.0.32.2	100.0.13.1	ESP	158	ESP (SPI=0x9c95cc97)
818	2694.364213	100.0.13.1	100.0.32.2	ESP	158	ESP (SPI=0x574084f3)
819	2694.368856	100.0.32.2	100.0.13.1	ESP	158	ESP (SPI=0x9c95cc97)
820	2695.373422	100.0.13.1	100.0.32.2	ESP	158	ESP (SPI=0x574084f3)
821	2695.379530	100.0.32.2	100.0.13.1	ESP	158	ESP (SPI=0x9c95cc97)

> Frame 817: 158 bytes on wire (1264 bits), 158 bytes captured (1264 bits) on interface 0
> Ethernet II, Src: 50:00:00:03:00:02 (50:00:00:03:00:02), Dst: 50:00:00:02:00:02 (50:00:00:02:00:02)
> Internet Protocol Version 4, Src: 100.0.32.2, Dst: 100.0.13.1
> Encapsulating Security Payload
  ESP SPI: 0x9c95cc97 (2627062935)
  ESP Sequence: 20

```
PC1> trace 192.168.2.1
trace to 192.168.2.1, 8 hops max, press Ctrl+C to stop
  1   192.168.1.254     2.876 ms   1.466 ms   1.971 ms
  2   192.168.1.254     1.249 ms   0.630 ms   4294967.218 ms
  3   172.16.10.2       3.948 ms   2.097 ms   0.482 ms
  4   *192.168.2.1      1.478 ms (ICMP type:3, code:3, Destination port unreachable)

PC1>
```

➤ 问题与思考

1. GRE Over IPSec VPN 与 IPSec VPN Over GRE 配置的区别？
2. IPSec VPN Over GRE 报文封装格式是什么？

第 5 章　IPv6 协议

5.1　IPv6 路由协议

5.1.1　IPv6 静态路由协议

> 原理

静态路由（Static Route）是指通过手动方式为路由器配置路由信息，可以简单地让路由器获知到达目标网络的路由。

静态路由具有配置简单、路由器资源负载小、可控性强等优点。缺点是不能动态反映网络拓扑，当网络拓扑发生变化时，网络管理员就必须手动配置改变路由表，因此静态路由不适合大型网络。

在静态路由中存在一种目的地/掩码为 "::/0" 的路由，称为默认路由（Default Route）。计算机或路由器的 IP 路由表中可能存在默认路由，也可能不存在默认路由。如果网络设备的 IP 路由表中存在默认路由，那么当一个待发送或待转发的 IP 报文不能匹配 IP 路由表中的任何非默认路由时，就会根据默认路由来进行发送或转发；如果网络设备的 IP 路由表中不存在默认路由，那么当一个待发送或待转发的 IP 报文不能匹配 IP 路由表中的任何路由时，该 IP 报文就会被直接丢弃。

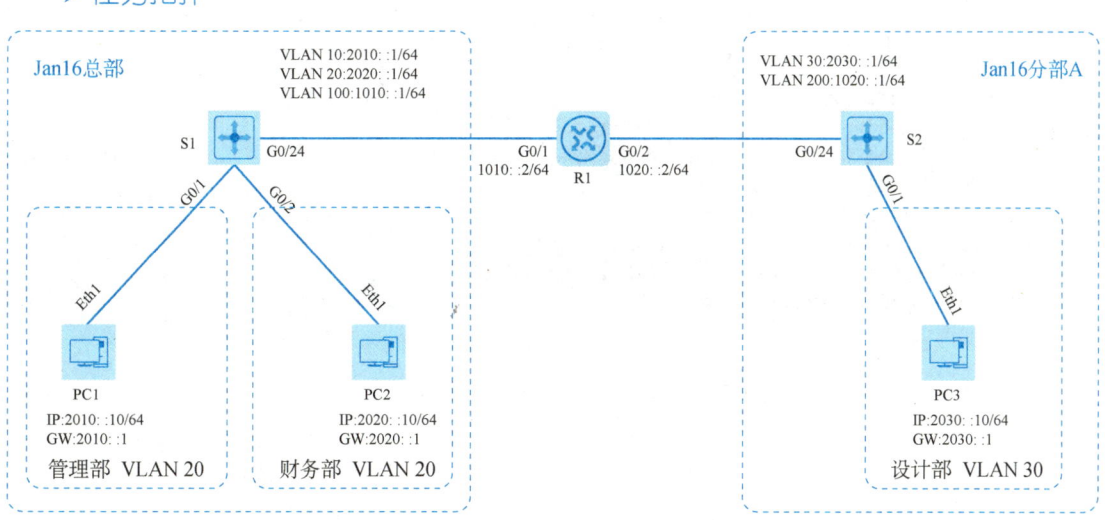

> 任务拓扑

> 实施步骤

1. 根据任务拓扑在 S1 创建部门 VLAN 10、VLAN 20 及互联 VLAN 100。

```
Ruijie>enable
Ruijie#configure terminal
Ruijie(config)#hostname S1
S1(config)#vlan 10
S1(config-vlan)#vlan 20
S1(config-vlan)#vlan 100
S1(config-vlan)#exit
S1(config)#
```

2. 根据任务拓扑在 S2 创建部门 VLAN 30 及互联 VLAN 200。

```
Ruijie>enable
Ruijie#configure terminal
Ruijie(config)#hostname S2
S2(config)#vlan 30
S2(config-vlan)#vlan 200
S2(config-vlan)#exit
S2(config)#
```

3. 根据任务拓扑在 S1 划分 VLAN,并将对应端口添加到部门 VLAN 中。

```
S1(config)#interface gigabitEthernet 0/1
S1(config-if-GigabitEthernet 0/1)#switchport mode access
S1(config-if-GigabitEthernet 0/1)#switchport access vlan 10
S1(config-if-GigabitEthernet 0/1)#eixt
S1(config)#interface gigabitEthernet 0/2
S1(config-if-GigabitEthernet 0/2)#switchport mode access
S1(config-if-GigabitEthernet 0/1)#switchport access vlan 20
S1(config-if-GigabitEthernet 0/1)#eixt
S1(config)#interface gigabitEthernet 0/24
S1(config-if-GigabitEthernet 0/24)#switchport mode access
S1(config-if-GigabitEthernet 0/24)#switchport access vlan 100
S1(config-if-GigabitEthernet 0/24)#eixt
S1(config)#
```

4. 根据任务拓扑在 S2 划分 VLAN,并将对应端口划分到部门 VLAN 中。

```
S2(config)#interface gigabitEthernet 0/1
S2(config-if-GigabitEthernet 0/1)#switchport mode access
S2(config-if-GigabitEthernet 0/1)#switchport access vlan 30
S2(config-if-GigabitEthernet 0/1)#eixt
S2(config)#interface gigabitEthernet 0/24
S2(config-if-GigabitEthernet 0/24)#switchport mode access
S2(config-if-GigabitEthernet 0/24)#switchport access vlan 200
S2(config-if-GigabitEthernet 0/24)#eixt
S2(config)#
```

5. 根据任务拓扑配置各部门 PC 的 IPv6 地址。

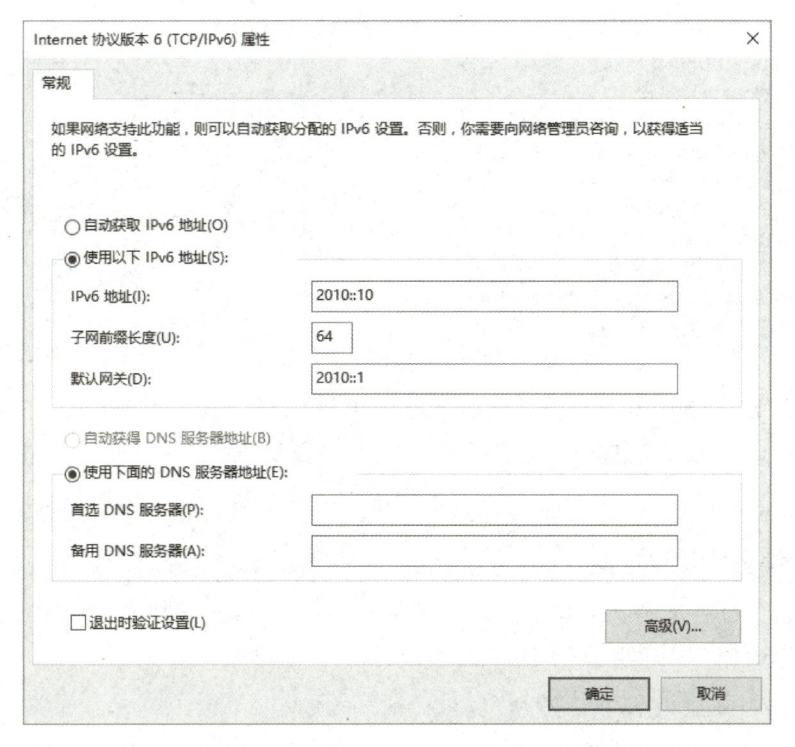

6. 在 S1 为各部门 VLAN 创建 VLANIF 接口并配置 IPv6 地址，作为部门的网关；为互联 VLAN 创建 VLANIF 接口并配置 IPv6 地址，作为与 R1 互联的地址。

```
S1(config)# interface vlan 10
S1(config-if-VLAN 10)#ipv6 enable
S1(config-if-VLAN 10)#ipv6 address 2010::1/64
S1(config-if-VLAN 10)#eixt
S1(config)# interface vlan 20
S1(config-if-VLAN 20)#ipv6 enable
S1(config-if-VLAN 20)#ipv6 address 2020::1/64
S1(config-if-VLAN 20)#eixt
S1(config)# interface vlan 100
S1(config-if-VLAN 100)#ipv6 enable
S1(config-if-VLAN 100)#ipv6 address 1010::1/64
S1(config-if-VLAN 100)#eixt
```

7. 在 S2 为部门 VLAN 创建 VLANIF 接口并配置 IPv6 地址，作为部门的网关；为互联 VLAN 创建 VLANIF 接口并配置 IPv6 地址，作为与 R1 互联的地址。

```
S2(config)# interface vlan 30
S2(config-if-VLAN 30)#ipv6 enable
S2(config-if-VLAN 30)#ipv6 address 2030::1/64
S2(config-if-VLAN 30)#exit
S2(config)#interface vlan 200
S2(config-if-VLAN 200)#ipv6 enable
S2(config-if-VLAN 200)#ipv6 address 1020::1/64
S2(config-if-VLAN 200)#exit
S2(config)#
```

8. 在 R1 为两个接口配置 IP 地址，作为与交换机 S1、S2 互联的地址。

```
Ruijie>enable
Ruijie#configure terminal
Ruijie(config)#hostname R1
R1(config)#interface gigabitEthernet 0/1
R1(config-if-GigabitEthernet 0/1)#ipv6 enable
R1(config-if-GigabitEthernet 0/1)#ipv6 address 1010::2/64
R1(config-if-GigabitEthernet 0/1)#exit
R1(config)#interface gigabitEthernet 0/2
R1(config-if-GigabitEthernet 0/2)#ipv6 enable
R1(config-if-GigabitEthernet 0/2)#ipv6 address 1020::2/64
R1(config-if-GigabitEthernet 0/2)#exit
R1(config)#
```

9．在总部 S1 配置默认路由，目标前缀为"::0"，下一跳地址为园区网路由器"1010::2"。

```
S1(config)#ipv6 route ::/0 1010::2
```

10．在分部 S2 配置默认路由，目标前缀为"::0"，下一跳地址为园区网路由器"1020::2"。

```
S2(config)#ipv6 route ::/0 1020::2
```

11．在园区 R1 配置静态路由，目标前缀为 Jan16 公司管理部网段："2010::64"下一跳地址为园区网路由器"1010::1"。

```
R1(config)#ipv6 route 2010::/64 1010::1
```

12．在园区 R1 配置静态路由，目标前缀为 Jan16 公司财务部网段："2020::64"下一跳地址为园区网路由器"101::1"。

```
R1(config)#ipv6 route 2020::/64 1010::1
```

13．在园区 R1 配置静态路由，目标前缀为 Jan16 公司设计部网段："2030::64"下一跳地址为园区网路由器"1020::1"。

```
R1(config)#ipv6 route 2030::/64 1020::1
```

➢ 任务验证

1．在 S1 使用 show vlan 命令验证 VLAN 的创建情况。

```
S1(config)#show vlan
VLAN Name                          Status    Ports
---- ------------------------------ --------- ---------------------------------
   1 VLAN0001                       STATIC    Gi0/3, Gi0/4, Gi0/5, Gi0/6
                                              Gi0/7, Gi0/8, Gi0/9, Gi0/10
                                              Gi0/11, Gi0/12, Gi0/13, Gi0/14
                                              Gi0/15, Gi0/16, Gi0/17, Gi0/18
                                              Gi0/19, Gi0/20, Gi0/21, Gi0/22
                                              Gi0/23, Gi0/25, Gi0/26, Gi0/27
                                              Gi0/28, Te0/29, Te0/30, Te0/31
                                              Te0/32
  10 VLAN0010                       STATIC    Gi0/1
  20 VLAN0020                       STATIC    Gi0/2
 100 VLAN0100                       STATIC    Gi0/24
```

2．在 S2 使用 show vlan 命令验证 VLAN 的创建情况。

```
S2(config)#show vlan
```

VLAN Name	Status	Ports
1 VLAN0001	STATIC	Gi0/2, Gi0/3, Gi0/4, Gi0/5
		Gi0/6, Gi0/7, Gi0/8, Gi0/9
		Gi0/10, Gi0/11, Gi0/12, Gi0/13
		Gi0/14, Gi0/15, Gi0/16, Gi0/17
		Gi0/18, Gi0/19, Gi0/20, Gi0/21
		Gi0/22, Gi0/23, Gi0/25, Gi0/26
		Gi0/27, Gi0/28, Te0/29, Te0/30
		Te0/31, Te0/32
30 VLAN0030	STATIC	Gi0/1
200 VLAN0200	STATIC	Gi0/24

3. 在 S1 使用 show interface switchport 命令验证链路配置情况。

```
S1(config)#show interface switchport
Interface              Switchport Mode      Access Native Protected VLAN lists
----------------------------------------------------------------------
GigabitEthernet 0/1    enabled     ACCESS    10    1    Disabled   ALL
GigabitEthernet 0/2    enabled     ACCESS    20    1    Disabled   ALL
……………
GigabitEthernet 0/24   enabled     ACCESS    100   1    Disabled   ALL
… …
```

4. 在 S2 使用 show interface switchport 命令验证链路配置情况。

```
S2(config)#show interface switchport
Interface              Switchport Mode      Access Native Protected VLAN lists
----------------------------------------------------------------------
GigabitEthernet 0/1    enabled     ACCESS    30    1    Disabled   ALL
… …
GigabitEthernet 0/24   enabled     ACCESS    200   1    Disabled   ALL
… …
```

5. 在 S1 使用 show ipv6 interface brief 命令验证 IPv6 地址配置情况。

```
S1(config)#show ipv6 interface brief

VLAN 10                         [up/up]
        FE80::274:9CFF:FECD:6922
        2010::1
VLAN 20                         [up/up]
        FE80::274:9CFF:FECD:6922
        2020::1
VLAN 100                        [up/up]
        FE80::274:9CFF:FECD:6922
        1010::1
```

6. 在 S2 使用 show ipv6 interface brief 命令验证 IPv6 地址配置情况。

```
S2(config)#show ipv6 interface brief

VLAN 30                         [up/up]
        FE80::274:9CFF:FE6B:A751
        2030::1
VLAN 200                        [up/up]
        FE80::274:9CFF:FE6B:A751
        1020::1
```

7. 在 R1 使用 show ipv6 interface brief 命令验证 IPv6 地址配置情况。

```
R1(config)#show ipv6 interface brief

GigabitEthernet 0/1              [up/up]
        FE80::8205:88FF:FED0:D8D3
        1010::2
GigabitEthernet 0/2              [up/up]
        FE80::8205:88FF:FED0:D8D2
        1020::2
```

8. 在 S1 使用 show ipv6 route 命令验证默认路由配置情况。

```
S1#show ipv6 route

IPv6 routing table name - Default - 12 entries
Codes:   C - Connected, L - Local, S - Static
         R - RIP, O - OSPF, B - BGP, I - IS-IS, V - Overflow route
         N1 - OSPF NSSA external type 1, N2 - OSPF NSSA external type 2
         E1 - OSPF external type 1, E2 - OSPF external type 2
         SU - IS-IS summary, L1 - IS-IS level-1, L2 - IS-IS level-2
         IA - Inter area

S       ::/0 [1/0] via 1010::2
                   (recursive via 1010::2, VLAN 100)
C       1010::/64 via VLAN 100, directly connected
L       1010::1/128 via VLAN 100, local host
C       2010::/64 via VLAN 10, directly connected
L       2010::1/128 via VLAN 10, local host
C       FE80::/10 via ::1, Null0
C       FE80::/64 via VLAN 10, directly connected
L       FE80::274:9CFF:FECD:6922/128 via VLAN 10, local host
C       FE80::/64 via VLAN 20, directly connected
L       FE80::274:9CFF:FECD:6922/128 via VLAN 20, local host
C       FE80::/64 via VLAN 100, directly connected
L       FE80::274:9CFF:FECD:6922/128 via VLAN 100, local host
```

9. 在 S2 使用 show ipv6 route 命令验证默认路由配置情况。

```
S2#show ipv6 route

IPv6 routing table name - Default - 6 entries
Codes:   C - Connected, L - Local, S - Static
         R - RIP, O - OSPF, B - BGP, I - IS-IS, V - Overflow route
         N1 - OSPF NSSA external type 1, N2 - OSPF NSSA external type 2
         E1 - OSPF external type 1, E2 - OSPF external type 2
         SU - IS-IS summary, L1 - IS-IS level-1, L2 - IS-IS level-2
         IA - Inter area

S       ::/0 [1/0] via 1020::2 (recursive via 1020::2, VLAN 200)
C       1020::/64 via VLAN 200, directly connected
L       1020::1/128 via VLAN 200, local host
C       FE80::/10 via ::1, Null0
C       FE80::/64 via VLAN 200, directly connected
L       FE80::274:9CFF:FE6B:A751/128 via VLAN 200, local host
```

10. 在 R1 使用 show ipv6 route 命令验证静态路由配置情况。

```
R1#show ipv6 route
IPv6 routing table name is - Default - 13 entries
Codes: C - Connected, L - Local, S - Static, R - RIP, B - BGP
       I1 - ISIS L1, I2 - ISIS L2, IA - ISIS interarea, IS - ISIS summary
       O - OSPF intra area, OI - OSPF inter area,   OE1 - OSPF external type 1, OE2 - OSPF external type 2
       ON1 - OSPF NSSA external type 1, ON2 - OSPF NSSA external type 2
L      ::1/128 via Loopback, local host
C      1010::/64 via GigabitEthernet 0/1, directly connected
L      1010::2/128 via GigabitEthernet 0/1, local host
C      1020::/64 via GigabitEthernet 0/2, directly connected
L      1020::2/128 via GigabitEthernet 0/2, local host
S      2010::/64 [1/0] via 1010::1 (recursive via 1010::1, GigabitEthernet 0/1)
S      2020::/64 [1/0] via 1010::1 (recursive via 1010::1, GigabitEthernet 0/1)
S      2030::/64 [1/0] via 1020::1 (recursive via 1020::1, GigabitEthernet 0/2)
L      FE80::/10 via ::1, Null0
C      FE80::/64 via GigabitEthernet 0/2, directly connected
L      FE80::8205:88FF:FED0:DC4B/128 via GigabitEthernet 0/2, local host
C      FE80::/64 via GigabitEthernet 0/1, directly connected
L      FE80::8205:88FF:FED0:DC4C/128 via GigabitEthernet 0/1, local host
```

11．使用管理部 PC1 Ping 财务部 PC2，能实现互通。

```
C:\Users\admin>ping 2020::10

正在 Ping 2020::10 具有 32 字节的数据:
来自 2020::10 的回复: 时间<1ms
来自 2020::10 的回复: 时间=2ms
来自 2020::10 的回复: 时间=1ms
来自 2020::10 的回复: 时间=1ms

2020::10 的 Ping 统计信息:
    数据包: 已发送 = 4, 已接收 = 4, 丢失 = 0 (0% 丢失),
往返行程的估计时间(以毫秒为单位):
    最短 = 0ms, 最长 = 2ms, 平均 = 1ms
```

12．使用管理部 PC1 Ping 设计部 PC3，能实现互通。

```
C:\Users\admin>ping 2030::10

正在 Ping 2030::10 具有 32 字节的数据:
来自 2030::10 的回复: 时间=5 ms
来自 2030::10 的回复: 时间=1 ms
来自 2030::10 的回复: 时间=1 ms
来自 2030::10 的回复: 时间=1 ms

2030::10 的 Ping 统计信息:
    数据包: 已发送 = 4, 已接收 = 4, 丢失 = 0 (0% 丢失),
往返行程的估计时间(以毫秒为单位):
最短 = 1 ms, 最长 = 5 ms, 平均 = 2 ms
```

➢ 问题与思考

1．IPv6 被请求节点组播组地址的产生过程和作用分别是什么？
2．根据任务拓扑描述 PC1 与 PC2 之间的通信过程（包含地址解析）。

5.1.2 OSPFv3 路由协议

➢ 原理

OSPF 是一种典型的链路状态路由协议。OSPFv3 在 IPv6 网络中提供路由功能，也是 IPv6 组网中的主流路由协议之一。OSPFv3 与 OSPFv2 的工作机制基本相同，但 OSPFv3 与 OSPFv2 之间不能互相兼容，因为 OSPFv3 与 OSPFv2 分别是根据 IPv6 网络和 IPv4 网络开发出来的。

OSPFv3 是运行在 IPv6 网络中的动态路由协议。运行 OSPFv3 的路由器使用物理接口链路本地地址为源地址来发送 OSPFv3 报文。在同一条链路上，路由器会学习其他路由器的链路本地地址，并在进行报文转发的过程中将这些地址当成下一跳地址使用。

➢ 任务拓扑

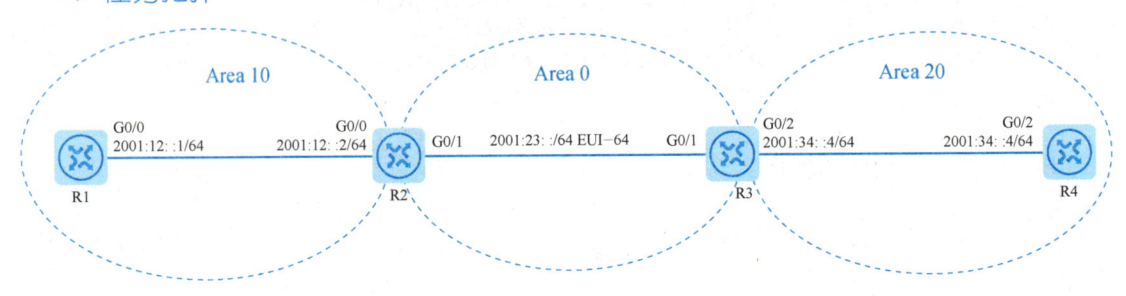

```
     Area 10                  Area 0                   Area 20
   G0/0       G0/0      G0/1              G0/1      G0/2       G0/2
 2001:12::1/64  2001:12::2/64   2001:23::/64 EUI-64    2001:34::4/64  2001:34::4/64
   R1           R2                           R3           R4
```

➢ 实施步骤

1. 根据任务拓扑配置各设备接口 IP 地址。

```
Ruijie>enable
Password:ruijie
Ruijie#configure terminal
Ruijie(config)#hostname   R1
R1(config)#interface    gigabitEthernet 0/0
R1(config-if-GigabitEthernet 0/0)#no   switchport
R1(config-if-GigabitEthernet 0/0)#ipv6    enable
R1(config-if-GigabitEthernet 0/0)#ipv6 address 2001:12::1/64
R1(config-if-GigabitEthernet 0/0)#exit
R1(config)#int loopback 0
R1(config-if-Loopback 0)#ipv6    enable
R1(config-if-Loopback 0)#ipv6    address    3000:1::1/128
R1(config-if-Loopback 0)#exit
R1(config)#

Ruijie>enable
Password:ruijie
Ruijie#configure terminal
Ruijie(config)#hostname    R2
R2(config)#interface    gigabitEthernet 0/0
R2(config-if-GigabitEthernet 0/0)#no   switchport
R2(config-if-GigabitEthernet 0/0)#ipv6 enable
R2(config-if-GigabitEthernet 0/0)#ipv6    address    2001:12::2/64
R2(config-if-GigabitEthernet 0/0)#exit
R2(config)#interface    gigabitEthernet 0/1
```

```
R2(config-if-GigabitEthernet 0/1)#no   switchport
R2(config-if-GigabitEthernet 0/1)#ipv6   enable
R2(config-if-GigabitEthernet 0/1)#ipv6   address   2001:23::/64 eui-64
R2(config-if-GigabitEthernet 0/1)#exit
R2(config)#

Ruijie>enable
Password:ruijie
Ruijie#configure terminal
Ruijie(config)#hostname   R3
R3(config)#interface   gigabitEthernet 0/1
R3(config-if-GigabitEthernet 0/1)#no   switchport
R3(config-if-GigabitEthernet 0/1)#ipv6   enable
R3(config-if-GigabitEthernet 0/1)#ipv6   address   2001:23::/64 eui-64
R3(config-if-GigabitEthernet 0/1)#exit
R3(config)#interface   gigabitEthernet 0/2
R3(config-if-GigabitEthernet 0/2)#no   switchport
R3(config-if-GigabitEthernet 0/2)#ipv6   enable
R3(config-if-GigabitEthernet 0/2)#ipv6   address   2001:34::4/64
R3(config-if-GigabitEthernet 0/2)#exit
R3(config)#

Ruijie>enable
Password:ruijie
Ruijie#configure terminal
Ruijie(config)#hostname   R4
R4(config)#interface   gigabitEthernet 0/2
R4(config-if-GigabitEthernet 0/2)#no   switchport
R4(config-if-GigabitEthernet 0/2)#ipv6 enable
R4(config-if-GigabitEthernet 0/2)#ipv6   address   2001:34::4/64
R4(config-if-GigabitEthernet 0/2)#exit
R4(config)#interface   loopback 0
R4(config-if-Loopback 0)#ipv6   enable
R4(config-if-Loopback 0)#ipv6   address 3000:2::2/128
R4(config-if-Loopback 0)#exit
R4(config)#
```

2．在各设备之间配置 OSPFv3 协议互通。

```
R1(config)#ipv6   router   ospf 10
R1(config-router)#router-id   1.1.1.1
*Jul 14 09:26:36: %OSPFV3-4-NORTRID: OSPFv3 process 10 failed to allocate unique router-id and cannot
start. //提示 OSPFv3 的 Router ID 必须通告手工配置
Change router-id and update OSPFv3 process! [yes/no]:y
R1(config-router)#exit
R1(config)#interface   gigabitEthernet 0/0
R1(config-if-GigabitEthernet 0/0)#ipv6   ospf   10 area   10
R1(config-if-GigabitEthernet 0/0)#exit
R1(config)#int loopback 0
R1(config-if-Loopback 0)#ipv6   ospf   10 area   10
R1(config-if-Loopback 0)#exit
R1(config)#

R2(config)#ipv6   router   ospf   10
R2(config-router)#router-id 2.2.2.2
Change router-id and update OSPFv3 process! [yes/no]:y
```

```
R2(config-router)#exit
R2(config)#interface    gigabitEthernet 0/0
R2(config-if-GigabitEthernet 0/0)#ipv6  ospf   10 area   10
R2(config-if-GigabitEthernet 0/0)#exit
R2(config)#interface    gigabitEthernet 0/1
R2(config-if-GigabitEthernet 0/1)#ipv6 ospf    10 area   0
R2(config-if-GigabitEthernet 0/1)#exit
R2(config)#

R3(config)#ipv6 router    ospf 10
R3(config-router)#router-id    3.3.3.3
Change router-id and update OSPFv3 process! [yes/no]:y
R3(config-router)#exit
R3(config)#interface    gigabitEthernet 0/1
R3(config-if-GigabitEthernet 0/1)#ipv6 ospf    10 area   0
R3(config-if-GigabitEthernet 0/1)#exit
R3(config)#interface    gigabitEthernet 0/2
R3(config-if-GigabitEthernet 0/2)#ipv6 ospf    10 area   20
R3(config-if-GigabitEthernet 0/2)#exit
R3(config)#

R4(config)#ipv6   router   ospf   10
R4(config-router)#router-id    4.4.4.4
Change router-id and update OSPFv3 process! [yes/no]:y
R4(config-router)#exit
R4(config)#interface    gigabitEthernet 0/2
R4(config-if-GigabitEthernet 0/2)#ipv6   ospf   10 area   20
R4(config-if-GigabitEthernet 0/2)#exit
R4(config)#interface    loopback 0
R4(config-if-Loopback 0)#ipv6   ospf   10 area   20
R4(config-if-Loopback 0)#exit
R4(config)#
```

➢ 任务验证

1．检查 IPv6 地址配置情况。

```
R1#show   ipv6 interface   brief

GigabitEthernet 0/0              [up/up]
        FE80::5200:FF:FE01:1
        2001:12::1
Loopback 0                       [up/up]
        FE80::5200:FF:FE01:1
        3000:1::1

R1#

R2#show   ipv6  interface   brief

GigabitEthernet 0/0              [up/up]
        FE80::5200:FF:FE02:1
        2001:12::2
GigabitEthernet 0/1              [up/up]
        FE80::5200:FF:FE02:2
        2001:23::5200:FF:FE02:2
```

R2#

R3#show ipv6 interface brief

GigabitEthernet 0/1   [up/up]
       FE80::5200:FF:FE03:2
       2001:23::5200:FF:FE03:2
GigabitEthernet 0/2   [up/up]
       FE80::5200:FF:FE03:3
       2001:34::4

R3#

R4#show ipv6 interface brief

GigabitEthernet 0/2   [up/up]
       FE80::5200:FF:FE04:3
       2001:34::4
Loopback 0   [up/up]
       FE80::5200:FF:FE04:3
       3000:2::2

R4#

2．检查 OSPFv3 邻居建立情况。

```
R2#show  ipv6  ospf  neighbor

OSPFv3 Process (10), 2 Neighbors, 2 is Full:
Neighbor ID       Pri    State       BFD State    Dead Time    Instance ID    Interface
1.1.1.1           1      Full/DR     -            00:00:39     0              GigabitEthernet
0/0
3.3.3.3           1      Full/DR     -            00:00:34     0              GigabitEthernet
0/1
R2#

R2#show  ipv6  ospf  neighbor

OSPFv3 Process (10), 2 Neighbors, 2 is Full:
Neighbor ID       Pri    State       BFD State    Dead Time    Instance ID    Interface
1.1.1.1           1      Full/DR     -            00:00:39     0              GigabitEthernet
0/0
3.3.3.3           1      Full/DR     -            00:00:34     0              GigabitEthernet
0/1
R2#
```

3．检查 OSPFv3 路由传递情况。

```
R1#show  ipv6  route

IPv6 routing table name - Default - 11 entries
Codes:   C - Connected, L - Local, S - Static
         R - RIP, O - OSPF, B - BGP, I - IS-IS, V - Overflow route
         N1 - OSPF NSSA external type 1, N2 - OSPF NSSA external type 2
         E1 - OSPF external type 1, E2 - OSPF external type 2
         SU - IS-IS summary, L1 - IS-IS level-1, L2 - IS-IS level-2
```

```
               IA - Inter area, EV - BGP EVPN, N - Nd to host

C       2001:12::/64 via GigabitEthernet 0/0, directly connected
L       2001:12::1/128 via GigabitEthernet 0/0, local host
O   IA  2001:23::/64 [110/2] via FE80::5200:FF:FE02:1, GigabitEthernet 0/0
O   IA  2001:34::/64 [110/3] via FE80::5200:FF:FE02:1, GigabitEthernet 0/0
LC      3000:1::1/128 via Loopback 0, local host
O   IA  3000:2::2/128 [110/3] via FE80::5200:FF:FE02:1, GigabitEthernet 0/0
C       FE80::/10 via ::1, Null0
C       FE80::/64 via GigabitEthernet 0/0, directly connected
L       FE80::5200:FF:FE01:1/128 via GigabitEthernet 0/0, local host
C       FE80::/64 via Loopback 0, directly connected
L       FE80::5200:FF:FE01:1/128 via Loopback 0, local host
R1#

R4#show  ipv6  route

IPv6 routing table name - Default - 11 entries
Codes:   C - Connected, L - Local, S - Static
         R - RIP, O - OSPF, B - BGP, I - IS-IS, V - Overflow route
         N1 - OSPF NSSA external type 1, N2 - OSPF NSSA external type 2
         E1 - OSPF external type 1, E2 - OSPF external type 2
         SU - IS-IS summary, L1 - IS-IS level-1, L2 - IS-IS level-2
         IA - Inter area, EV - BGP EVPN, N - Nd to host

O   IA  2001:12::/64 [110/3] via FE80::5200:FF:FE03:3, GigabitEthernet 0/2
O   IA  2001:23::/64 [110/2] via FE80::5200:FF:FE03:3, GigabitEthernet 0/2
C       2001:34::/64 via GigabitEthernet 0/2, directly connected
L       2001:34::4/128 via GigabitEthernet 0/2, local host
O   IA  3000:1::1/128 [110/3] via FE80::5200:FF:FE03:3, GigabitEthernet 0/2
LC      3000:2::2/128 via Loopback 0, local host
C       FE80::/10 via ::1, Null0
C       FE80::/64 via Loopback 0, directly connected
L       FE80::5200:FF:FE04:3/128 via Loopback 0, local host
C       FE80::/64 via GigabitEthernet 0/2, directly connected
L       FE80::5200:FF:FE04:3/128 via GigabitEthernet 0/2, local host
R4#
```

4. 测试 R1 设备和 R4 设备的 loopback 0 接口互通情况。

```
R1#ping  ipv6 3000:2::2
Sending 5, 100-byte ICMP Echoes to 3000:2::2, timeout is 2 seconds:
   < press Ctrl+C to break >
!!!!!
Success rate is 100 percent (5/5), round-trip min/avg/max = 3/19/80 ms.
R1#

R4#ping  ipv6   3000:1::1
Sending 5, 100-byte ICMP Echoes to 3000:1::1, timeout is 2 seconds:
   < press Ctrl+C to break >
!!!!!
Success rate is 100 percent (5/5), round-trip min/avg/max = 3/3/4 ms.
R4#
```

> 问题与思考

1. OSPFv3 路由协议，增加了几个特有的 LSA 类型，为什么需要增加这几个特有的 LSA 类型，描述新增 LSA 类型的作用。
2. 描述 OSPFv3 与 OSPFv2 的相同点与不同点。
3. 在以上配置的基础上将 Area 20 改为 Stub 区域。

5.1.3 BGP4+路由协议

> 原理

传统 BGP-4 只管理 IPv4 路由信息，对于使用其他网络程协议（如 IPv6 等）的应用未给予支持。

IETF 对 BGP-4 扩展，提出 BGP4+，可以提供对 IPv6、IPX 和 MPLS VPN 的支持（简单说：扩展 IPv6 协议栈支持）。

BGP4+针对 IPv6 网络层协议的信息反映到 NLRI 及 Next_Hop 属性中，新增扩展属性：

MP_REACH_NLRI：Multiprotocol Reachable NLRI，多协议可达 NLRI。用于发布可达路由及下一跳地址信息。

MP_UNREACH_NLRI：Multiprotocol Unreachable NLRI，多协议不可达 NLRI。用于撤销不可达路由。

Next-Hop 属性信息用 IPv6 地址来表示，可以是 IPv6 全球单播地址或者下一跳的链路本地地址。BGP4+利用 BGP 的多协议扩展属性来达到在 IPv6 网络中应用的目的，BGP 协议原有的消息机制和路由机制没有改变。

> 任务拓扑

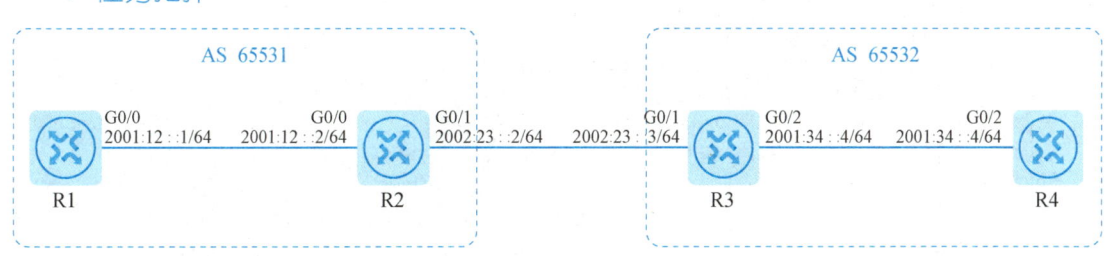

```
       AS 65531                                AS 65532
  G0/0         G0/0    G0/1         G0/1    G0/2         G0/2
R1 2001:12::1/64  2001:12::2/64 R2  2002:23::2/64  2002:23::3/64 R3  2001:34::4/64  2001:34::4/64 R4
```

> 实施步骤

1. 根据任务拓扑配置各设备接口 IP 地址。

```
Ruijie>enable
Password:ruijie
Ruijie#configure terminal
Ruijie(config)#hostname   R1
R1(config)#interface   gigabitEthernet 0/0
R1(config-if-GigabitEthernet 0/0)#no   switchport
R1(config-if-GigabitEthernet 0/0)#ipv6 enable
R1(config-if-GigabitEthernet 0/0)#ipv6   address    2001:12::1/64
R1(config-if-GigabitEthernet 0/0)#exit
R1(config)#interface   gigabitEthernet 0/1
```

```
R1(config-if-GigabitEthernet 0/1)#no    switchport
R1(config-if-GigabitEthernet 0/1)#ipv6    enable
R1(config-if-GigabitEthernet 0/1)#ipv6    address    3000:10::1/64
R1(config-if-GigabitEthernet 0/1)#exit
R1(config)#


Ruijie>enable
Password:ruijie
Ruijie#configure terminal
Ruijie(config)#hostname   R2
R2(config)#interface    gigabitEthernet 0/0
R2(config-if-GigabitEthernet 0/0)#no    switchport
R2(config-if-GigabitEthernet 0/0)#ipv6 enable
R2(config-if-GigabitEthernet 0/0)#ipv6    address    2001:12::2/64
R2(config-if-GigabitEthernet 0/0)#exit
R2(config)#interface    gigabitEthernet 0/1
R2(config-if-GigabitEthernet 0/1)#no    switchport
R2(config-if-GigabitEthernet 0/1)#ipv6    enable
R2(config-if-GigabitEthernet 0/1)#ipv6    address    2002:23::2/64
R2(config-if-GigabitEthernet 0/1)#exit
R2(config)#


Ruijie>enable
Password:ruijie
Ruijie#configure terminal
Ruijie(config)#hostname   R3
R3(config)#interface    gigabitEthernet 0/1
R3(config-if-GigabitEthernet 0/1)#no    switchport
R3(config-if-GigabitEthernet 0/1)#ipv6    enable
R3(config-if-GigabitEthernet 0/1)#ipv6    address    2002:23::3/64
R3(config-if-GigabitEthernet 0/1)#exiti
R3(config)#interface    gigabitEthernet 0/2
R3(config-if-GigabitEthernet 0/2)#no    switchport
R3(config-if-GigabitEthernet 0/2)#ipv6    enable
R3(config-if-GigabitEthernet 0/2)# ipv6    address    2001:34::3/64
R3(config-if-GigabitEthernet 0/2)#exit
R3(config)


Ruijie>enable
Password:ruijie
Ruijie#configure terminal
Ruijie(config)#hostname   R4
R4(config)#interface    gigabitEthernet 0/2
R4(config-if-GigabitEthernet 0/2)#no switchport
R4(config-if-GigabitEthernet 0/2)#ipv6    enable
R4(config-if-GigabitEthernet 0/2)#ipv6    address    2001:34::4/64
R4(config-if-GigabitEthernet 0/2)#exit
R4(config)#interface    gigabitEthernet 0/1
R4(config-if-GigabitEthernet 0/1)#no    switchport
R4(config-if-GigabitEthernet 0/1)#ipv6 enable
R4(config-if-GigabitEthernet 0/1)#ipv6    address    3002:20::1/64
R4(config-if-GigabitEthernet 0/1)#exit
R4(config)#
```

2. BGP4+配置 AS65531 内部和 AS65532 内部 IBGP 邻居关系，在 AS65531 和 AS65532 之间建立 EBGP 邻居关系。

```
R1(config)#router   bgp   65531
R1(config-router)#bgp    router-id   1.1.1.1
R1(config-router)#neighbor    2001:12::2 remote-as    65531
R1(config-router)#address-family ipv6
R1(config-router-af)#neighbor    2001:12::2 activate
R1(config-router-af)#exit

R2(config)#router   bgp   65531
R2(config-router)#bgp    router-id   2.2.2.2
R2(config-router)#neighbor    2001:12::1 remote-as    65531
R2(config-router)#neighbor    2002:23::3 remote-as    65532
R2(config-router)#address-family ipv6
R2(config-router-af)#neighbor    2001:12::1 activate
R2(config-router-af)#neighbor    2001:12::1 next-hop-self
R2(config-router-af)#neighbor    2002:23::3 activate
R2(config-router-af)#exit
R2(config-router)#exit
R2(config)#

R3(config)#router   bgp   65532
R3(config-router)#bgp    router-id 3.3.3.3
R3(config-router)#neighbor    2002:23::2 remote-as    65531
R3(config-router)#address-family ipv6
R3(config-router-af)#neighbor    2002:23::2 activate
R3(config-router-af)#exit
R3(config-router)#neighbor    2001:34::4 remote-as    65532
R3(config-router)#address-family ipv6
R3(config-router-af)#neighbor    2001:34::4 activate
R3(config-router-af)#neighbor    2001:34::4 next-hop-self
R3(config-router-af)#exit
R3(config-router)#exit
R3(config)#

R4(config)#router   bgp   65532
R4(config-router)#bgp    router-id   4.4.4.4
R4(config-router)#neighbor    2001:34::3 remote-as    65532
R4(config-router)#address-family ipv6
R4(config-router-af)#neighbor    2001:34::3 activate
R4(config-router-af)#exit
R4(config-router)#exit
R4(config)#
```

3. 通告 IPv6 网段到 BGP4+中。

```
R1(config)#router   bgp   65531
R1(config-router)#address-family ipv6
R1(config-router-af)#network    3000:10::/64
R1(config-router-af)#exit
R1(config-router)#exit
R1(config)#


R4(config)#router   bgp 65532
```

```
R4(config-router)#address-family ipv6
R4(config-router-af)#network 3002:20::/64
R4(config-router-af)#exit
R4(config-router)#exit
R4(config)#
```

> 任务验证

1. 检查 IPv6 地址配置情况。

```
R1#show  ipv6  interface  brief

GigabitEthernet 0/0            [up/up]
        FE80::5200:FF:FE01:1
        2001:12::1
GigabitEthernet 0/1            [up/up]
        FE80::5200:FF:FE01:2
        3000:10::1

R1#


R2#show  ipv6  interface  brief

GigabitEthernet 0/0            [up/up]
        FE80::5200:FF:FE02:1
        2001:12::2
GigabitEthernet 0/1            [up/up]
        FE80::5200:FF:FE02:2
        2002:23::2

R2#


R3#show   ipv interface   brief

GigabitEthernet 0/1            [up/up]
        FE80::5200:FF:FE03:2
        2002:23::3
GigabitEthernet 0/2            [up/up]
        FE80::5200:FF:FE03:3
        2001:34::3

R3#


R4#show   ipv interface   brief

GigabitEthernet 0/1            [up/up]
        FE80::5200:FF:FE04:2
        3002:20::1
GigabitEthernet 0/2            [up/up]
        FE80::5200:FF:FE04:3
        2001:34::4

R4#
```

2. 检查 BGP4+邻居关系建立情况。

```
R2#show    bgp   ipv6 unicast    summary
For address family: IPv6 Unicast
BGP router identifier 2.2.2.2, local AS number 65531
BGP table version is 3
0 BGP AS-PATH entries
0 BGP Community entries
0 BGP Prefix entries (Maximum-prefix:4294967295)

Neighbor          V        AS MsgRcvd MsgSent    TblVer    InQ OutQ Up/Down    State/PfxRcd
2001:12::1        4     65531      10       9         2      0    0 00:06:21         0
2002:23::3        4     65532       9       8         3      0    0 00:05:24         0

Total number of neighbors 2, established neighbors 2

R2#


R3#show    bgp   ipv6 unicast summary
For address family: IPv6 Unicast
BGP router identifier 3.3.3.3, local AS number 65532
BGP table version is 3
0 BGP AS-PATH entries
0 BGP Community entries
0 BGP Prefix entries (Maximum-prefix:4294967295)

Neighbor          V        AS MsgRcvd MsgSent    TblVer    InQ OutQ Up/Down    State/PfxRcd
2001:34::4        4     65532       7       9         3      0    0 00:04:56         0
2002:23::2        4     65531       9       7         3      0    0 00:05:52         0

Total number of neighbors 2, established neighbors 2

R3#
```

3. 检查 BGP4+路由通告的路由信息。

```
R1#show    bgp ipv6 unicast
BGP table version is 5, local router ID is 1.1.1.1
Status codes: s suppressed, d damped, h history, * valid, > best, i - internal,
              S Stale, b - backup entry, m - multipath, f Filter, a additional-path
Origin codes: i - IGP, e - EGP, ? - incomplete

     Network           Next Hop         Metric     LocPrf      Weight Path
*>   3000:10::/64      ::                  0                   32768     i
*>i 3002:20::/64       2001:12::2          0        100         0 65532 i

Total number of prefixes 2
R1#



R1#show    bgp   ipv6 unicast    3002:20::/64
BGP routing table entry for 3002:20::/64(#0x7f70993a28d0)
Paths: (1 available, best #1, table Default-IP-Routing-Table)
    Not advertised to any peer
```

```
                65532
           2001:12::2(fe80::5200:ff:fe02:1) from 2001:12::2 (2.2.2.2)
           (fe80::5200:ff:fe02:1)
             Origin IGP, metric 0, localpref 100, valid, internal, best
             Last update: Fri Jul 14 09:55:06 2023
             RX ID: 0,TX ID: 0
R1#


R4#show    bgp ipv6 unicast
BGP table version is 2, local router ID is 4.4.4.4
Status codes: s suppressed, d damped, h history, * valid, > best, i - internal,
              S Stale, b - backup entry, m - multipath, f Filter, a additional-path
Origin codes: i - IGP, e - EGP, ? - incomplete

     Network              Next Hop              Metric       LocPrf      Weight Path
*>i 3000:10::/64          2001:34::3            0            100         0 65531 i
*>  3002:20::/64          ::                    0                        32768      i

Total number of prefixes 2
R4#
```

> 问题与思考

在 BGP 路由器上想要实现 BGP4+路由进行路由聚合，如何操作？

5.2 IPv6 过渡技术

5.2.1 IPv6 手工隧道（GRE）

> 原理

隧道（Tunnel）技术也是 IPv6 过渡技术的一种。隧道技术是一种数据封装技术，它利用一种网络传输协议，将其他协议产生的数据作为数据载荷，然后封装在自己的报文中在网络中进行传输（当边界路由器转发数据查找路由表时，发现数据的出口为隧道接口，便会进行数据封装）。当 IPv6 网络 A 的数据要穿越 IPv4 网络到达 IPv6 网络 B 时，因为 IPv6 与 IPv4 互不兼容，所以需要在 RA 和 RB（两端的设备都要支持双栈协议）上配置隧道技术，通过隧道技术将 IPv6 数据作为 IPv4 的数据载荷，封装在 IPv4 报文中，使数据通过 IPv4 网络传输到 IPv6 网络 B 中。这便是隧道技术的关键。

隧道需要有一个起点和一个终点，起点和终点确定了以后，隧道也就可以确定了。IPv6 Over IPv4 隧道起点的 IPv4 地址必须为手工配置，而终点的 IPv4 地址有手工配置和自动获取两种配置方式。

（1）手动隧道：边界设备不能自动获得隧道终点的 IPv4 地址，需要手工配置隧道终点的 IPv4 地址，报文才能正确发送至隧道终点。

（2）自动隧道：边界设备可以自动获得隧道终点的 IPv4 地址，所以不需要手动配置终点的 IPv4 地址，一般做法是隧道两个接口的 IPv6 地址采用内嵌 IPv4 地址的特殊 IPv6 地址形式，这样路由设备可以从 IPv6 报文中的目的 IPv6 地址中提取出 IPv4 地址。

IPv6 Over IPv4 GRE 隧道的工作原理

通用路由协议封装（Generic Routing Encapsulation，GRE）隧道是一种手动隧道，在 IPv6 Over IPv4 GRE 应用中，GRE 隧道把 IPv6 协议称为乘客协议，把 GRE 称为承载协议。当 GRE 隧道封装数据时，IPv6 数据报文先被封装为 GRE 数据报文，再封装为 IPv4 数据报文，封装之后的数据结构如图 5-1 所示。GRE 可以支持多种网络层乘客协议，如：IP、IPX、Apple Talk。

IPv4 头部	GRE 头部	IPv6 头部	IPv6 数据

图 5-1　封装之后的数据结构图

IPv6 Over IPv4 GRE 隧道的特点

GRE 隧道通用性好，原理简单，易于配置。但作为手动隧道，每个隧道都需要手动配置 IPv4 地址，在 IPv6 网络过渡的前期阶段，随着互联网中需要互联的 IPv6 网络数量不断增加，需要配置的隧道数量以及维护和管理的难度也会随之增加。

> **任务拓扑**

> **实施步骤**

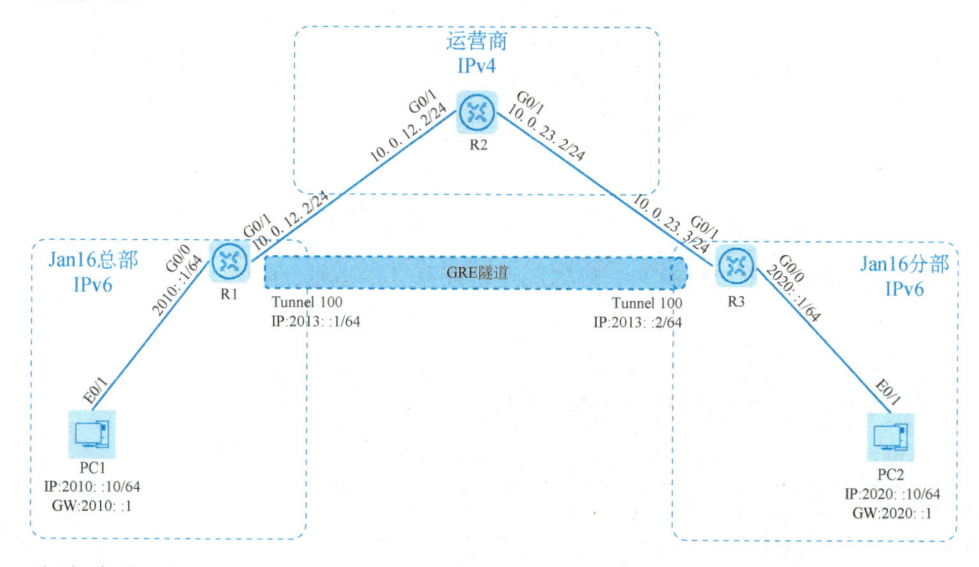

1. 根据任务拓扑配置各设备接口 IP 地址。
2. 在路由器 R2 配置 IPv4 地址，作为与总部路由器、分部 A 路由器互联的地址。

```
Ruijie>enable
Ruijie#configure terminalfigure terminal
Ruijie(config)#hostname R2
R2(config)#interface GigabitEthernet 0/1
R2(config-if-GigabitEthernet 0/1)#ip address 10.1.12.2 255.255.255.0
R2(config-if-GigabitEthernet 0/1)#exit
R2(config)#interface gigabitEthernet 0/2
R2(config-if-GigabitEthernet 0/2)#ip address 10.1.23.2 255.255.255.0
R2(config-if-GigabitEthernet 0/2)#exiti
```

3．PC1 的 IPv6 地址配置结果如图 5-2 所示，同理完成 PC2 的 IPv6 地址配置。

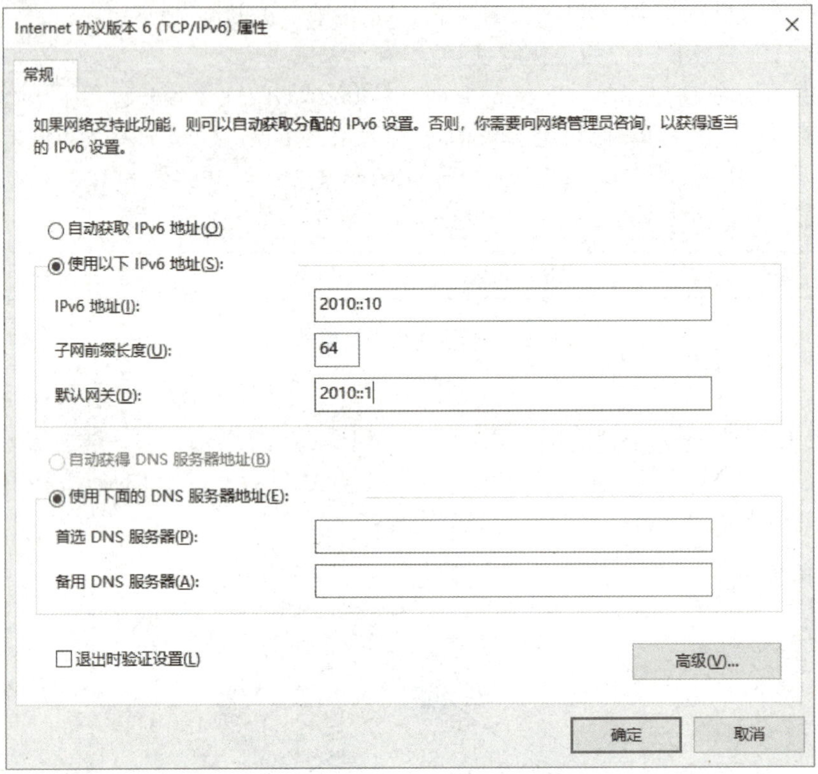

图 5-2　PC1 的 IPv6 地址配置结果

4．在路由器 R1 配置 IPv4 地址，作为与运营商互联的地址，配置 IPv6 地址，作为总部的网关。

```
Ruijie>enable
Ruijie#configure terminalfigure terminal
Ruijie(config)#hostname R1
R1(config)#interface GigabitEthernet 0/1
R1(config-if-GigabitEthernet 0/1)#ip address 10.1.12.1 255.255.255.0
R1(config-if-GigabitEthernet 0/1)#exit
R1(config)#interface gigabitEthernet 0/0
R1(config-if-GigabitEthernet 0/0)# ipv6 address 2010::1/64
R1(config-if-GigabitEthernet 0/0)#exit
```

5．在路由器 R3 配置 IPv4 地址，作为与运营商互联的地址，配置 IPv6 地址，作为分部 A 的网关。

```
Ruijie>enable
Ruijie#configure terminalfigure terminal
Ruijie(config)#hostname R3
R3(config)#interface GigabitEthernet 0/2
R3(config-if-GigabitEthernet 0/2)#ip address 10.1.23.3 255.255.255.0
R3(config-if-GigabitEthernet 0/2)#exit
R3(config)#interface gigabitEthernet 0/0
R3(config-if-GigabitEthernet 0/0)# ipv6 address 2020::1/64
R3(config-if-GigabitEthernet 0/0)#exit
```

6. 在路由器 R1 配置默认路由，下一跳地址为运营商路由器 R2。

R1(config)#ip route 0.0.0.0 0.0.0.0 10.1.12.2

7. 在路由器 R3 配置默认路由，下一跳地址为运营商路由器 R2。

R3(config)#ip route 0.0.0.0 0.0.0.0 10.1.23.2

8. 配置路由器 R1 的 IPv6 Over IPv4 GRE 隧道，在路由器 R1 创建 IPv6 Over IPv4 GRE 隧道，并配置去往分部 A 的 IPv6 静态路由，下一跳地址为隧道接口。

```
R1(config)#interface Tunnel 100
R1(config-if-Tunnel 100)#ipv6 enable
R1(config-if-Tunnel 100)#ipv6 address 2013::1/64
R1(config-if-Tunnel 100)#tunnel source 10.1.12.1
R1(config-if-Tunnel 100)#tunnel destination 10.1.23.3
R1(config-if-Tunnel 100)#exit
R1(config)#ipv6 route 2020::/64 tunnel 100
```

9. 在路由器 R3 创建 IPv6 Over IPv4 GRE 隧道，并配置去往总部的 IPv6 静态路由，下一跳地址为隧道接口。

```
R3(config)#interface Tunnel 100
R3(config-if-Tunnel 100)#ipv6 enable
R3(config-if-Tunnel 100)#ipv6 address 2013::2/64
R3(config-if-Tunnel 100)#tunnel source 10.1.23.3
R3(config-if-Tunnel 100)#tunnel destination 10.1.12.1
R3(config-if-Tunnel 100)#exit
R3(config)#ipv6 route 2010::/64 tunnel 100
```

➢ 任务验证

1. 查看 IPv4 地址配置情况。

```
R2(config)#show ip interface brief
Interface           IP-Address(Pri)    IP-Address(Sec)   Status    Protocol   Description
… …
GigabitEthernet 0/1   10.1.12.2/24      no address        up        up
GigabitEthernet 0/2   10.1.23.2/24      no address        up        up
… …
R2(config)#
```

2. 在 R1 使用 show ip interface brief 和 show ipv6 interface brief 命令验证 IP 地址配置情况。

```
R1(config)#show ip interface brief
Interface           IP-Address(Pri)    IP-Address(Sec)   Status    Protocol   Description
… …
GigabitEthernet 0/1   10.1.12.1/24      no address        up        up
… …
R1(config)#

R1(config)#show ipv6 interface brief

GigabitEthernet 0/0              [up/up]
       FE80::8205:88FF:FED0:D8D4
       2010::1
```

R1(config)#

3. 在 R3 使用 show ip interface brief 和 show ipv6 interface brief 命令验证 IP 地址配置情况。

```
R3(config)#show ip interface brief
Interface              IP-Address(Pri)    IP-Address(Sec)    Status    Protocol    Description
… …
GigabitEthernet 0/1    10.1.23.3/24       no address         up        up
… …
R3(config)#show ipv6 interface brief

GigabitEthernet 0/0               [up/up]
         FE80::8205:88FF:FED0:DC4D
         2020::1
R3(config)#
```

4. 在 R1 使用 show ip route 命令验证默认路由配置情况。

```
R1(config)#show ip route
… …
Gateway of last resort is 10.1.12.2 to network 0.0.0.0
S*    0.0.0.0/0 [1/0] via 10.1.12.2
C     10.1.12.0/24 is directly connected, GigabitEthernet 0/1
C     10.1.12.1/32 is local host.
C     192.168.1.0/24 is directly connected, VLAN 1
C     192.168.1.1/32 is local host.
R1(config)#
```

5. 在 R3 使用 show ip route 命令验证默认路由配置情况。

```
R3(config)#show ip route
… …
Gateway of last resort is 10.1.23.2 to network 0.0.0.0
S*    0.0.0.0/0 [1/0] via 10.1.23.2
C     10.1.23.0/24 is directly connected, GigabitEthernet 0/2
C     10.1.23.3/32 is local host.
C     192.168.1.0/24 is directly connected, VLAN 1
C     192.168.1.1/32 is local host.
R3(config)#
```

6. 在 R1 使用 show ipv6 route 命令验证 IPv6 静态路由配置情况。

```
R1(config)#show ipv6 route
… …
L       ::1/128 via Loopback, local host
C       2010::/64 via GigabitEthernet 0/0, directly connected
L       2010::1/128 via GigabitEthernet 0/0, local host
C       2013::/64 via Tunnel 100, directly connected
L       2013::1/128 via Tunnel 100, local host
S       2020::/64 [1/0] via Tunnel 100, directly connected
L       FE80::/10 via ::1, Null0
C       FE80::/64 via GigabitEthernet 0/0, directly connected
L       FE80::8205:88FF:FED0:D8D4/128 via GigabitEthernet 0/0, local host
C       FE80::/64 via Tunnel 100, directly connected
L       FE80::A01:C01/128 via Tunnel 100, local host
R1(config)#
```

7. 在 R3 使用 show ipv6 route 命令验证 IPv6 静态路由配置情况。

```
R3(config)#show ipv6 route
… …
L       ::1/128 via Loopback, local host
S       2010::/64 [1/0] via Tunnel 100, directly connected
C       2013::/64 via Tunnel 100, directly connected
L       2013::2/128 via Tunnel 100, local host
C       2020::/64 via GigabitEthernet 0/0, directly connected
L       2020::1/128 via GigabitEthernet 0/0, local host
L       FE80::/10 via ::1, Null0
C       FE80::/64 via GigabitEthernet 0/0, directly connected
L       FE80::8205:88FF:FED0:DC4D/128 via GigabitEthernet 0/0, local host
C       FE80::/64 via Tunnel 100, directly connected
L       FE80::A01:1703/128 via Tunnel 100, local host
R3(config)#
```

8. 以 R1 为隧道起点，尝试 Ping 通隧道终点 R3 的隧道接口地址 2013::2，能成功 Ping 通，表示隧道建立成功。

```
R1#ping ipv6 2013::2
Sending 5, 100-byte ICMP Echoes to 2013::2, timeout is 2 seconds:
   < press Ctrl+C to break >
!!!!!
Success rate is 100 percent (5/5), round-trip min/avg/max = 1/1/1 ms
R1#
```

9. 使用 PC1 Ping PC2 的 IPv6 地址（2020::10）。

```
C:\Users\admin>ping 2020::10

正在 Ping 2020::10 具有 32 字节的数据:
来自 2020::10 的回复: 时间=1ms
来自 2020::10 的回复: 时间=1ms
来自 2020::10 的回复: 时间=1ms
来自 2020::10 的回复: 时间=1ms

2020::10 的 Ping 统计信息:
    数据包: 已发送 = 4，已接收 = 4，丢失 = 0 (0% 丢失)，
往返行程的估计时间(以毫秒为单位):
    最短 = 1ms，最长 = 1ms，平均 = 1ms
```

> 问题与思考

1. IPv6 过渡隧道的手工隧道和自动隧道的区别？
2. IPv6 手工隧道和 GRE 隧道的区别？
3. IPv6 手工隧道无法建立的原因？

5.2.2 IPv6 自动隧道（6to4）

> 原理

6to4 隧道是一种自动隧道，要求站点内网络设备使用特殊的 IPv6 地址——6to4 地址。

6to4 地址格式：

6to4 地址将 IPv4 地址嵌入到 IPv6 地址中。

FP（3 位）	TLA ID（13 位）	IPv4 地址（32 位）	SLA ID（16 位）	接口 ID（64 位）

（1）FP：可聚合全球单播地址的格式前缀（Format Prefix），其值为 001（固定的）。
（2）TLA ID：顶级聚合标识符（Top Level Aggregator Identifier），其值为 0x0002（固定的）。
（3）IPv4 地址：隧道起点 IPv4 地址，使用时需将 32 位 IPv4 地址转换为十六进制形式。
（4）SLA ID：站点级聚合标识符（Site Level Aggregator Identifier），其值由用户自定义。
（5）接口 ID：IPv6 接口 ID，其值由用户自定义。

> 任务拓扑

> 实施步骤

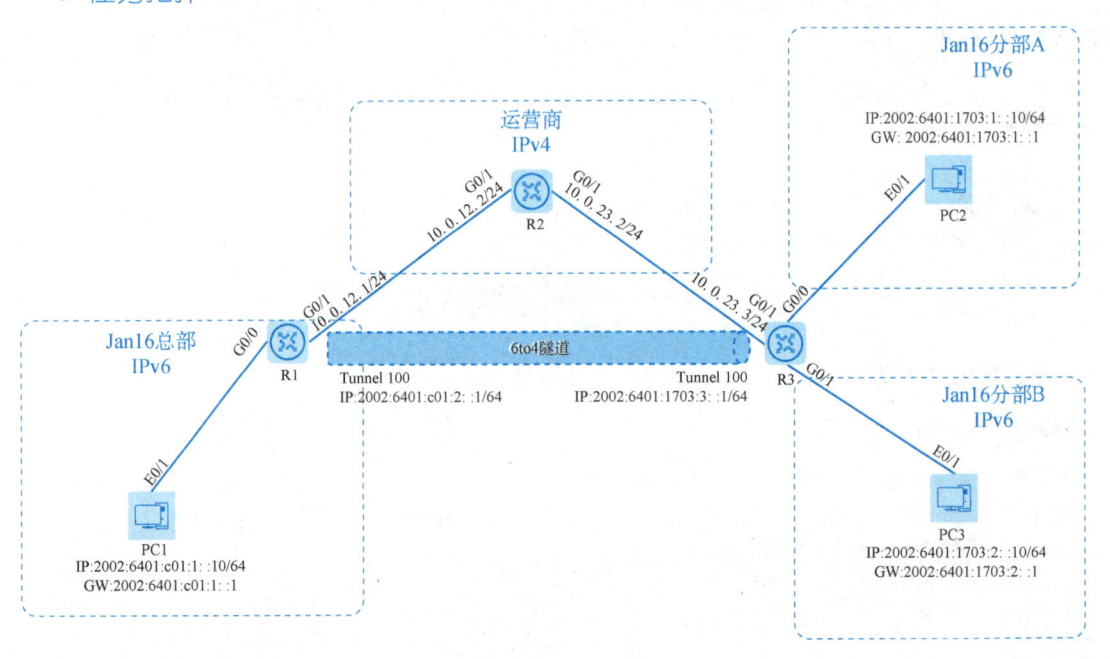

1. 根据任务拓扑在路由器 R2 配置 IPv4 地址，作为与总部路由器、分部 A 路由器互联的地址。

```
Ruijie>enable
Ruijie#configure terminal
Ruijie(config)#hostname R2
R2(config)#interface GigabitEthernet 0/1
R2(config-if-GigabitEthernet 0/1)#ip address 100.1.12.2 255.255.255.0
R2(config-if-GigabitEthernet 0/1)#exit
R2(config)#interface gigabitEthernet 0/0
R2(config-if-GigabitEthernet 0/0)#ip address 100.1.23.2 255.255.255.0
R2(config-if-GigabitEthernet 0/0)#exit
```

2. 根据任务拓扑在路由器 R1 配置 IPv4 地址，作为与运营商互联的地址，配置 IPv6 地址，作为总部的网关。

```
Ruijie>enable
Ruijie#configure terminal
Ruijie(config)#hostname R1
R1(config)#interface GigabitEthernet 0/1
R1(config-if-GigabitEthernet 0/1)#ip address 100.1.12.1 255.255.255.0
R1(config-if-GigabitEthernet 0/1)#exit
R1(config)#interface gigabitEthernet 0/0
R1(config-if-GigabitEthernet 0/0)# ipv6 address 2002:6401:c01:1::1/64
R1(config-if-GigabitEthernet 0/0)#exit
```

3．在路由器 R3 配置 IPv4 地址，作为与运营商互联的地址，配置 IPv6 地址，作为总部的网关。

```
Ruijie>enable
Ruijie#configure terminal
Ruijie(config)#hostname R3
R3(config)#interface GigabitEthernet 0/2
R3(config-if-GigabitEthernet 0/2)#ip address 100.1.23.3 255.255.255.0
R3(config-if-GigabitEthernet 0/2)#exit
R3(config)#interface gigabitEthernet 0/0
R3(config-if-GigabitEthernet 0/0)# ipv6 address 2002:6401:1703:1::1/64
R3(config-if-GigabitEthernet 0/0)#exit
R3(config)#interface gigabitEthernet 0/1
R3(config-if-GigabitEthernet 0/1)# ipv6 address 2002:6401:1703:2::1/64
R3(config-if-GigabitEthernet 0/1)#exit
```

4．配置 PC1 的 IPv6 地址，同理完成 PC2 和 PC3 的 IPv6 地址配置。

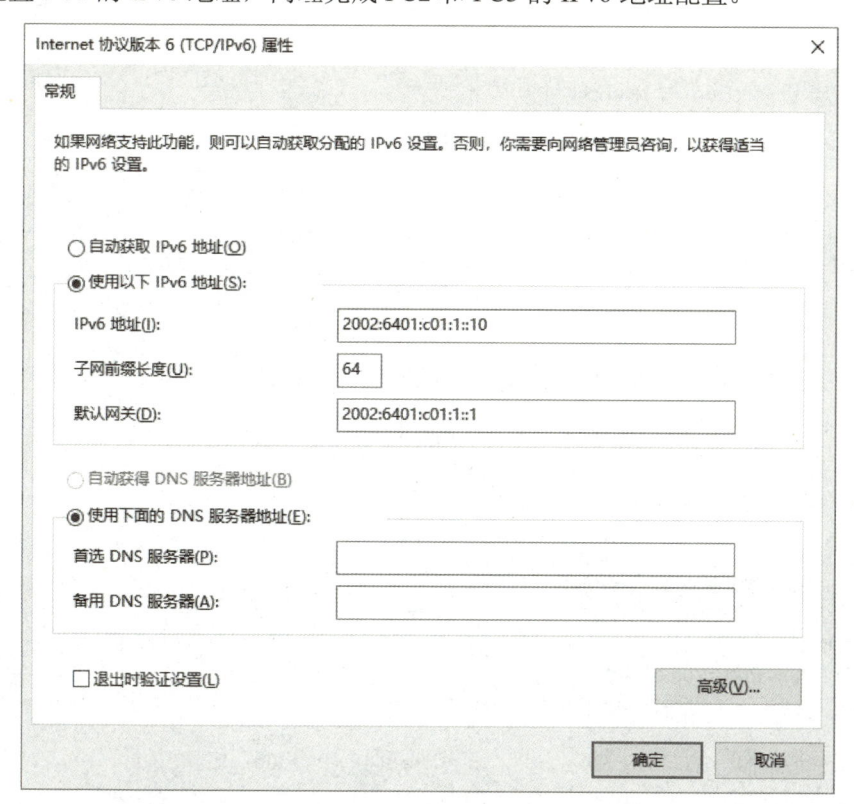

5．在总部路由器 R1 配置默认路由，下一跳地址指向运营商路由器 R2。

```
R1(config)#ip route 0.0.0.0 0.0.0.0 100.1.12.2
```

6. 在路由器 R3 配置默认路由，下一跳地址指向运营商路由器 R2。

```
R3(config)#ip route 0.0.0.0 0.0.0.0 100.1.23.2
```

7. 在 R1 路由器创建 6to4 隧道，并配置去往分部 A 的 IPv6 静态路由，下一跳地址为隧道接口。

```
R1(config)#interface Tunnel 100
R1(config-if-Tunnel 100)#ipv6 enable
R1(config-if-Tunnel 100)#ipv6 address 2002:6401:c01:2::1/64
R1(config-if-Tunnel 100)#tunnel mode ipv6ip 6to4
R1(config-if-Tunnel 100)# tunnel source 100.1.12.1
R1(config-if-Tunnel 100)#exit
R1(config)#ipv6 route 2002::/16 Tunnel 100
```

8. 在 R3 路由器创建 6to4 隧道，并配置去往总部的 IPv6 静态路由，下一跳地址为隧道接口。

```
R3(config)#interface Tunnel 100
R3(config-if-Tunnel 100)#ipv6 enable
R3(config-if-Tunnel 100)#ipv6 address 2002:6401:1703:3::1/64
R3(config-if-Tunnel 100)#tunnel mode ipv6ip 6to4
R3(config-if-Tunnel 100)# tunnel source 100.1.23.3
R3(config-if-Tunnel 100)#exit
R3(config)#ipv6 route 2002::/16 Tunnel 100
```

➢ 任务验证

1. 检查各设备 IPv4 与 IPv6 地址配置情况。

```
R2#show ip interface brief
Interface              IP-Address(Pri)   IP-Address(Sec)   Status    Protocol   Description
… …
GigabitEthernet 0/0    100.1.23.2/24     no address        up        up
GigabitEthernet 0/1    100.1.12.2/24     no address        up        up
… …
R2#
R1(config)#show ip interface brief
Interface              IP-Address(Pri)   IP-Address(Sec)   Status    Protoco Description
… …
GigabitEthernet 0/1    100.1.12.1/24     no address        up        up
… …
R1(config)#show ipv6 interface brief

GigabitEthernet 0/0                  [up/up]
       FE80::8205:88FF:FED0:D8D4
       2002:6401:C01:1::1
R1(config)#
R3(config)#show ip interface brief
Interface              IP-Address(Pri)   IP-Address(Sec)   Status    Protoco Description
… …
GigabitEthernet 0/0    100.1.23.3/24     no address        up        up
… …
R3(config)#show ipv6 interface brief
```

```
GigabitEthernet 0/1.20          [up/down]
        2002:6401:1703:2::1
        FE80::5A69:6CFF:FE6A:3926
GigabitEthernet 0/1.10          [up/down]
        2002:6401:1703:1::1
        FE80::5A69:6CFF:FE6A:3926
R3(config)#
```

2．检查出口路由配置情况。

```
R1(config)#show ip route
… …
Gateway of last resort is 100.1.12.2 to network 0.0.0.0
S*    0.0.0.0/0 [1/0] via 100.1.12.2
C     100.1.12.0/24 is directly connected, GigabitEthernet 0/1
C     100.1.12.1/32 is local host.
C     192.168.1.0/24 is directly connected, VLAN 1
C     192.168.1.1/32 is local host.
R1(config)#
R3(config)#show ip route
… …
Gateway of last resort is 100.1.23.2 to network 0.0.0.0
S*    0.0.0.0/0 [1/0] via 100.1.23.2
C     100.1.23.0/24 is directly connected, GigabitEthernet 0/2
C     100.1.23.3/32 is local host.
C     192.168.1.0/24 is directly connected, VLAN 1
C     192.168.1.1/32 is local host.
R3(config)#
```

3．检查隧道静态路由配置情况。

```
R1#show ipv6 route
… …
S     2002::/16 [1/0] via Tunnel 100, directly connected
… …

R1(config)#

R3#show ipv6 route
… …
S     2002::/16 [1/0] via Tunnel 100, directly connected
… …
R3(config)#
```

4．以 R1 设备为隧道起点，尝试 ping 通隧道终点 R3 的隧道接口地址 2002:6401:1703:3::1。

```
R1#ping ipv6 2002:6401:1703:3::1
Sending 5, 100-byte ICMP Echoes to 2002:6401:1703:3::1, timeout is 2 seconds:
  < press Ctrl+C to break >
!!!!!
Success rate is 100 percent (5/5), round-trip min/avg/max = 1/1/1 ms
R1#
```

5．总部 PC1 测试与分部 A PC2 的 IPv6 地址（2002:a01:1703:1::10）。

```
C:\Users\admin>ping 2002:a01:1703:1::10
```

正在 Ping 2002:a01:1703:1::10 具有 32 字节的数据:
来自 2002:a01:1703:1::10 的回复: 时间=1ms
来自 2002:a01:1703:1::10 的回复: 时间=2ms
来自 2002:a01:1703:1::10 的回复: 时间=2ms
来自 2002:a01:1703:1::10 的回复: 时间=2ms

2002:a01:1703:1::10 的 Ping 统计信息:
    数据包: 已发送 = 4，已接收 = 4，丢失 = 0 (0% 丢失)，
往返行程的估计时间(以毫秒为单位):
    最短 = 1ms，最长 = 2ms，平均 = 1ms

6．总部 PC1 测试与分部 A PC2 的 IPv6 地址（2002:a01:1703:2::10）。

C:\Users\admin>ping 2002:a01:1703:2::10

正在 Ping 2002:a01:1703:2::10 具有 32 字节的数据:
来自 2002:a01:1703:2::10 的回复: 时间=1ms
来自 2002:a01:1703:2::10 的回复: 时间=2ms
来自 2002:a01:1703:2::10 的回复: 时间=2ms
来自 2002:a01:1703:2::10 的回复: 时间=1ms

2002:a01:1703:2::10 的 Ping 统计信息:
    数据包: 已发送 = 4，已接收 = 4，丢失 = 0 (0% 丢失)，
往返行程的估计时间(以毫秒为单位):
    最短 = 1ms，最长 = 2ms，平均 = 1ms

7．使用分部 A PC2 ping 分部 B PC3 的 IPv6 地址（2002:a01:1703:2::10）。

C:\Users\admin>ping 2002:a01:1703:2::10

正在 Ping 2002:a01:1703:2::10 具有 32 字节的数据:
来自 2002:a01:1703:2::10 的回复: 时间=1ms
来自 2002:a01:1703:2::10 的回复: 时间=1ms
来自 2002:a01:1703:2::10 的回复: 时间=1ms
来自 2002:a01:1703:2::10 的回复: 时间=1ms

2002:a01:1703:2::10 的 Ping 统计信息:
    数据包: 已发送 = 4，已接收 = 4，丢失 = 0 (0% 丢失)，
往返行程的估计时间(以毫秒为单位):
    最短 = 1ms，最长 = 1ms，平均 = 1ms

> 问题与思考

1．IPv6 自动隧道的地址结构是怎样的？
2．根据以上配置完成 IPv6 自动隧道的 6to4 中继配置。

5.2.3 IPv6 自动隧道（ISATAP）

> 原理

站点内自动隧道寻址协议（Intra-Site Automatic Tunnel Addressing Protocol，ISATAP）隧道是一种自动隧道技术，多用于实现站点内被 IPv4 网络分隔的 IPv6 设备之间的通信。

SATAP 地址也是一种使用内嵌 IPv4 地址的特殊 IPv6 地址，它将 IPv4 地址嵌入 ISATAP 地址的接口 ID 部分。在 ISATAP 地址中，对于前缀部分并没有特殊要求，前缀可以是本地链路、本地站点、6to4 前缀。

➢ 任务拓扑

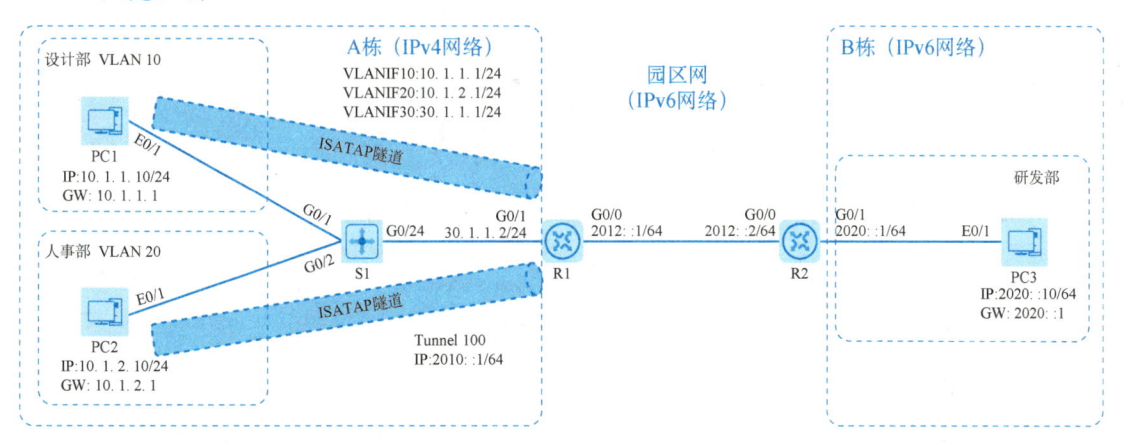

```
设计部 VLAN 10            A栋（IPv4网络）              园区网             B栋（IPv6网络）
                         VLANIF10:10.1.1.1/24        (IPv6网络)
    E0/1                 VLANIF20:10.1.2.1/24
    PC1                  VLANIF30:30.1.1.1/24                              研发部
IP:10.1.1.10/24              ISATAP隧道
GW:10.1.1.1
                    G0/1         G0/1      G0/0      G0/0     G0/1    E0/1
                    G0/24   30.1.1.2/24  2012::1/64  2012::2/64  2020::1/64
人事部 VLAN 20      G0/2    S1                R1              R2           PC3
                                                                      IP:2020::10/64
    E0/1                    ISATAP隧道                                  GW:2020::1
    PC2
IP:10.1.2.10/24                     Tunnel 100
GW:10.1.2.1                         IP:2010::1/64
```

➢ 实施步骤

1. 根据任务拓扑在 S1 创建设计部 VLAN 10、人事部 VLAN 20 和通信 VLAN 30。

```
Ruijie>enable
Ruijie#configure terminal
Ruijie(config)#hostname S1
S1(config)# vlan 10
S1(config-vlan)#vlan 20
S1(config-vlan)#vlan 30
S1(config-vlan)#exit
```

2. 根据任务拓扑在 S1 划分 VLAN，并将对应端口添加到部门 VLAN 中。

```
S1(config)#interface GigabitEthernet0/1
S1(config-if-GigabitEthernet 0/1)#switchport mode access
S1(config-if-GigabitEthernet 0/1)#switchport access vlan 10
S1(config-if-GigabitEthernet 0/1)#exit
S1(config)#interface GigabitEthernet0/2
S1(config-if-GigabitEthernet 0/2)#switchport mode access
S1(config-if-GigabitEthernet 0/2)#switchport access vlan 20
S1(config-if-GigabitEthernet 0/2)#exit
S1(config)#interface GigabitEthernet0/24
S1(config-if-GigabitEthernet 0/24)#switchport mode access
S1(config-if-GigabitEthernet 0/24)#switchport access vlan 30
S1(config-if-GigabitEthernet 0/24)#exit
S1(config)#
```

3. 在主机 PC1 和 PC2 配置 IPv4 地址。

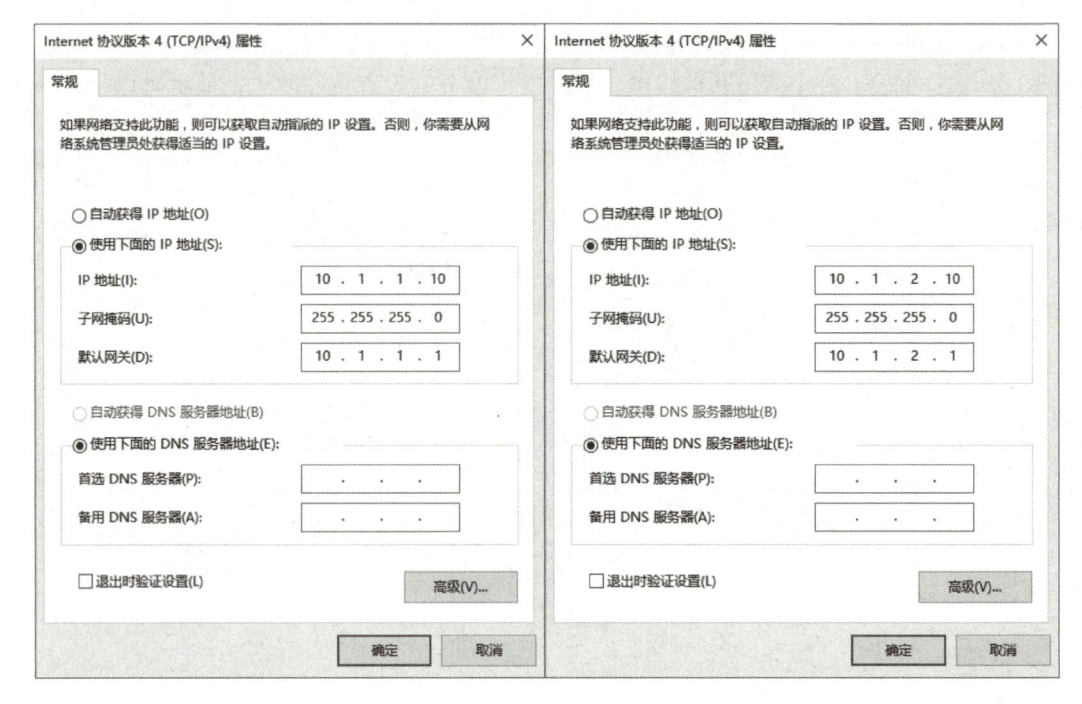

4. 主机 PC3 配置 IPv6 地址，PC1 与 PC2 的 IPv6 地址配置为自动获取。

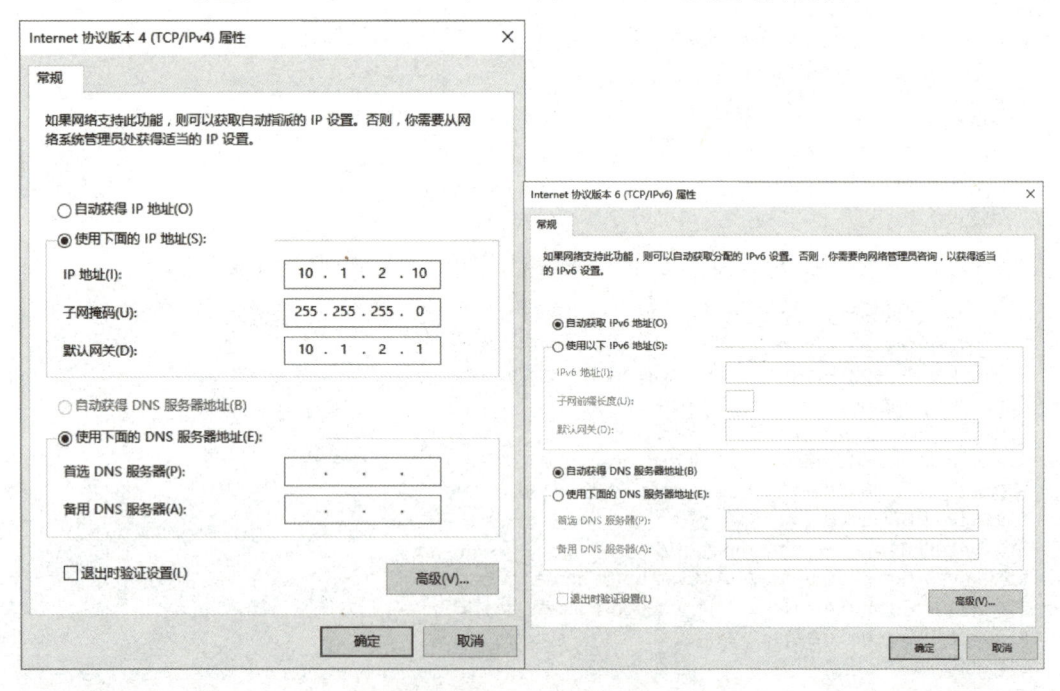

5. 在路由器 R1 配置 IPv4 和 IPv6 地址，作为与网关交换机 S1 和园区网路由器 R2 互联的地址。

```
Ruijie>enable
Ruijie#configure terminal
Ruijie(config)#hostname R1
R1(config)#interface GigabitEthernet 0/1
```

```
R1(config-if-GigabitEthernet 0/1)#ip address 30.1.1.2 255.255.255.0
R1(config-if-GigabitEthernet 0/1)#exit
R1(config)# interface GigabitEthernet 0/0
R1(config-if-GigabitEthernet 0/0)#ipv6 enable
R1(config-if-GigabitEthernet 0/0)#ipv6 address 2012::1 64
R1(config-if-GigabitEthernet 0/0)#exit
R1(config)#
```

6. 在路由器 R2 配置 IPv6 地址，作为与园区网路由器 R1 互联的地址及研发部的网关。

```
Ruijie>enable
Ruijie#configure terminal
Ruijie(config)#hostname R2
R2(config)#interface GigabitEthernet 0/1
R2(config-if-GigabitEthernet 0/1)#ipv6 enable
R2(config-if-GigabitEthernet 0/1)# ipv6 address 2020::1 64
R2(config-if-GigabitEthernet 0/1)#exit
R2(config)# interface GigabitEthernet 0/0
R2(config-if-GigabitEthernet 0/0)#ipv6 enable
R2(config-if-GigabitEthernet 0/0)#ipv6 address 2012::2 64
R2(config-if-GigabitEthernet 0/0)#exit
R2(config)#
```

7. 在 S1 配置 IPv4 地址，作为设计部与人事部的网关，以及与园区路由器 R1 互联的地址。

```
S1(config)#interface vlan 10
S1(config-if)#ip address 10.1.1.1 255.255.255.0
S1(config-if)#exit
S1(config)#interface vlan 20
S1(config-if)#ip address 10.1.2.1 255.255.255.0
S1(config-if)#exit
S1(config)#interface vlan 30
S1(config-if)#ip address 30.1.1.1 255.255.255.0
S1(config-if)#exit
```

8. 在路由器 R1 配置设计部与人事部的 IPv4 静态路由，下一跳地址为网关交换机 S1。

```
R1(config)#ip route 10.1.1.0 255.255.255.0 30.1.1.1
R1(config)#ip route 10.1.2.0 255.255.255.0 30.1.1.1
```

9. 在路由器 R1 配置通往研发部的 IPv6 静态路由，下一跳地址为园区网路由器 R2。

```
R1(config)#ipv6 route 2020::/64 2012::2
```

10. 在路由器 R2 配置通往 ISATP 隧道前缀的 IPv6 静态路由，下一跳地址为园区网路由器 R1。

```
R2(config)#ipv6 route 2010::/64 2012::1
```

11. 在路由器 R1 创建 ISATAP 隧道接口，配置 IPv6 地址并开启 RA 报文发送功能。

```
R1(config)#interface Tunnel 100
R1(config-if-Tunnel 100)#ipv6 enable
R1(config-if-Tunnel 100)#ipv6 address 2010::1/64 eui-64
R1(config-if-Tunnel 100)#tunnel mode ipv6ip isatap
R1(config-if-Tunnel 100)#tunnel source 30.1.1.2
R1(config-if-Tunnel 100)#no ipv6 nd    suppress-ra
R1(config-if-Tunnel 100)#exit
R1(config)#
```

12. 在 PC1 指定 ISATAP 路由器 IPv4 地址为 30.1.1.2，以管理员身份运行 CMD 命令提示符窗口进行配置。

注：此配置以 Windows 10 系统进行试验，不同操作系统的命令可能不同，set router 30.1.1.2 命令用于指定 ISATAP 路由器。set state enable 命令用于启用 ISATAP 隧道。

```
C:\Users\Administrator>netsh
netsh>interface ipv6
netsh interface ipv6 >isatap
netsh interface ipv6 isatap>set router 30.1.1.2
确定。
netsh interface ipv6 isatap>set state enable
确定。
netsh interface ipv6 isatap>
```

13. 在 PC2 指定 ISATAP 路由器 IPv4 地址为 30.1.1.2，以管理员身份运行 CMD 命令提示符窗口进行配置。

```
C:\Users\Administrator>netsh
netsh>interface ipv6
netsh interface ipv6 >isatap
netsh interface ipv6 isatap>set router 30.1.1.2
确定。
netsh interface ipv6 isatap>set state enable
确定。
netsh interface ipv6 isatap>
```

➢ 任务验证

1. 在 S1 使用 show vlan 命令验证 VLAN 的创建情况。

```
S1(config)#show vlan
VLAN Name                          Status    Ports
---- -------------------------------- --------- ----------------------------------
   1 VLAN0001                       STATIC    Gi0/3, Gi0/4, Gi0/5, Gi0/6
                                              Gi0/7, Gi0/8, Gi0/9, Gi0/10
                                              Gi0/11, Gi0/12, Gi0/13, Gi0/14
                                              Gi0/15, Gi0/16, Gi0/17, Gi0/18
                                              Gi0/19, Gi0/20, Gi0/21, Gi0/22
                                              Gi0/23, Gi0/25, Gi0/26, Gi0/27
                                              Gi0/28, Te0/29, Te0/30, Te0/31
                                              Te0/32
  10 VLAN0010                       STATIC    Gi0/1
  20 VLAN0020                       STATIC    Gi0/2
  30 VLAN0030                       STATIC    Gi0/24
 100 VLAN0100                       STATIC
 300 VLAN0300                       STATIC
S1(config)#
```

2. 在 S2 使用 show vlan 命令验证 VLAN 的创建情况。

```
S1(config)#show interface switchport
Interface               Switchport Mode      Access Native Protected VLAN lists
----------------------- ---------- --------- ------ ------ --------- ----------------------
GigabitEthernet 0/1     enabled    ACCESS    10     1      Disabled  ALL
GigabitEthernet 0/2     enabled    ACCESS    20     1      Disabled  ALL
```

```
… …
GigabitEthernet 0/24              enabled    ACCESS    30    1    Disabled    ALL
… …
S1(config)#
```

3. 在 R1 使用 show ip interface brief 和 show ipv6 interface brief 命令验证 R1 的 IP 地址配置情况。

```
R1(config)#show ip interface brief
Interface              IP-Address(Pri)    IP-Address(Sec)    Status        Protocol    Description
… …
GigabitEthernet 0/1                30.1.1.2/24            no address                  up            up
… …
R1(config)#
R1(config)#show ipv6 interface brief

GigabitEthernet 0/0           [up/up]
       FE80::8205:88FF:FED0:D8D4
       2012::1
R1(config)#
```

4. 在 R2 使用 show ipv6 interface brief 命令验证 R2 的 IP 地址配置情况。

```
R2(config)#show ipv6 interface brief

GigabitEthernet 0/0           [up/up]
       FE80::8205:88FF:FED0:D848
       2012::2
GigabitEthernet 0/1           [up/up]
       FE80::8205:88FF:FED0:D847
       2020::1
R2(config)#
```

5. 在 S1 使用 show ip interface brief 命令验证 S1 的 IP 地址配置情况。

```
S1(config)#show ip interface brief
Interface       IP-Address(Pri)    IP-Address(Sec)    Status    Protocol
VLAN 10         10.1.1.1/24        no address         up        up
VLAN 20         10.1.2.1/24        no address         up        up
VLAN 30         30.1.1.1/24        no address         up        up
… …
S1(config)#
```

6. 在 R1 使用 show ip route show ipv6 route 命令验证 R1 的静态路由配置情况。

```
R1(config)#show ip route
… …
Gateway of last resort is no set
S    10.1.1.0/24 [1/0] via 30.1.1.1
S    10.1.2.0/24 [1/0] via 30.1.1.1
… …
R1(config)#
R1(config)#show ipv6 route
… …
S       2020::/64 [1/0] via 2012::2 (recursive via 2012::2, GigabitEthernet 0/0)
```

……
R1(config)#

7. 在 R2 使用 show ipv6 route 命令验证 R2 的静态路由配置情况。

R2(config)#show ipv6 route
……
S   2010::/64 [1/0] via 2012::1 (recursive via 2012::1, GigabitEthernet 0/0)
……
R2(config)#

8. 在 PC1 使用 ipconfig/all 命令查看 ISATAP 接口信息。

C:\Users\Administrator>ipconfig /all
……
隧道适配器 isatap.{51E1BCF0-D1AC-44F3-A68F-98CEE44E0E53 }:

    连接特定的 DNS 后缀 ……… :
    描述………………: Microsoft ISATAP Adapter
    物理地址…………: 00-00-00-00-00-00-00-E0
    DHCP 已启用 …………: 否
    自动配置已启用………: 是
    本地链接 IPv6 地址……: fe80::5efe:10.1.1.10%26(首选)
    默认网关…………:
    DHCPv6 IAID …………: 436207616
    DHCPv6 客户端 DUID ……: 00-01-00-01-27-A2-C5-DD-00-0C-29-7D-4B-AC
    DNS 服务器 …………: fec0:0:0:ffff::1%1
                                  fec0:0:0:ffff::2%1
                                  fec0:0:0:ffff::3%1
    TCPIP 上的 NetBIOS ……: 已禁用
C:\Users\Administrator>

9. 在 PC2 使用 ipconfig/all 命令查看 ISATAP 接口信息。

C:\Users\Administrator>ipconfig /all
……
隧道适配器 isatap.{FE9C6DDF-91F8-4D1B-9482-AEABF330F047 }:

    连接特定的 DNS 后缀 ……… :
    描述………………: Microsoft ISATAP Adapter
    物理地址…………: 00-00-00-00-00-00-00-E0
    DHCP 已启用 …………: 否
    自动配置已启用………: 是
    IPv6 地址 …………: 2010::5efe:10.1.2.10(首选)
    本地链接 IPv6 地址……: fe80::5efe:10.1.2.10%25(首选)
    默认网关…………: fe80::5efe:30.1.1.2%2
    DHCPv6 IAID …………: 419430400
    DHCPv6 客户端 DUID ……: 00-01-00-01-27-A2-CE-F0-00-0C-29-4A-80-34
    DNS 服务器 …………: fec0:0:0:ffff::1%1
                                  fec0:0:0:ffff::2%1
                                  fec0:0:0:ffff::3%1
    TCPIP 上的 NetBIOS ……: 已禁用

10. 使用设计部 PC1 ping 研发部 PC3 的 IPv6 地址 2020::10。

C:\Users\Administrator>ping 2020::10

正在 Ping 2020::10 具有 32 字节的数据:
来自 2020::10 的回复: 时间=1ms
来自 2020::10 的回复: 时间=1ms
来自 2020::10 的回复: 时间=1ms
来自 2020::10 的回复: 时间=1ms

2020::10 的 Ping 统计信息:
    数据包: 已发送 = 4，已接收 = 4，丢失 = 0 (0% 丢失)，
往返行程的估计时间(以毫秒为单位):
    最短 = 1ms，最长 = 1ms，平均 = 1ms

11. 使用人事部 PC2 ping 研发部 PC3 的 IPv6 地址 2020::10。

C:\Users\Administrator>ping 2020::10

正在 Ping 2020::10 具有 32 字节的数据:
来自 2020::10 的回复: 时间=1ms
来自 2020::10 的回复: 时间<1ms
来自 2020::10 的回复: 时间=2ms
来自 2020::10 的回复: 时间=1ms

2020::10 的 Ping 统计信息:
    数据包: 已发送 = 4，已接收 = 4，丢失 = 0 (0% 丢失)，
往返行程的估计时间(以毫秒为单位):
    最短 = 1ms，最长 = 2ms，平均 = 1ms

➤ 问题与思考

1. ISATAP 自动隧道的地址结构？
2. 在隧道接口上配置 no ipv6 nd suppress-ra 命令的作用是什么？